Selected papers from the Third Conference on Experimental Medicine and Surgery in Primates, Lyon, June 1972

Medical Primatology 1972

Part I

General Primatology, Reproduction and Perinatal Studies, Genetics, Phylogenetics and Evolution

Edited by

EDWARD I. GOLDSMITH
Cornell University Medical College, New York, N.Y.

J. MOOR-JANKOWSKI
New York University School of Medicine, New York, N.Y.

In collaboration with

C. R. Austin
H. Balner
N. Barnicot
R. Cortesini
F. Deinhardt
N. Gengozian
G. A. Gresham
J. E. Hamner III
S. S. Kalter
R. Naquet
M. Panigel
F. T. Perkins
J. Ruffié
A. Schrier
M. Vastesaeger
G. Zbinden
E. Zuckerkandl

With 132 figures and 50 tables

S. Karger · Basel · München · Paris · London · New York · Sydney 1972

Medical Primatology 1970
XXII+997 p., 284 fig., 174 tab., 1971
ISBN 3-8055-1163-9

S. Karger · Basel · München · Paris · London · New York · Sydney
Arnold-Böcklin-Strasse 25, CH–4011 Basel (Switzerland)

Printed in Switzerland by City-Druck AG, Glattbrugg
ISBN 3-8055-1486-7

To Frank Glenn, MD

Professor of Surgery Emeritus, Cornell University Medical College, whose pioneering efforts on behalf of Medical Primatology have guided and encouraged the editors of these volumes

Contents

Part I

General Primatology. Husbandry, Management, Breeding Choice of Species Conservation. Communication Media in Medical Primatology

Part II

Experimental Surgery

Transplantation Immunology

Part III

Infectious Diseases

Foreword

The three volumes of 'Medical Primatology 1972' consist of papers selected from communications presented at the Third Conference on Experimental Medicine and Surgery in Primates, held in Lyon, France, 21st–23rd June 1972. The program of the Conference was designed to gather new information to complement the former volume *Medical Primatology 1970*, which resulted from the Second Conference.

Papers chosen for the present volumes passed two selection procedures: (1) 246 abstracts were submitted of which 176 were accepted to be presented as papers at the Third Conference; (2) these complete papers were then evaluated for validity and relevance of the results presented. With the exception of review articles, only papers containing new information were accepted. The selection process led to some difficult decisions, as no time was available for resubmission of rejected papers. The chosen 120 papers were edited for style, correct simian taxonomy, bibliography, and appropriate cross references.

Despite the rigorous standards applied, we expect that shortcomings will still be found since little time was available for editing in order to obtain publication within seven months after the Conference. We hope, however, that *Medical Primatology 1972* will satisfy the objectives set at the First Conference in 1967, namely, to provide periodically an up-to-date and complete handbook and reference source covering the broad field of Medical Primatology.

Mr. THOMAS KARGER has continued to back up the new field of medical primatology despite the commercial risk involved. Miss DENISE GREDER and Mr. ROLF STEINEBRUNNER of the Karger firm, have assisted us at the Conference and during the following months of hard editorial labor. Miss

GETTA SKOTCHDOPOLE, Assistant Editor, attended most ably to the coordination of editorial efforts and to proof-reading. The Editors alone, being nonsimian primates, must accept the blame for inadequacies.

EDWARD I. GOLDSMITH, MD
Professor of Surgery
Cornell University Medical College, New York

J. MOOR-JANKOWSKI, MD
Research Professor
New York University School of Medicine, New York

Laboratory for Experimental Medicine and Surgery in Primates (LEMSIP) New York, November 1972

Organization Committee of the Third Conference on Experimental Medicine and Surgery in Primates 1972

WHO Delegates	W. I. B. Beveridge, Geneva Z. Fejfar, Geneva
National Delegates	
Belgium	Dr. Robert Gortz University of Bruxelles Dr. J. Mortelmans Institut de Médecine tropicale, Antwerp
Canada	Dr. Samuel Solomon Royal Victoria Hospital, Montreal
France	Dr. Charles Merieux Institut Merieux, Lyon Dr. Robert Naquet Institut de Neuro-physiologie du CNRS, Marseille Dr. René Triau Fondation Merieux, Lyon
Germany	Dr. Hans-Jürg Kuhn Universität Frankfurt
Great Britain	Mr. K. R. Hobbs Medical Research Council, Carshalton Dr. F. T. Perkins National Institute for Medical Research, London
Holland	Dr. Hans Balner Radiobiological Institute TNO, Rijswijk
Israel	Dr. Uri Marinov National Council for Research and Development, Jerusalem

Italy	Dr. RAFFAELLO CORTESINI Università di Roma
Norway	Professor KARE BERG University of Oslo
Sweden	Professor ERNST BARANY Uppsala Universitet
Switzerland	Dr. HEINZ WEBER Sandoz AG, Basel Professor G. ZBINDEN Universität Zürich
United States	Dr. JULES S. CASS Veterans Administration, Washington, D.C. Dr. S. S. KALTER Southwest Foundation for Research and Education San Antonio, Texas Dr. C. H. KRATOCHVIL The Upjohn Company Kalamazoo
Scientific Secretary	Dr. J. RUFFIÉ Collège de France, Paris
Co-Chairmen	Dr. E. I. GOLDSMITH Cornell University Medical College, New York Dr. J. MOOR-JANKOWSKI New York University School of Medicine, New York

Sponsors and Benefactors of the Third Conference on Experimental Medicine and Surgery in Primates 1972

The Conference participants are greatly indebted to Dr. PIERRE DOROLLE, Deputy Director, World Health Organization, for the fruitful collaboration of WHO in the Conference.

The Conference was sponsored by the following international and national agencies:

International	European Society for the Study of Drug Toxicity International Association of Biologic Standardization
Belgium	Ministère de la Politique et Programmation Scientifique
Canada	McGill University, Montreal
France	Ministère de la Santé Publique et de la Sécurité Sociale Centre National de la Recherche Scientifique (CNRS) Institut National de la Santé et de la Recherche Médicale (INSERM) Université Paul Sabatier de Toulouse Université Claude Bernard de Lyon Fondation Merieux
Germany	German Pharmacological Society
Great Britain	The Royal College of Surgeons of England The Zoological Society of London
Holland	Netherlands Organization for Health Research
Israel	The Weizmann Institute

Italy	Italian Ministry of Health
Sweden	The Royal Swedish Academy of Sciences
Switzerland	Union of Swiss Societies of Experimental Biology
USA	New York University School of Medicine American Association for Laboratory Animal Science

The Conference Organization Committee is grateful to the governments and agencies which provided travel grants for their nationals to attend the Conference.

Our sincere thanks are due to Dr. CHARLES MERIEUX, the Merieux Foundation and the Merieux Institute. Their staff provided valuable scientific contributions to the Conference, assumed responsibility for the logistics, and offered warm hospitality to the participants, far beyond the usual role of hosts. The Merieux Foundation provided generously for the considerable expenditures of the local organization and helped many speakers to attend the Conference by defraying their expenses.

We wish to thank the following organizations, listed in alphabetical order, which contributed substantially to cover the expenses of the American participation in the Conference.

Major Contributors	*Baylor University* *Ciba-Geigy* *Generics Corporation of America* *Hoffmann-La Roche* *Schering Corporation* *The Upjohn Company*
Contributors	*Ortho Pharmaceutical Corporation* *Riker Laboratories, Inc.* *Sandoz Ltd. Basel* *Sherwood Medical Industries Inc.*

General Primatology

Husbandry, Management, Breeding Choice of Species Conservation
Communication Media in Medical Primatology

Medical Primatology 1972. Proc. 3rd Conf. exp. Med. Surg. Primates, Lyon 1972, part I, pp. 2–4 (Karger, Basel 1972)

Primate Animals in Biomedical Research

Z. FEJFAR

Cardiovascular Diseases, WHO, Geneva

Although medical research has long made use of various species of animals, it is only comparatively recently that monkeys and apes have been widely used. The close phylogenetic relationship of these animals to man means that their anatomic, physiological and psychological characteristics and disease patterns are in many respects more similar to those of man than are those of other animals; therefore, they offer special advantages for investigation of a number of human health problems. Their potential for medical research has not yet been fully explored and every year new uses for them are being found.

The World Health Organization has undertaken coordinated international research in comparative medicine since 1959, particularly into the etiology, epidemiology and pathogenesis of malignant growth and of cardiovascular diseases in a number of species [1].

WHO has initiated and assisted in the organization of an international symposium on the use of primates in medicine outlining more specifically the problems of husbandry and technology [2]; a symposium on infections and immunosuppression in subhuman primates [3]; a symposium on breeding primates for laboratory use [4]; and most recently a symposium on the use of primates for research on problems of human reproduction [5]. Health aspects of the supply and use of human primates for biomedical purposes were outlined at a meeting of a WHO scientific group on Health Aspects of the Supply and Use of Nonhuman Primates for Biomedical Purposes [6]. Recommendations made by the group include such problems as those connected with holding and exportation, transportation by air, quarantine, health aspects related to biomedical use, and national control of the above. The Organization also designated a WHO regional reference laboratory for simian viruses.

The present conference with the programme embracing such a broad spectrum of topics contains many items which advance the above-mentioned WHO interest. We see in this a good example of a convergence between the international governmental organization and a scientific community both aiming to promote knowledge by profound investigations. Studies on primates are not a transient fashion. Their increasing importance is documented among others, by the exponential rise of publications on the subject and by the present volume which continues and completes the previous published in 1969 [7] and 1971 [8].

As a physician, I have been constantly faced with the problem of how to relate experimental results to the conditions in man. We have tried to use to the maximum observations on man before seeking the proof in experiments on animals. More and more, however, investigations on etiology of chronic diseases require studies at the level of the cell on the one hand, and those on the effect of the environment on the individual in the community on the other. Greater emphasis is placed upon early diagnosis of a chronic condition and on studies to assure the optimum physical and mental development of man, well adapted to his civilization. These aspects may well be explored by studying systematically the nonhuman primates to approximate the health and disease conditions of man.

Volumes such as this are valuable in providing a handbook and guidelines for scientists working with primate animals in a variety of disciplines.

References

1 The Medical Research Programme of the World Health Organization, 1964–1968: Report of the Director-General (WHO, Geneva 1969).

2 BEVERIDGE, W.I.B. (ed.): In GOLDSMITH and MOOR-JANKOWSKI Primates in Medicine, vol. 2 and 3. Using primates in medical research: I. Husbandry and technology. II. Recent comparative research (Karger, Basel 1969).

3 BALNER, H. and BEVERIDGE, W.I.B. (ed.): Infections and immunosuppression in subhuman primates. Proc. Int. Symp. Infections and Immunosuppression in Subhuman Primates, Rijswijk 1969 (Munksgaard, Copenhagen 1970).

4 BEVERIDGE, W.I.B. (ed.): Breeding primates. Proc. Int. Symp. Breeding Nonhuman Primates for Laboratory Use, Berne 1971 (Karger, Basel 1972).

5 Symposium on the Use of Nonhuman Primates for Research on Problems of Human Reproduction, Sukhumi 1971 (1972).

6 Health Aspects of the Supply and Use of Nonhuman Primates for Biomedical Purposes, Report of a WHO Scientific Group. Wld Hlth Org. Techn. Rep. Ser., vol. 490 (1971).

7 GOLDSMITH, E. I. and MOOR-JANKOWSKI, J. (ed.): Experimental medicine and surgery in primates. Ann. N.Y. Acad. Sci. *162:* 1–702 (1969).
8 GOLDSMITH, E. I. and MOOR-JANKOWSKI, J. (ed.): Medical primatology 1970 (Karger, Basel 1971).

Author's address: Dr. Z. FEJFAR, Chief, Cardiovascular Diseases, WHO, *CH-1211 Geneva* (Switzerland)

Medical Primatology 1972. Proc. 3rd Conf. exp. Med. Surg. Primates, Lyon 1972, part I, pp. 5–14 (Karger, Basel 1972)

Nonhuman Primates, a Vulnerable Resource[1]

Barbara Harrisson

Deaprtment of Antropology, Cornell University, Ithaca, N.Y.

Introduction

The exploitation of primate animals is one of the most serious causes for their depletion. It is due to the unawareness of the presently-occurring rapid changes by the institutions which regulate the acquisition, trade and usage of the animals. Existing controls pertain to a situation before the advent of animal traffic by air, and of the world wide demand for captive simians to serve industry, zoos, pet trade and research. Moreover, resource nations have seen rapid developments in vast land areas which rendered them accessible to international communication. New technologies fostered wide-scale destruction and alteration of habitats, especially of lowland tropical forests.

Most influenced by these events were specialized primate animals with limited distribution. More generalized forms, with wider distribution, were also affected by uncontrolled killing as agricultural pests and by extensive harvesting for exportation. Rare and endangered species entered the international commerce in spite of protection laws because of the high prices offered by the purchasers. New species were listed as endangered, as time went on. But the controls at point of initial resource remained ineffective.

1 This paper, which evaluates current activities supporting the long-term availability of nonhuman primates, is based on a comprehensive analysis of the 'Conservation of Nonhuman Primates in 1970' [1] where an extensive list of references is included. It is also based in part on unpublished material and discussion of issues within the Survival Service Commission of the International Union for Conservation of Nature and Natural Resources (IUCN, CH-1110 Morges, Switzerland) where the author is Secretary of the Primate Specialist Group.

The improvements which are necessary require investments and procedures beyond the means of resource nations. These insist upon the international character of the problem and the obligation of users to bring about the improvements. Researchers clearly stated their interest in conservation: Nonhuman primates are unique because of their close relationship to man. The disappearance of any such species is a loss for biomedical research [2]. But the legal and administrative research framework is not designed to support and protect the resources. New legislation cannot proceed because basic information is lacking. Existing controls cannot operate because they do not apply internationally. Veterinary and health controls are not adequate to the increased usage and movements of captive simians. The only effectively regulated aspect is the international transportation. Airline carriers now provide facilities to avoid excessive loss and inhumane conditions of animals en route. But major areas of initial and subsequent wastage and abuse are not under control. The consequence is serious overexploitation and a general rapist attitude to the resource.

This paper identifies users and the trade in broad terms. It then describes the main activities in support of conservation, of resource management and of trade control. It finally advocates how these may be improved.

Users and Traders of Primate Animals

The users of primate animals may be separated into three groups. In or near the natural habitats, there is firstly a native interest in simians as a source of food and cash or status value in local terms. This group of users connects with the second, that of traders. Linking supply and demand, traders are concerned mainly with captive animals on national and international levels. They make use of transporters, a subsidiary group. The third group combines ultimate users, researchers, biomedical industry, zoological parks, entertainment industry and pet owners. All three groups operate within framework of national interest: governments who derive economic benefits from the exportation or from the ultimate usage of simians.

The large group of ultimate users makes possible the overexploitation. Far removed from the resource, they inadvertently prompt irresponsible activities contrary to the long-term interests of their own and of the resource countries. The majority of users is, unfortunately, satisfied to exercise demand without enquiry into the status of the resource and the methods by which traders produce and distribute supplies.

The trade in simians is complex and only a few generalizations are possible. The majority of trappers are individual entrepreneurs. Many operate only incidentally or seasonally. Some countries license their activities; others do not. A few trappers are organized in commercial business with substantial investments. The majority of trapped animals percolate through many hands to a local dealer who has marketing capability, and to an exporter or wholesaler. Exporters and wholesalers assemble their stock from many dealers. Some of the larger institutions and industries buy directly from foreign wholesalers. Most buy from dealers in their own countries, many of whom perform only brokerage functions.

An unknown portion of the trade is illicit. Protected species are captured, illegally transported over national borders and only then exported from a neighbouring country, or state in the case of federal nations where protection laws may differ from one state to another. Protected species are also exported directly with false declarations of identification. The commercial incentive is highest for the most vulnerable and valuable species: for apes, for rare species with limited distributions, for species that do not establish well in captivity. These operations succeed because of ineffective law enforcement and limited capacity of customs to identify species and subspecies.

The interest in rare and vulnerable species is, by tradition, a major feature of trade promotion. Wholesalers and brokers invite interest by advertising since this has shown to foster demands. The demands, in turn, increase the pressure on species most in need of protection. Another type of advertising is addressed to researchers, laboratories and industries. Here the capability to deliver disease-free specimens of desirable genera and species with determined ages, sexes and weights is stressed. This produces the impression that advertisers exercise controls at point of initial supply which, generally is not the case. The specimens which cannot enter the market because they do not meet the established criteria are wasted. Stress, disease and neglect take additional tolls in resource countries during trapping, transportation and holding by chains of unskilled individuals; all losses remain unaccounted for in trade statistics. Dealers buying stock on speculation take additional high losses if they cannot find immediate buyers. Strong competition among dealers intensifies abuses of which most customers are unaware.

Prices for imports increase as species become overexploited and wasted. Thus, point when imports are as costly as captive-bred specimens may be that of no return for a number of species in their natural environment.

Activities Supporting Resource Management and Control

A variety of programs have evolved. The main issues are summarized below in the order of importance accorded them by user groups.

Captive Breeding

Captive breeding, acclaimed as the best conservation measure, is valid if it represents a long-term concern and supports species most vulnerable in their natural habitats.

Importers primarily promote captive breeding of simians most used in research and industry, and which establish reasonably well in captivity. But these genera and species are also under least pressure. A recent study which assesses the long-term breeding potential in zoos (inclusive of eight primate species [3]) indicates grave concern. The situation in research laboratories is more favorable but laboratory breeding is not yet adequate. The Institute of Laboratory Animal Resources (ILAR), National Academy of Sciences, provides reference data and description of breeding colonies in the US.[2] The existing breeding programs are handicapped by fragmentation and by the lack of data on the natural environments and behavior of species used.

Recovery of Information in Resource Countries

The information needed for laboratory breeding is also needed by the habitat countries for protection measures. They need basic information on the distribution and coexistence of genera and species; their ranges, troup and group sizes, food habits, life statistics and adaptive properties, to enable administrators to promote formation of parks and preserves and to establish ceilings of exploitation. For apes and for forest-dwelling simians in general, including those used extensively in research, this information is spotty. The ecological studies made [4] have identified a few species in few limited ranges. They indicate the impossibility of generalization, considerable differences for the same species living under different environmental conditions, and variation in group and individual adaption.

2 The program is summarized in Lab. Prim. Newsl. No. 11, pp. 20 and 21 (1972).

The drastic upsets of simian populations through timber felling, hunting, trapping and killing at forest edges where simian populations concentrate in prime habitat, is known. STRUHSAKER [5], after a recent field investigation in Africa, emphasizes the great scientific interest in the rain forests and the need of support for appropriate studies.

Resource nations have an economic interest in their rain forests which they exploit for timber and agriculture in accordance with established traditions, policies and expertise. This is detrimental to simian populations. Their survival requires the replacement of man's present economic stake in rain forests by an alternative. STRUHSAKER [5] calls this an environmental problem of international proportions. 'It seems apparent', he states, 'that the precarious position of Africa's rain forests necessitates a carefully planned and ecologically founded multi-purpose use, which, by definition, would include areas completely protected against human exploitation.'

The facilities of ongoing UNESCO, IBP and FAO programs of the United Nations related to tropical forestry[3] present opportunities for studies and surveys of nonhuman primates, which would contribute to habitat preservation and fresh concepts of resource management.

International, national and institutional biological field stations where facilities and expertise can be shared, have proven advantageous. But some are now too far removed from the areas where scientific presence is most needed. If the protection of habitat and controlled acquisition in strategic resource areas now under pressure of exploitation is intended, the initiation of ecological surveys is urgently needed to provide basic information.

One vital area, the upper Amazon, developed such activities over the past two years with the cooperative support of user interests. Aware of the dangers of consumer demands which may drain the upper Amazon, US researchers obtained grants for biological and status surveys. The Colombian government, through the auspices of the Instituto de Desarrollo de los Recursos Naturales Renovables (INDERENA) is accordingly to be placed in a position where it can recover the information it needs for the promotion of habitat preservation and resource management. Supported by the Smithsonian Institution, fieldwork has gone ahead. A biological station near Leticia is planned. The urgency of many tasks, including that of employee

3 The UNESCO and FAO Headquarters in Paris and Rome for general information. UNESCO's Field Science Office for Africa (P.O. Box 30592, Nairobi) and the Pacific Science Association (B.P. Bishop Museum, P.O. Box 6037, Honolulu, Hawaii) which connects with regional committees on conservation and tropical forestry in Southeast Asia can also provide links and information.

training and public education which Colombia faces, require outside participation and support.[4]

The Old World regions, from where endangered and vulnerable species are exported, are widely scattered. Ecological studies proceed in forests of Southeast Asia and tropical Africa. But they are isolated and cannot fulfill the needs the rapidly-deteriorating situation demands. Lack of funds is the main handicap. A cooperative user concern by importing nations outside the US is urgently needed. These countries accounted for 41 % of total trade in simians in 1971. Current user-responsibility is reliably documented in the US [6]. That of Eurasia has been computed from some government statistics from airline carriers and from research opinions (see table I).

Trade Controls

The 3rd Congress of Primatology held in Switzerland in 1970, recommended that *Pan troglodytes* be declared endangered in the Red Data Book of the International Union for Conservation of Nature and Natural Resources (IUCN) [7]. The listing of *P. troglodytes* as endangered could have introduced import restrictions in countries which follow IUCN recommendations, irrespective of export restrictions applicable in resource countries.

However, for listing in the Red Data Book proof is required that survival of *P. troglodytes* is unlikely without early implementation of protective measures, and that numbers are reduced to a critically low level. Such data is available for parts of its range. But because *P. troglodytes* has a very wide distribution and remains secure, or unidentified, in other parts of its range, its present status was declared as *depleted*—that is, endangered if further deterioration takes place. If the data is evaluated for each subspecies of

Table I. Imports of nonhuman primates during 1971

	New World species		Old World species		Total	
	numbers	%	numbers	%	numbers	%
USA	54,500	83	32,000	39	86,500	59
Eurasia	10,500	17	50,000	61	60,500	41
Total	65,000		82,000		147,000	

4 HERNANDEZ J., Asesor de Investigacion de Fauna Silvestrel, Instituto de Desarrollo de los Recursos Naturales Renovables, D.E. Correro 14, No. 25A–66, Apdo. Aereo 13458, Bogota.

P. troglodytes separately, it turns out that *P. troglodytes verus,* the western subspecies, is endangered. But the Red Data Book limits breakdown into subspecies because listing of subspecies extends and complicates data to a point where the information becomes counter-productive. A young captive chimpanzee cannot be readily identified through customs or veterinary inspection into any one subspecies. False identification is a common feature in illicit trade operations.

A recent example may serve to illustrate another complicating aspect. *Leontideus,* listed as endangered in the Red Data Book, occurs only in Brazil where its protection receives intensive support through the Brasilian government and the Instituto de Desinvolvimento Florestal[5]. The captive population in the world's zoos numbered 84 in 1970, including 34 captive-born specimens [3]. Golden marmosets continue to be offered on overseas markets with false claims that they are captive-bred. Recently 37 specimens were illegally exported to Paraguay from where they may enter the international trade. They may be distributed to countries which do not require documentation of legal origin (i.e. a Brazilian export permit) or appropriate verification of claims that an animal was bred outside Brazil. Parallel cases involving other genera and species may be complicated further if they occur in several countries where protection laws differ and if their propagation capacity is less well ascertained[6].

IUCN's Draft Convention on the Export, Import and Trade of Certain Species of Wild Animals and Plants[7] proposes a treaty which coordinates national controls on an international platform. It suggests a central authority to monitor the trade on the basis of current information it receives. The Convention distinguishes two categories of wildlife; species which are already threatened by worldwide extinction, and species which are approaching that condition. It excludes specimens which were bred in captivity.

The species are listed in two appendices to the main text. Appendix I defines species to be used only under exceptional circumstances. Appendix II lists species not yet threatened, but the export, import and transit of which must be controlled to avoid exploitation incompatible with survival. All species of primate animals, excluding those listed in Appen-

5 Dr. A. MAGNANINI, Jardin Botanica and Instituto Brasileira de Desinvolvimento Florestal, Rio de Janeiro. A Golden Marmoset Conference was held in Washington in February, 1972. Its results are in preparation. For summary of background, see reference [8] and DUMOND's Comments on minimum requirements of the husbandry of the golden marmoset, in Lab. Prim. Newsl., vol. 10, April 1971.

6 A recent case involved advertisement of 'orangutan twins born and bred in captivity', from Taiwan. Enquiries resulted in photographs showing animals of different sizes betrween 1 and 3 years old, and a 'birth certificate' from a local zoo. The gradual accumulation of stud-books greatly facilitates tracing of information.

7 International Union for Conservation of Nature and Natural Resources, review from February, 1971, distributed internationally to governments through embassies, consulates and commissions from its headquarter at CH-1110 Morges (Switzerland).

dix I are in this category. They are to be marketed under the Convention subject to regulation primarily at the point of origin, through the institution of standard export permits. Permits are to be issued by the national authority only for specimens lawfully captured or, in the case of re-export, after proof of lawful import has been provided. Description and marks must enable the permit to be linked with the animal subject to trade. Copies of all permits are to be submitted to a central authority. For the primate animals listed in Appendix I, two controls are performed. First, the prospective importer has to apply for a standard import permit from a national scientific authority. This authority may grant permits only for adequately justified scientific and educational purposes. The next step is lawful capture on the basis of the import permit, provided the approval by authorities in the country of origin. The animal is then traded with import and export permits attached, copies being passed to the central authority.

It is obvious that the Convention can only succeed if supported by the majority of exporting countries, and if the importing countries, conscious of their responsibility as users, also support it. A large number of states have agreed with the fundamental principle of the Convention. Exporting countries favor the international controls. But in some major importing countries objections persist, mainly because of doubts about the effectiveness of the required difficult legal and administrative mechanisms.

International treaties are slow to come alive. The Convention, which involves complex issues and commands low priorities in human and economic terms, has been in circulation since 1964. Various national mechanisms have proceeded in the meantime, from conservation and public health point of view. Most comprehensive coverage exists to date in the USA. The Endangered Species Act [9] uses Red Data Book guidance combined with the requirement of proof of legal acquisition from the country of origin in accordance with the Lacey Act (18 US Code 43 and 44 as amended) which prohibits the importation of any wildlife that was collected, killed or exported illegally from its country of origin. These national measures provide important relief, and assist exporting countries. But the fact remains that such shipments as cannot now enter the US may be redirected to major European markets where less comprehensive restrictions apply, none from a regional point of view. This would be impossible with an international treaty.

Rationalisation of Usage

Scientists recommend a code of ethics for themselves which requires a judicious choice of species and maximum usage of individual animals [2]. Scientists who take only a limited interest in the continued availability of the species they use, or in nonhuman primates as a whole, can be guided through policies which funding institutions, industries and associations may adopt. Policies which require research proposals to include detailed justifications for the choice of endangered and vulnerable species, and for the numbers of simians to be used, can be beneficial, particularly if combined with a requirement to identify how the acquisition of test animals is compatible with the current status of the species in its country of origin. Policies which restrict acquisition of primate animals to traders who supply proof of legal origin and disclose the methods by which specimens are obtained, could have excellent effects. The requirement that breeding projects provide assurance of long-term maintenance and for the accommodation of surplus, as well as pooling of resources with existing projects, would eliminate damaging fragmentation and short-term concerns.

A proposal which suggests the elimination of one large group of users, the pet owners, appears likely to be most effective. A significant number of primate animals are pets.[8] Most pets are wild-caught and lead short, single lives; a majority are vulnerable New World species. Moreover, uncontrolled private possession, especially in urban and suburban areas, involve hazards to public health which remain unaccounted for in the present system of trade and distribution of imports.

Conservation in Practice

As mentioned previously, conservation requires surveys and baseline research to obtain information needed for preservation of habitat and controlled exploitation of animals for export.

The manpower and techniques for such fieldwork are available, but not funds. In the resource countries prices for wildlife are generally based on overexploitation, rising in accordance with diminishing availability. The levy of dues to pay for the establishment of effective management of the resources would price wildlife out of international markets. Therefore, an investment from outside sources is necessary to make a beginning. Unfortunately, research which enquires into the availability of primate animals takes very low priority. Most research is funded for specialized topics, primarily relevant to human health, where the emphasis is on laboratory

8 The percentage of imports destined for pet markets varies greatly from country to country. It is largest in the US, between 40 and 50%. See footnote 1.

experimentation. Although the argument is accepted that laboratory studies are enriched by data on ecology and behavior of simians in natural habitats, no funds are available for such investigations.

International agencies which support the development in the resource countries are presently oriented towards improvements of established systems of exploitation, such as timber industry and agriculture. Research foundations and pharmaceutical industries do not recognize their own long-term interest in field projects. Therefore, reorientation is necessary to ensure the continued availability of the resource.

References

1 HARRISSON, B.: Conservation of nonhuman primates in 1970; in GOLDSMITH and MOOR-JANKOWSKI Primates in medicine, vol. 5, pp. 1–99 (Karger, Basel 1971).

2 GOLDSMITH, E. I. and MOOR-JANKOWSKI, J. (ed.): Experimental medicine and surgery in primates. Ann. N.Y. Acad. Sci. *162:* 1–704 (1969).

3 PERRY, J.; BRIDGWATER, D. D., and HORSEMEN, D. L.: Captive propagation, a progress report. Zoologica (in press).

4 CROOK, J. H.: The socioecology of primates; in CROOK Social behavior in birds and mammals (Academic Press, London 1970).

5 STRUHSAKER, T. T.: Rainforest conservation in Africa. Primates (in press).

6 US Department of the Interior, Fish and Wildlife Service, Bureau of Sport, Fisheries and Wildlife, vol. WL-498 (1972).

7 International Union for Conservation of Nature and Natural Resources: Red data book. Mammalia, vol. 1, rev. (Arts Graphiques Heliographica, Lausanne 1966).

8 PERRY, J.: The golden lion marmoset. Oryx *11:* 1 (1971).

9 US Department of the Interior, Fish and Wildlife Service: Endangered species conservation. Notice of proposed rule making. Fed. Regist., vol. 35, pp. 6069–6075 (1970).

Author's address: Mrs. BARBARA HARRISSON, Department of Anthropology, McGraw Hall, Cornell University, *Ithaca, NY 14850* (USA)

Medical Primatology 1972. Proc. 3rd Conf. exp. Med. Surg. Primates, Lyon 1972, part I, pp. 15–29 (Karger, Basel 1972)

Development of Biomedical Research in Nonhuman Primates in the United States

W. J. Goodwin

Primate Research Centers, Section Animal Resources, Branch Divisions of Research Resources, National Institutes of Health, Bethesda, Md.

The nonhuman primates, being biologically and behaviorally closer to man than other animals, are becoming important in many areas of biomedical research. The use of these animals in research requires specialized facilities, carefully trained personnel and an adequate number of the appropriate species. This paper will review the development of primate research facilities in the United States and trends in the use of nonhuman primates in biomedical research during recent years.

The first important primate research facility established in the US was the Yerkes Laboratory of Primate Biology established in 1928 by Dr. Robert Yerkes. This laboratory and several other early primate research facilities were devoted primarily to behavioral studies. In addition, primate animals were used in studies of reproductive biology and of infectious diseases such as malaria, yellow fever, tuberculosis, and poliomyelites.

In the late 1940's and early 1950's, several groups of scientists expressed concern regarding the lack of adequate primate research facilities in the United States. The concerned scientists included at least three National Institutes of Health (NIH) advisory groups and the National Academy of Science – National Research Council – Institute of Laboratory Animal Resources (ILAR). The efforts of these groups to establish primate resources were not immediately successfull; however, during the late 1950's a series of events took place that ultimately resulted in the establishment of the Regional Primate Research Centers Program in the early 1960's. These events have been documented in the NIH publication entitled 'Regional Primate Research Centers – The Creation of a Program' [6].

The development of any major Government-supported scientific program in the United States requires the interest and efforts of many individuals including university scientists, Federal scientists and administrators, advisory groups and the Congress. Although a number of individuals played a role in the development of this program, there are two former members of the NIH staff that deserve special recognition. Dr. JAMES WATT, then Director of the National Heart Institute, was largely responsible for the initiation of the Primate Centers program, and Dr. W. H. EYESTONE was the staff member chiefly responsible for establishment and development of the seven Centers.

The National Advisory Heart Council recommended the development of a primate colony on the mainland of the US, near a university, which would serve as a site for long-term multidisciplinary, multiuniversity research programs on cardiovascular diseases. In planning for this center two major issues were raised, namely: (a) Should the proposed primate colony be devoted solely to cardiovascular research, or should its research program be broader to include other biomedical research areas? (b) Should this facility be primarily for research, or for primate breeding to supply a number of investigators in the United States, or for both purposes? The concensus at that time was that the primate facility should have a broad research program and that it should be limited to research, not to the supplying of primate animals for all investigators. The original plan called for the establishment of one large primate center; however, when this plan was considered at higher governmental echelons an alternate plan for the establishment of several regional centers was proposed. In developing this plan two major proposals were considered, namely, (a) the establishment of a central large primate research station plus several smaller subsidiary national research centers to be operated by the Federal Government, and (b) the establishment of a number of regional centers, funded by Federal funds and operated as an integral part of universities. The latter course of action was undertaken.

When the Primate Research Centers were initially established, a number of US scientists opposed the center concept and felt funds should be made available to universities for the establishment of smaller primate resources. The primate centers concept prevailed. At the stage of development of primate research in the early 1960's, large primate centers were desirable to develop the specialized facilities and obtain the necessary highly trained professional and technical personnel. The Primate Centers were instrumental in establishing some of the technology of primate experimentation, in increasing the visibility of primate research, and in introducing many investigators to the use of primates. The existence of the Centers has, in turn, provided the impetus for the establishment of a number of small primate research facilities in the United States and in several other countries.

The Primate Research Centers by their very nature were established on a long-term basis thus involving long-term moral and financial commitments on the part of the Federal Government. The Centers were established by grant funds from the NIH, and their operation is supported by NIH grant funds awarded annually. The grant support is renewed at intervals no greater than every five years by submission of a renewal grant application for consideration by the NIH advisory groups and staff.

The administrative relationship between the Federal Government and the seven host universities is unusual in many respects. Basically, since the Federal Government has a long-term commitment to support the Centers, it must assume a major share of the responsibility for assuring scientific excellence in the Centers. On the other hand, the host universities are also responsible for maintaining scientific excellence in the Centers and are sensitive to interference by Federal officials. The Primate Centers Program was established to meet a national need; therefore, the host universities and the Federal agency consider the operation of the Centers as mutually advantageous and work co-operatively to assure success. This partnership has been quite successful.

For each Center the host university has designated a senior official with overall administrative responsibility and a director responsible to him and for the scientific direction and management of the Center. The director remains in close contact with the responsible NIH staff. A successful director must be a highly competent biomedical scientist and administrator, and must be capable of directing research in several broad areas. Experience has shown that a good director is the key to the operation of a successful center. The scientific staff are appointed by the host institution and usually hold joint appointments in the appropriate academic departments of the university. In addition to the core scientific staff, the research teams of each Center include visiting scientists, collaborative investigators and graduate students. Visiting scientists have formal appointments and usually remain for one or two years. The collaborative investigators work on a more informal basis, usually supported through their home institution, and the length of stay depends primarily on the nature of the research. The supporting staff includes specialists and technicians in many fields such as pathology, veterinary medicine, animal husbandry, clinical chemistry, microbiology, public information, medical illustrations, and administration.

The Primate Centers are playing an ever increasing role in the training of scientists and technicians interested in specific aspects of primate research. The program annually sponsors a number of symposia in selected areas. In addition, visitors from other countries stay at the Centers for varying periods of time in order to pursue their research interests. Thus, the Centers offer a national and international resource for specialized training in the use of primates in biomedical research.

The research program in each Center has developed as a joint effort of the host institution and the NIH. Initially, the research programs reflected the interest and research capabilities of the host institutions, and the Centers were permitted to develop their own research programs without direct influence by the NIH. However, the NIH did insist that high caliber research programs be conducted and reviewed the activities on a regular basis. In the early years, the grants supporting the operation of the Centers funded the majority of the research activities; however, experience has shown that it is more desirable for the Center core scientists to compete for project grant support on an equal basis with other university scientists. As the research programs developed, it became apparent that certain research areas were receiving considerably more attention than others, and that several areas of national importance were not being adequately developed. Currently the NIH, in co-operation with the host universities, is co-ordinating the total program and establishing major research missions for each Center.

The statistics on the personnel, colonies and scientific production of the seven Centers are shown in table I.

A recent publication 'NIH Primate Research Centers – A Major Scientific Resource' [5] presents a complete description of each Center including research programs, administration, facilities, animal colonies and services. This publication is available upon request; therefore, this paper will be limited to a brief discussion of the unique features of each Center.

Table I. Primate research centers statistical data

Scientists with doctoral degree	120
Supporting staff	637
Total core staff	757
Collaborative scientists	334
Visiting scientists	11
Graduate students	172
Scientific papers published	653
Biologic specimens provided for scientists	325
Scientists visiting centers	1,525
Nonhuman primates maintained	7,340

A. Oregon

The major research missions of this Center are reproductive biology, cardiovascular and metabolic diseases, immune diseases, and cutaneous biology. The principal primates species is *Macaca mulatta*. The unique features include (a) an outstanding graphic arts unit, (b) a two-acre enclosure containing a troop of *Macaca fuscata* for behavioral studies, and (c) an excellent colony of prosimians including four species of the genus Lemur.

B. Washington

This Center is particularly noted for its research on the central nervous system control of the cardiovascular system. In addition to the core research program, the Center has a large collaborative research effort in several areas. The unique features include (a) the physical location as an integral part of the health sciences complex of the Washington University, (b) the Primate Information Center which supplies bibliographic information on biomedical research utilizing primate animals, and (c) the field station located at Medical Lake, Washington, specialized in breeding of *Macaca nemestrina* [1]. The principal species are *M. nemestrina* and *Papio anubis*.

C. Delta

This Center is located in a semitropical setting (Louisiana) and contains a number of outdoor areas where primate animals can be housed on a year-round basis. The principal research programs are infectious diseases, nutrition, and reproductive biology. A radiation facility containing a cobalt-60 radiation source is another feature of this Center. The colony contains a large number of species including *Erythrocebus patas, M. mulatta, Saimiri sciureus* and *Pan troglodytes*.

D. Wisconsin

This Center has a long history in behavioral research which continues to be one of its major missions. Reproductive physiology and neurophysiology are other major research areas. The Center has an excellent simian metabolic nursery and a unique holding facility at the local zoo in Madison. *M. mulatta* is the principal species.

E. Yerkes

The original Yerkes primate laboratory, founded in 1928 by Dr. ROBERT YERKES, was located in Florida, and was moved to Atlanta, Georgia in 1965, to become one of the NIH Primate Research Centers. This Center has the largest collection of great apes of any primate facility in the world. The research mission is neurological and behavioral sciences and experimental pathology, including neoplastic and degenerative diseases. In addition to the main facility, a 115-acre field station located 30 miles from Atlanta has a number of outdoor facilities for great apes and monkeys.

F. New England

This Center is located 30 miles from Boston, Massachusetts, and its primary research efforts are in microbiology and pathology. The Center has a relatively small core staff; however, it participates in a large collaborative program involving some 70 investigators in a variety of research areas. The Center is well designed, and laboratories are built on a ten-foot modular system with all utilities on the outside walls. This permits flexibility as research needs develop. The principal species are *M. mulatta, M. fascicularis, Aotus trivirgatus,* and *S. sciureus.*

G. California

This Center was formerly named the National Center for Primate Biology. It was established to conduct studies on primate biology including husbandry, medicine, and baseline studies. The mission of this Center has been changed recently to a broad program in environmental health sciences including perinatal biology, respiratory diseases and oncology. The reasons for this change are as follows: (a) Experience has shown that biological baseline studies on primate species can best be conducted in Centers when these species are used in regular research programs. (b) In recent years a great deal of baseline data has been accumulated in the Primate Research

Centers and in other major primate laboratories in this country. (c) The Center, in co-operation with the host university, has the capability of developing an outstanding program in the environmental health sciences.

Primate Laboratories

A number of major primate research laboratories have been established in the United States during the past decade. They are supported in a variety of ways; however, encouragement and financial support from several Federal agencies have contributed significantly to their development. The primate research laboratories vary in size, research programs, and administration. Some of the larger laboratories are listed in tables II–IV. In addition, the pharmaceutical and chemical industries maintain large numbers of primate animals for use in their research and development programs.

A long-established primate laboratory is the Caribbean Primate Research Center. Its interesting history began in 1938 with the release of a number of *M. mulatta* onto the island of Cayo Santiago, off the southeast

Table II. Multidisciplinary primate laboratories in the US 1972

Laboratory	Research areas	Primate species
University		
Caribbean Primate Research Center, University of Puerto Rico, San Juan, P.R.	behavior and reproductive biology	890 *M. mulatta*
Holloman Primate Facility, Holloman AFB, N.M.	environmental toxicology	300 *M. mulatta* 95 *P. troglodytes*
Laboratory for Experimental Medicine and Surgery in Primates, New York University, Sterling Forest, N.Y.	immunogenetics, reproductive biology and experimental surgery	180 *Papio* spp. 100 *M. mulatta* 75 *P. troglodytes*
Nonprofit Foundations		
Southwest Foundation for Research and Education, San Antonio, Tex.	biochemistry, endocrinology and microbiology	619 *Papio* spp. 84 *Saquinus* spp. 40 *P. troglodytes*
Gulf South Research Foundation, New Iberia, La.	virology and toxicology	200 *M. mulatta* 200 *S. sciureus*
Commercial		
Bionetics Research Laboratories Inc., Kensington, Md.	virology and toxicology	2,100 *M. mulatta* 352 *M. fascicularis*
Hazleton Laboratories Inc., Falls Church, Va.	oncology and toxicology	525 *M. mulatta* 425 *M. fascicularis*

coast of Puerto Rico. These free-ranging troops of *M. mulatta* have been maintained through the years and are used primarily for behavioral observations. Additional facilities include two islands off the southwest coast where monkeys can be studied under corral and free-ranging conditions, and the central facility near San Juan, which will probably be used for the production of certain New World species. The responsibility for this Center has recently been transferred to the Division of Research Resources, and closer co-ordination between this Center and the Primate Research Centers is anticipated.

The Holloman Primate Laboratory was formerly operated by the US Air Force and is located at Holloman Air Force Base near Alamogordo, New Mexico. The facilities include seven buildings with 69,000 ft^2 of space on 37 acres. Albany Medical College assumed responsibility for the operation of this laboratory on July 1, 1971, with the basic support being provided by an NIH contract administered by the Division of Research Resources. Research programs in environmental toxicology, reproductive biology, and

Table III. Specialized primate laboratories in the US 1972

Laboratory	Research areas	Primate species
Department of Physiology, University of Pittsburgh Medical School, Pittsburgh, Pa.	reproductive physiology	220 *M. mulatta*
Division of Reproductive Physiology, University of Pennsylvania Medical School, Philadelphia, Pa.	reproductive physiology	200 *M. mulatta*
Population Council, Rockefeller University, New York, N.Y.	reproductive biology	400 *M. mulatta*
Department of Laboratory Animal Medicine, Bowman Gray Medical School, Winston Salem, N.C.	cardiovascular diseases	600 *S. sciureus*
Dental Sciences Institute, University of Texas, Houston, Tex.	dental research	400 *Saguinus* spp.
Departments of Medicine and Microbiology, Rush Presbyterian-St. Luke Medical Center, Chicago, Ill.	viral oncology	400 *Saguinus* spp.
Oak Ridge Association University, Oak Ridge, Tenn.	immunology	500 *Saguinus* spp.

behavioral sciences are currently underway. This laboratory is well equipped for research activities and currently maintains approximately 100 chimpanzees in its colony.

Another major university-sponsored primate laboratory is the Laboratory for Experimental Medicine and Surgery in Primates (LEMSIP). This laboratory is administered by New York University Medical School; however, it is sponsored by Associated Medical Schools of New York and New Jersey and provides primate research facilities and expert services for the 13 affiliated medical schools and a number of investigators from other institutions. This primate facility consists of some 30 prefabricated laboratory modules $50 \times 10 \times 7.5'$ that house the animal colony and supporting services. It is currently divided into two compounds, one for chimpanzees (5,000 ft^2) and one for other species (12,000 ft^2), each compound being self-supporting with respect to staff and essential facilities. There are currently some 120

Table IV. Primate research laboratories – US Government 1972

Agency	Research areas	Primate genera
National Institutes of Health, Bethesda, Md.	general and biomedical	9,500 of many species
Center for Disease Control, Atlanta, Ga. and Phoenix, Ariz.	malaria and hepatitis	895 *P. troglodytes, Macaca* spp., *Saguinus* spp. *A. trivirgatus*
Environmental Protection Administration, Perrine, Fla.	toxicology	320 *Macaca* spp., *S. sciureus*
Walter Reed Army Institute of Research, Washington, D.C.	neural sciences and infectious diseases	865 *Macaca* spp., *Papio* spp., *A. trivirgatus, S. sciureus*
Medical Laboratory, Edgewood, Arsenal, Md.	toxicology and behavior	365 *Macaca* spp., *S. sciureus*
Aerospace Medical Laboratory Wright-Patterson AFB, Ohio	toxicology and preventive medicine	265 *Macaca* spp., *Papio* spp.
School of Aerospace Medicine, Brooks AFB, Tex.	physiology and radiobiology	1,330 *Macaca* spp.
Naval Medical Institute, Bethesda, Md.	immunology and dental	150 *Macaca* spp., *Papio* spp., *S. sciureus*
Naval Aerospace Medical Research Laboratories, Pensacola, Fla.	behavior	95 *Macaca* spp., *S. sciureus*
Naval Biomedical Research Laboratories, Oakland, Calif.	immunology and microbiology	90 *Macaca* spp., *Cercopithecus* spp.
Armed Forces Institute of Pathology, Washington, D.C.	experimental surgery	80 *Macaca* spp.

active research projects in a number of research areas underway, using primarily the baboon and the chimpanzee.

Another type of multidisciplinary primate laboratory is the Southwest Foundation for Research and Education (SFRE) located in San Antonio, Texas. This nonprofit research foundation became operational in 1950 and is supported by endowment funds and Federal grants and contracts. This is an extensive facility consisting of a number of research buildings and animal quarters. A unique feature of this laboratory is the outdoor facilities for housing baboons. This laboratory, known as the 'baboon center' in the US, has conducted extensive studies on the biologic profiles of the baboon and has been responsible for much of the information now available on this species. This laboratory has maintained a trapping facility in Kenya for a number of years where animals are trapped, culled and conditioned prior to shipment to the US. The research program at SFRE is divided into biochemistry, endocrinology and microbiology. This laboratory also administers the Simian Virus Reference Center sponsored by WHO and supported by NIH.

Bionetics Research Laboratories, Inc., is a private commercial research and development laboratory established in 1961 and located in Maryland. This is an example of an independent laboratory being developed to conduct biomedical research under contract support from government and industry. The laboratory has a staff of some 25 scientists with expertise in virology, toxicology, immunology, and primate medicine. This is one of the largest primate breeding facilities in the world. During the past ten years, Bionetics has produced some 4,000 newborn of several species, including *M. mulatta, Cercopithecus aethiops,* and *Papio papio.* There are currently some 2,100 primate animals housed in this laboratory.

Universities with biomedical research programs in the United States usually maintain one or more laboratory animal resources. In nearly all cases nonhuman primates can be found in these colonies. Accurate statistics on the numbers of currently used are difficult to obtain; however, in 1967 a total of 193 university animal resources reported a daily inventory of 10,300 simians being used in research.

There are a large number of government laboratories using primate animals in their research programs. The NIH uses more nonhuman primates in research than any other governmental agency. A survey conducted in 1970 indicated a total of some 9,300 simians being used by over 60 investigators representing essentially all aspects of biomedical research. Animal facilities at the NIH house approximately 2,400 of these, with the remaining 6,900 being housed under contracts by commercial concerns.

Trends in Use of Primates in Research

The use of nonhuman primates in biomedical research in the US has increased rapidly during the past decade. Currently they are being used in essentially all major biomedical research areas and are increasingly being identified as the only suitable animal model in many research projects. This diversity of the use is well documented by the papers listed in the Current Primate References published by the Primate Information Center of the Washington University Primate Research Center. An examination of the research papers presented at major scientific meetings also indicates the increasing emphasis on the use of these animals in many areas of biomedi-research. It is not feasible to discuss each of these areas; however, certain major areas such as neurophysiology and behavior, reproductive biology, and infectious diseases bear mentioning.

The earlier behavioral research involved naturalistic studies of simians in their native habitat and social studies under semicontrolled conditions. These observational studies were directed toward understanding the mental capabilities of the animals and their social interactions as a basis for better understanding human behavioral characteristics. More recently, the perfecting of specialized techniques in neuroanatomy, neurophysiology, and neurochemistry have made it possible to relate brain function more directly with behavior. In addition, studies on the relationship between endocrine functions and behavior are yielding promising results. Hopefully, these modern techniques will have direct application to the understanding of neurological processes underlying the mental health problems in man.

The similarity between the reproductive system of the nonhuman primate and man has resulted in ever increasing use of primate animals in reproductive biology studies. Their use in reproductive biology would be even more widespread if sufficient numbers of well-defined animals were available. Currently there is an urgent need for pregnant, fetal and newborn animals. The NIH is supporting a number of breeding colonies for use in reproductive biology research, and it is anticipated that additional ones will be established. In the future one would expect nonhuman primates to be increasingly used in the following research areas: gametogenesis, maturation, capacitation and fertilization, implantation and early development, hormonal control, reproductive behavior, ovulation control, placental function, intrauterine devices, and congenital anomalies.

Primate animals have proven to be excellent research models in the study of many infectious diseases. The classical studies on yellow fever, polio, and tuberculosis are well known and proved the usefulness of simians in microbiology. More recently New World monkeys have been used extensively for malaria chemotherapy studies. The owl monkey can be infected with *Plasmodium vivax,* and this parasite can now be maintained indefinitely in this species. In addition the Gorgas Memorial Laboratory in Panama has found the squirrel monkey and titi marmoset are also proving to be good hosts for vivax malaria. Therefore, the use of nonhuman primates as models to study human malaria appears increasingly promising. The historical development of the use of these animals in viral research has recently been reviewed by some authors in the book 'Pathology of Simian Primates' [2]. Currently, increasing emphasis is being placed on the importance of simians in viral oncology, infectious hepatitis, and neurological disorders caused by 'slow' viruses.

In order to illustrate the trends in the numbers of nonhuman primates used in the United States, the annual laboratory animal surveys conducted by the Institute of Laboratory Animal Resources [4] have been summarized in table V. The total numbers used increased rapidly until 1968 and since have decreased gradually. The increase until 1968 was due to the increasing recognition of the value of these animals in biomedical research, to the

Table V. Trends in the use of primates in the US 1966–1970

Primates	1966	1967	1968	1969	1970
New World	8,731	12,398	32,683	16,532	15,282
Cebidae	8,062	11,619	28,825	14,117	12,480
Callithrichidae	669	779	3,858	2,415	2,802
Old World	50,105	47,256	50,004	46,748	37,161
Cercopithecus spp.	7,522	7,200	5,262	5,852	3,017
M. mulatta	28,307	30,620	37,290	34,665	27,092
M. fascicularis	12,044	6,565	4,372	3,836	3,487
M. nemestrina	264	337	603	353	830
M. arctoides	557	679	1,174	486	863
Papio spp.	1,158	1,472	985	1,254	1,651
Pan spp.	253	383	378	302	221
Total[1]	62,783	70,416	85,283	68,002	54,437[2]
Questionnaires returned	1,442	1,366	1,493	2,258	1,523
Percent returned	61	53	63	69	49

1 Includes undesignated species.
2 Only 49% of questionnaires returned.

establishment of Primate Research Centers and primate laboratories, and to the increase of Federal funds for biomedical research. The subsequent decrease since 1968 is probably related to changes in commercial production of poliomyelitis vaccine and to the decrease in research funds, but also to better utilization of animals by investigators. In spite of the decrease in the numbers of simians used in research, the number of research projects using nonhuman primates has increased significantly. The number of projects reported by the Science Information Exchange increased from 666, in 1965, to 951 in 1968 [3]. The Science Information Exchange reports that in 1971 simians were employed in 1183 research projects. The major areas of research were reproductive biology, behavioral and neurological research, and infectious diseases. As one would expect, the rhesus monkey was the species most commonly used.

The decrease in research funds has probably resulted in better use being made of simians in general. It has encouraged investigators to make optimum use of these expensive animals and has helped to prevent their use in terminal research where other species would suffice. Moreover, investigators realize that the supply of simians from the wild is becoming a critical problem and are attempting to conserve them whenever possible.

An examination of table V reveals some interesting trends in the research usage of certain species. The large increase in the use of Cebidae in 1969 is probably due to importation of owl monkeys for malaria research and the squirrel monkey for viral and cardiovascular studies. The Callitrichidae likewise increased significantly in 1968 because of increased needs for marmosets mainly in viral oncology, dental research and immunology.

In reviewing the trends in the use of Old World monkeys, the rhesus is still by far the species most used in research. It is interesting to note, however, that some of the other macaque species, namely *M. arctoides, M. nemestrina* and *M. radiata,* may be gaining in popularity. These macaque species have been found to be better models than the rhesus for a number of research purposes and are slowly replacing it in a number of laboratories.

There has been a noticeable decrease in the number of Cercopithecus, probably due to the Marburg virus incident, and because of the decrease in demand for monkey kidneys to be used for cell culture.

In recent years the use of baboons has increased. In many ways the baboon is bridging the gap between the rhesus monkey and the greater apes. As the chimpanzee and other great apes become more difficult to obtain, one would expect even greater usage of baboons especially for research involving surgical procedures.

The availability of nonhuman primates for research needs has been discussed on numerous occasions. It is generally agreed that the establishment of breeding colonies for a number of important species will be essential within the next ten years in order to provide simians required for research in the United States. Currently the Primate Research Centers are breeding many of the animals used in their research programs, and several Centers are rapidly becoming self-sufficient in this respect. Recently the NIH implemented a contract program to determine the commercial feasibility of producing rhesus monkeys in outdoor environments.

A recent survey conducted by THORINGTON [7] listed 125 breeding colonies in the US with a total of 10,293 animals and a production of 2,271 live births during 1970. These figures are encouraging; however, this is a very small percentage of the number required for research use.

In my opinion it is essential that more breeding colonies be established, not only in the United States, but in all countries using significant numbers of nonhuman primates in research. The feasibility of establishing breeding colonies in the countries of origin should also be carefully investigated.

Another critical factor that needs to be carefully examined and controlled ist the use of simians for other than research purposes. THORINGTON [8] has compared the numbers of New World primates imported into the US with the numbers used in research. In 1969 he found that only 24% of the approximately 72,000 New World primates imported were used for biomedical purposes. The remainder can be accounted for largely by the pet trade, and mortality in dealer compounds.

Since the research use of various species is discussed elsewhere in this volume, it may be of interest to present data on the species most commonly used in the US during 1970. The statistics from the 1970 ILAR survey reveal the following, in decreasing order of number maintained: *M. mulatta,* 23,000; *S. sciureus,* 3,941; *M. fascicularis,* 2,689; *P. cynocephalus,* 1,535; *Saginus nigricollis*, 1,359; *M. nemestrina,* 1,327; and *A. trivirgatus*, 1,061. Other species used in significant numbers were *M. arctoides, Saguinus mystax, Galago crassicaudatus, P. troglodytes* and *M. radiata.*

It is anticipated that new biomedical research areas requiring nonhuman primates as models will continue to be identified and the numbers required to satisfy the research needs in the United States will increase annually. As the supply of primate animals from the countries of origin will continue to decrease, the establishing of breeding colonies will become essential in order to provide sufficient animals for research purposes.

References

1 BLAKLEY, G. A.; MORTON, W. R., and SMITH, O. A.: Husbandry and breeding of *Macaca nemestrina;* in GOLDSMITH and MOOR-JANKOWSKI Medical primatology 1972, part I, pp. 61–72 (Karger, Basel 1972).
2 FIENNES, R. N. T-W (ed.): Infectious and parasitic diseases; in Pathology of simian primates, vol. 2, pp. 382–747 (Karger, Basel 1972).
3 GOODWIN, W. J.: Current use of primates in biomedical research. Lab. anim. Sci. *20:* 329–332 (1970).
4 Institute of Laboratory Animal Resources. ILAR News *12/1* (1968), *13/1* (1969), *14/1* (1970), *15/1* (1971).
5 National Institutes of Health Primate Research Centers – A major scientific resource (US Department of Health, Education and Welfare, NIH, Bethesda, Md. 1972).
6 Regional Primate Research Centers – The creation of a program (US Department of Health, Education and Welfare, NIH, Bethesda, Md. 1968).
7 THORINGTON, R. W.: Survey of nonhuman primates being maintained on January 1, 1971. ILAR News *15/1* (1971).
8 THORINGTON, R. W.: Importation, breeding, and mortality of New World primates in the United States. Int. Zool. Yb. (in press).

Author's address: WILLIAM J. GOODWIN, PhD, Chief, Primate Research Centers, Section Animal Resources, Branch Division of Research Resources, National Institute of Health, *Bethesda, MD* (USA)

Medical Primatology 1972. Proc. 3rd Conf. exp. Med. Surg. Primates, Lyon 1972, part I, pp. 30–36 (Karger, Basel 1972)

Primate Animals in Medical Research

Recent Achievements and the Economics of Experimentation

J. MOOR-JANKOWSKI[1] and E. I. GOLDSMITH

Laboratory for Experimental Medicine and Surgery in Primates, New York University School of Medicine and Department of Surgery, Cornell University Medical College, New York, N.Y.

The unique importance of nonhuman primates for medical research has been proved most strikingly to our generation by the virtual eradication of poliomyelitis by vaccines prepared in monkey tissues and controlled through bioassays in monkeys. The initial discovery which, ultimately, led to the development of the vaccines now in use, was made as early as 1909 when LANDSTEINER and POPPER demonstrated that poliomyelitis can be transmitted from man to monkeys and apes. LANDSTEINER's desire to expand and continue this investigation was frustrated, however, when the high costs of primate animals precluded their use in his studies. Thus polio vaccine was produced half a century later when financial support became finally available to provide monkeys for the studies of ENDERS, SALK and SABIN.

Another important advance directly related to the use of monkeys occurred in 1937, when experiments with rhesus monkey blood led to the discovery of the Rh factor by LANDSTEINER and WIENER. This work opened completely new horizons in preventive perinatal medicine, in understanding of human blood grouping and of human genetics in general, and resulted in spectacular expansion of blood transfusion therapy which, in turn, made possible many of the achievements of modern surgery.

Despite these great scientific and therapeutic advances resulting directly from the use of nonhuman primates, and despite the obviousness of their value for medical experimentation to approximate the anatomy, physiology and pathology of man, the availability of apes and monkeys for experimental use is still severely restricted in most countries. The United States, where several primate facilities have been established during the last decade and,

1 Supported in part by grant RR 00316.

more recently, Holland, represent notable exceptions. In Soviet Russia a primate center has existed since 1927, but we know little about the availability of simians for experimentation in that country. It should be noted that in Russia dogs preceded man in the space flight, while simians were used in the United States. It is interesting to speculate on how much the rapid advances of the American manned space program owe to simian experimentation which has accompanied all its major developments.

However, it is in the literature survey by our Russian colleague, Dr. FRIDMAN from the Sukhumi Primate Information Center that we find the most striking proof of the importance and of the rapid growth in the use of simians in biomedical research [8]. His data demonstrate that while the number of publications in medicine and biology has doubled in the last 12–15 years, biomedical publications on primates have grown five times during the seven years from 1964 to 1971.

In view of such demonstrable success, the question arises why primate animals are not more accessible to scientists, especially in countries other than the USA and Holland. What are the underlying reasons for the delays in plans to create primate facilities in several countries despite ambitious plans drawn several years ago [2, 24, 26, 31]?

We believe that the reasons are twofold:

(1) insufficient dissemination of information on the importance of recent scientific progress and on advances in clinical medicine achieved by experimentation in simians;

(2) misconception that simians are too expensive for medical experimentation. This presentation intends to deal with both these problems.

Recent Progress in Medical Primatology

It is not possible to review all the important findings reported in what FRIDMAN [8] estimates to have been 5,000 publications in 1971 alone. Volumes resulting from the Conferences on Experimental Medicine and Surgery [11, 12] are intended to provide the more detailed surveys at two- or three-year intervals. Therefore, we shall limit ourselves here to pointing out the highlights of the most important achievements.

(1) Similarity of cellular antigens, serum components and organ functions of man and of apes and monkeys has been used to prolong patients' lives for many months by kidney heterotransplantation. The success obtained indicates the possibility of transplantation therapy by replacing human organs

with those of simians, once the exact histocompatibility pattern is determined [1, 29].

The discovery in simians of blood groups homologous (A-B-O, M-N, Rh-system) or analogous to those of man [28] underscored their similarity to man and opened new areas for experimentation. The observation that simians can survive when completely exchanged with compatible human blood (while nonprimate animals die during the exchange procedure) has led to the cross-circulation procedure for patients in hepatic failure [13].

(2) Only nonhuman primates share with man viruses which cannot be transmitted to other animals [7]. The exclusive use of monkeys for preparation and testing of polio vaccine has already been mentioned. Recently, the existence in man of the so-called 'slow virus' disease was for the first time demonstrated by experimental transmission to chimpanzees of the fatal neuropathy 'kuru' of New Guinea aborigines [9]. Many consider cancer, arthritis, multiple sclerosis and several other incurable human diseases to be caused by slow viruses; investigations are now being carried out on transmission of these diseases from human patients to monkeys and apes.

(3) Recently, the Australia antigen, associated with the virus of serum hepatitis of man, has been found in chimpanzees, orangutans and gibbons [21, 25, 30]. As a result, chimpanzees are now being used in large-scale studies of epidemiology, pathology, diagnosis and therapy of this disease.

(4)The chimpanzee is the only animal in which naturally occurring streptococcal pharyngitis has been found [32]. A similar condition, though not naturally occurring, may be induced in baboons [32], but not in nonprimate animals.

(5) Primates are the only animals with menstrual cycles; thus they are particularly suitable for reproduction studies such as the testing of fertility and antifertility drugs and methods applicable to man. The similarity of the fetoplacental anatomy, physiology and of intrauterine development allows for surgical experimentation in conditions shared with man by nonhuman primates only [14–16]. These studies have led, among others, to the development of the intra-uterine transfusion technique now used in erythroblastotic human fetuses.

In teratological studies, thalidomide given to most pregnant nonprimate animals has had no effect, but caused pregnant rhesus monkeys [6] and baboons [20] to produce malformed offspring. In another series of experiments, compression of the umbilical cord was shown to cause in newborn monkeys a syndrome very similar to cerebral palsy in man [34].

(6) So far two genetically transmitted diseases of man, not found in other

species, have been demonstrated in the chimpanzee: mongolism (Down's syndrome) [27] and alcaptonuria [33]. Epilepsy has been observed in baboons [23].

Certain surgical experimental procedures, particularly those involving the heart and the lungs, and those dealing with the liver, cannot for anatomical and physiological reasons be performed successfully in the formerly standard experimental animal, the dog [17, 19, 22]. Therefore, surgeons are increasingly moving to the use of primate animals, especially since blood grouping information on several primate species is by now far more nearly complete than that on dogs, and permits extracorporeal circulation using the heart-lung machine primed with compatible blood.

(7) Another rewarding aspect of the use of primates in medical experimentation is the speed with which experimental findings made in simians may be applied to diagnosis or therapy in human patients. The method of surgical removal of *Schistosoma mansoni* was introduced in Brazil only six months after the initial experiment was performed in a baboon [10].

The examples listed above are those which appear to us as the most spectacular and in the forefront of biomedical experimentation. However, statistics of animals imported and maintained in the US indicate that by far the largest number of primate animals is used in direct testing of newly developed procedures and products by the pharmaceutical industry [18].

Economics of Medical Experimentation in Simians

It has been mentioned previously that, because of their high cost, simians were not available to LANDSTEINER to pursue his poliomyelitis studies in the early nineteen-hundreds. Had his brilliant mind been allowed the opportunity to continue his work, the polio vaccine might have been developed earlier and billions of dollars would not have had to be spent for treatment, rehabilitation and chronic care of victims of the disease.

Despite these considerations, simians are still widely reputed to be too expensive for medical experimentation. The unit price of a monkey or ape is compared to that of a rabbit, rat or mouse, without evaluating the final cost-effectiveness of a study from the viewpoint of similarity of man and simians and man's dissimilarity from rabbits, rats and mice. Between LANDSTEINER's induction of polio in primate animals in 1909 and ENDERS' observation of polio virus growth in monkey tissue cells in 1945, vast sums were expended in vain on polio research, using inappropriate species of animals [18].

Another common mistake which results in considering research in primates to be especially expensive is the computation of their costs when maintained under conditions designed for other laboratory animals or for zoological gardens. Much lower costs are achieved by using modern methods of simian husbandry pioneered by laboratories aimed specifically at the needs of biomedical experimentation and applying labor-saving management techniques as well as appropriate accounting techniques [3–5]. Under such conditions, the costs of maintaining medium-sized primate animals [baboons and rhesus monkeys) compares favorably with those of a well-maintained laboratory dog [17]. This is due on the one hand to the lower cost of a rationally maintained simian and, on the other hand, to the increase in the price of dog experimentation brought about by the relatively recent understanding that all animal experimentation is valid only if performed in healthy and well-maintained animals. This has ended the practice of bringing the 'cheap' unexamined and untreated stray dog directly from the pound to the laboratory and has, at the same time, increased the expense of the experimentation in dogs to the level equal to or exceeding that in simians.

We have chosen as an example for this comparison the situation in the animal house of the New York Hospital-Cornell Medical Center which employs the most modern husbandry techniques for dogs as well as the LEMSIP 'dry method' for simians. At this writing the *per diem* charge for rhesus monkeys is $1.10, for baboons $1.40 and for dogs $1.75 [EWALD, personal commun.] In New York, the purchase price for a conditioned mature mongrel dog, wormed, vaccinated and observed for two weeks, is $70, a beagle may cost $100 and a foxhound $200. On the other hand a young female rhesus monkey costs $70 and a baboon $130. Thus, the actual cost of experimentation in a rhesus monkey, which is usually purchased after a two-week dealer's quarantine, becomes cheaper than in a dog after the first day spent in the animal house, while the cost for a baboon becomes equal to that for a dog after one year of experimentation, and less expensive thereafter.

In those European countries where dog prices tend to be higher than in the United States [18], experimentation in simians, if properly managed, should prove to have an additional economic advantage.

References

1 BALNER, H. and ROOD, J. J. VAN (guest ed.): Proc. Workshop and Symposium on Transplantation Genetics of Primates, Rijswijk 1971. Transplant. Proc. *4:* 1–147 (1972).

2 BOIRON, M.: Remarks about the French national primate center; in GOLDSMITH and MOOR-JANKOWSKI Experimental medicine and surgery in primates. Ann. N.Y. Acad. Sci. *162:* 390 (1969).

3 BOORMAN, G. A.; SPELTIE, T.; GOOSEN, C., and BALNER, H.: Chimpanzee breeding and management at the primate center TNO; in GOLDSMITH and MOOR-JANKOWSKI Medical primatology 1972, part I, pp. 73–80 (Karger, Basel 1972).

4 BOORMAN, G. A.; SPELTIE, T.; GOODEN, C., and BALNER, H.: Rhesus monkey breeding husbandry and cost accounting at the primate center TNO; in GOLDSMITH and MOOR-JANKOWSKI Medical primatology 1972, part I, pp. 73–80 (Karger, Basel 1972).

5 DAVIS, J. H. and FITZGERALD, T. A.: Fiscal management of a primate facility for biomedical research; in GOLDSMITH and MOOR-JANKOWSKI Medical primatology 1972, part I, pp. 37–49 (Karger, Basel 1972).

6 DELAHUNT, C. S. and LASSEN, J.: Thalidomide syndrome in monkeys. Science *146:* 1300–1305 (1964).

7 FIENNES, R. N., T-W (ed.): Pathology of simian primates. Viruses and small organisms of indeterminate position, vol. 2, pp. 382–747 (Karger, Basel 1972).

8 FRIDMAN, E. P.: Recent developments of biomedical research in primates with a short historical review. J. med. Prim. *1:* 220–229 (1972).

9 GAJDUSEK, D. C. and ALPERS, M. P.: Bibliography of kuru (National Institute of Neurological Diseases and Stroke, NIH, Bethesda, Md. 1970).

10 GOLDSMITH, E. I. and KEAN, B. H.: Surgical treatment of schistosomiasis by extracorporeal hemofiltration in baboons and in man: *S. mansoni* and *S. haematobium;* in GOLDSMITH and MOOR-JANKOWSKI Experimental medicine and surgery in primates. Ann. N.Y. Acad. Sci. *162:* 453–458 (1969).

11 GOLDSMITH. E. I. and MOOR-JANKOWSKI (ed.): Experimental medicine and surgery in primates. Ann. N.Y. Acad. Sci. *162* (1969).

12 GOLDSMITH, E. I. and MOOR-JANKOWSKI, J. (ed.): Medical primatology 1970 (Karger, Basel 1971).

13 GOLDSMITH, E. I. and MOOR-JANKOWSKI, J. (ed.): Cross-circulation between man and simians, Medical primatology 1970, pp. 52–91 (Karger, Basel 1971).

14 GOLDSMITH, E.I. and MOOR-JANKOWSKI, J. (ed.): Experimental medicine and surgery in primates. Contribution of nonhuman primates to the study of perinatal problems, gynecology and obstetrics; Ann. N.Y. Acad. Sci. *162:* 225–300 (1969).

15 GOLDSMITH, E.I. and MOOR-JANKOWSKI, J. (ed.): Medical primatology 1970. Reproduction, perinatal growth and development studies; pp. 472–645 (Karger, Basel 1971).

16 GOLDSMITH, E.I. and MOOR-JANKOWSKI, J. (ed.): Medical primatology 1972. Reproduction and perinatal studies, art I, pp. 202–324 (Karger, Basel 1972).

17 GOLDSMITH, E. I.: Surgical experiences with nonhuman primate animals 1963–1971. Proc. National Conf. on Research Animals in Medicine, Washington, D.C. 1972 (in press).

18 GOLDSMITH, E. I. and MOOR-JANKOWSKI, J.: Report on the projected Israeli laboratory for production of conditioned nonhuman primates for science (Life Sciences Division of the National Council for Research and Development, Prime Minister's Office, Jerusalem 1972).

19 HAGLIN, J.; TELANDER, R. L., and MUZZAL, R. E.: Comparison of lung autotransplantation in the primate and dog. Surg. Forum *14:* 196–198 (1963).

20 HENDRICKX, A. G.; AXELROD, L. R., and CLAYBORN, L. D.: Thalidomide syndrome in baboons. Nature, Lond. *210:* 958–959 (1966).

21 HIRSCHMAN, R. J.; SHULMAN, H. R.; BARKER, L. F., and SMITH, K. O.: Virus-like particles in sera of patients with infectious and serum hepatitis. J. amer. med. Ass. *208:* 1667–1670 (1969).

22 HITCHCOCK, C. R.: Experimental surgery in primates and in standard laboratory animals. A comparative survey; in GOLDSMITH and MOOR-JANKOWSKI Experimental medicine and surgery in primates. Ann. N.Y. Acad. Sci. *162:* 393–403 (1969).

23 KILLAM, K. F.: Experimental epilepsy in primates; in GOLDSMITH and MOOR-JANKOWSKI Experimental medicine and surgery in primates. Ann. N.Y. Acad. Sci. *162:* 610–616 (1969).

24 KUHN, H.-J.: Denkschrift zur Situation der Primatenforschung (Steiner, Wiesbaden 1970).

25 LICHTER, E. A.: Chimpanzee antibodies to Australia antigen. Nature, Lond. *224:* 810–811 (1969).

26 MAHOUY, G.; FRIEDMANN, J. C., and MENDELSSON, R.: Architectural conception and building design of the French primate center; in GOLDSMITH and MOOR-JANKOWSKI Medical primatology 1970, pp. 961–969 (Karger, Basel 1971).

27 MCCLURE, H. M.; BELDEN, K. H.; PIEPER, W. A.; JACOBSON, C. B., and PICCIANO, D.: Cytogenetic studies and observations in the Yerkes great ape colony; in GOLDSMITH and MOOR-JANKOWSKI Medical primatology 1970, pp. 281–296 (Karger, Basel 1971).

28 MOOR-JANKOWSKI, J. and WIENER, A. S.: Red cell antigens of primates; in FIENNES Pathology of primates, pp. 270–317 (Karger, Basel 1972).

29 MURPHY, G. P.: Transplantation in primates; in GOLDSMITH and MOOR-JANKOWSKI Primates in medicine, vol. 7, (Karger, Basel 1972).

30 PRINCE, A. M.: Hepatitis associated antigen. Long-term persistence in chimpanzees; in GOLDSMITH and MOOR-JANKOWSKI Medical primatology 1970, pp. 731–739 (Karger, Basel 1971).

31 SPIEGEL, A.: Use of primate animals in biological and medical research in Germany; in GOLDSMITH and MOOR-JANKOWSKI Experimental medicine and surgery in primates. Ann. N.Y. Acad. Sci. *162:* 391 (1969).

32 TARANTA, A.; GOLDSTEIN, G.; SPAGNUOLO, M.; DAVIDSON, M., and UHR, J.: Experimental streptococcal infections in nonhuman primates; in GOLDSMITH and MOOR-JANKOWSKI Medical primatology 1970, pp. 748–760 (Karger, Basel 1971).

33 WATKINS, S.; BINLEY, H., and SHULMAN, N. R.: Alkaptonuria in a chimpanzee; in GOLDSMITH and MOOR-JANKOWSKI Medical primatology 1970, pp. 297–299 (Karger, Basel 1971).

34 WINDLE, W. F.: Brain damage at birth. Functional and structural modifications with time. J. amer. med. Ass. *206:* 1967–1972 (1968).

Authors' addresses: Dr. J. MOOR-JANKOWSKI, Laboratory for Experimental Medicine and Surgery in Primates, New York University Medical Center, 550 First Avenue, *New York, NY 10016;* and Dr. E.I. GOLDSMITH, The New York Hospital, Cornell Medical Center, 525 East 68th Street, Room 2006, *New York, NY 10021* (USA)

Medical Primatology 1972. Proc. 3rd Conf. exp. Med. Surg. Primates, Lyon 1972, part I, pp. 37–49 (Karger, Basel 1972)

Fiscal Management of a Primate Facility for Biomedical Research

J. DAVIS and T. FITZGERALD

Laboratory for Experimental Medicine and Surgery in Primates and Controller's Division, New York University Medical Center, New York, N.Y.

This paper is concerned with fiscal and cost-accounting methods developed for efficient management of an interinstitutional laboratory for biomedical experimentation in primate animals. The management techniques developed by us over several years may be useful to other facilities serving a single institution. A short description will present the administrative characteristics of our laboratory which served as the model in this presentation.

The Laboratory for Experimental Medicine and Surgery in Primates (LEMSIP) was originally established within New York University School of Medicine to accommodate the immunological studies of Drs. J. MOOR-JANKOWSKI and A. S. WIENER who shared their resources with several other investigators. Their sharing of animals, know-how and facilities provided a stimulus for the growth of LEMSIP which, in 1967, was transformed into a community resource and obtained sponsorship of the Associated Medical Schools of Greater New York.[1] At present, LEMSIP serves as a regional primate animal resource for research and education for the New York-New Jersey region and is also used, though to a lesser extent, by investigators from other areas of the United States and from abroad. At this time, 126 ongoing research projects are accommodated at LEMSIP.

Broad policy guidelines for LEMSIP are provided by the Associated Medical Schools of New York and New Jersey; however, daily administrative, fiscal and accounting management functions under the jurisdiction of New York University School of Medicine (see table I). Administrative efforts are geared toward providing optimum support for the work of

1 Since 1971, Associated Medical Schools of New York and New Jersey, Inc., including 13 medical schools in both states.

Table I. Community organization chart[1]

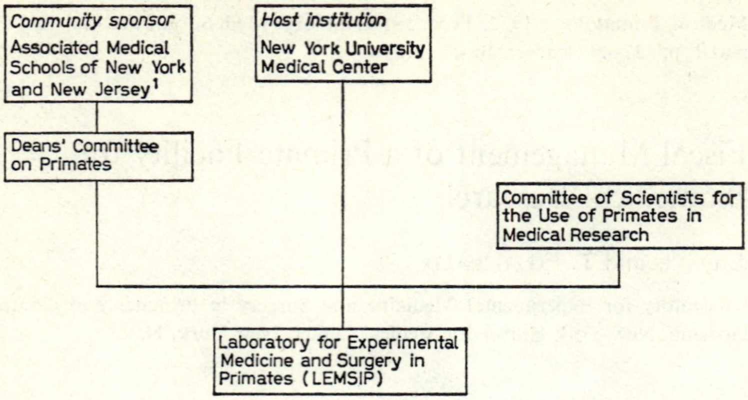

scientists. Their extensive utilization of the Laboratory for many different research and teaching objectives required the development of more adequate administrative and financial management techniques, which are the subject of the present paper.

LEMSIP was developed by physicians in need of an animal model most closely approximating man. Therefore, the internal organization, approach to health care of the animals, services to investigators and clinical capabilities have been patterned after the practice of human medicine and hospital operations. The organization of the Laboratory is very similar to that of a voluntary hospital.

One of the crucial problems which LEMSIP shares with hospitals, preventive medicine clinics, teaching institutions and other public service organizations is the problem of how to fund costs involved in operating the facility but not easily recognized as a part of the service rendered. Such costs, usually identified as the 'core operation', generally consist of a part of professional and administrative salaries, plant maintenance costs, utilities, etc., and, in the case of a primate laboratory, also include the cost of maintaining valuable primate animals while they are not being used between experiments. The decision to be reached on funding the 'core operation'

1 Albany Medical College; Albert Einstein College of Medicine of Yeshiva University; Columbia University-College of Physicians and Surgeons; Cornell University Medical College; Mount Sinai School of Medicine of City University of New York; Colleges of Medicine and Dentistry of New Jersey-Newark and Rutgers; New York Medical College; New York University School of Medicine; School of Medicine and Dentistry, University of Rochester; State University of New York-Downstate Medical Center, Stony Brook, and Upstate Medical Center.

Table II. A comparison of functions

Hospital for experimental medicine and surgery in primates	The hospital
Cage complement	Bed complement
general care cages	general and surgical beds
breeding cages	maternity beds
intensive care cages	intensive care beds
isolation cages	isolation beds
Building services	Building services
operation and maintenance of plant	operation and maintenance of plant
housekeeping services	housekeeping services
General and administrative services	General and administrative services
professional care administration	professional care administration
business services	business services
medical records and statistics	medical records and statistics
communications	communications
employee health care	employee health care
Animal husbandry services	Patient care services
professional services	professional services
house staff – M.D. and D.V.M.	house officers – interns and residents
nursing services	nursing services
sr. technicians	registered nurses
technicians	licensed practical nurses
animal caretakers	nurses' aids
Dietary service	Dietary service
Laundry service	Laundry service
Transportation service	Ambulance service
Clinical laboratories	Clinical laboratories
hematology	hematology
pathology and histology	pathology and histology
microbiology	bacteriology
biochemistry	chemistry etc.
Blood Storage	Blood bank
Surgery	Surgery
X-ray	X-ray
Pharmacy	Pharmacy
Teaching programs	Teaching programs
Research programs	Research programs

(core support) is to decide whether the investigators should pay from their research funds their fair share of the core operation or whether the core support should be funded through special grants and gifts.

It would appear fair to expect that investigators' research support should contain provisions for the *full cost* of their resource utilization. How-

Table III. Laboratory for Experimental Medicine and Surgery in Primates, New York University School of Medicine. Exhibit I: Schedule of total costs

	Total costs $	Salaries and wages $	Employee fringe benefits $	Supplies materials and service $
Operation and maintenance of plant	163,275	56,619	7,360	99,296
General and administrative services	270,822	147,718	19,203	103,901
Ancillary departments				
laundry	14,560	1,850	240	12,470
dietary	18,258	4,260	554	13,444
transportation	10,329			10,329
equipment development	14,087	11,051	1,437	1,599
blood storage	7,498	6,544	850	104
pharmacy	7,248	3,490	454	3,304
Animal husbandry				
chimpanzee compound				
general care cages	28,435	21,394	2,781	4,260
breeding cages	6,998	5,349	696	953
intensive care cages	7,069	5,349	696	1,024
isolation cages	29,315	21,394	2,781	5,140
monkey compound				
general care cages	52,376	43,484	5,653	3,239
breeding cages	12,549	10,521	1,368	660
intensive care cages	2,573	2,106	274	193
isolation cages	17,105	14,027	1,823	1,255
Research services				
intramural				
chimpanzee	12,653	9,330	1,213	2,110
monkey	41,743	31,235	4,060	6,448
extramural[1]				
chimpanzee	4,279	3,674	478	127
monkey	12,749	11,023	1,433	293
Clinical laboratories				
hematology	7,989	6,576	855	558
pathology and histology	24,481	20,170	2,622	1,689
microbiology	11,946	5,556	722	5,668
biochemistry	10,327	7,742	1,007	1,578
Surgery	23,035	16,434	2,136	4,465
X-ray	4,504	3,921	510	73
Total	816,203	470,817	61,206	284,180

1 Special projects.

Table IV. Laboratory for Experimental Medicine and Surgery in Primates, New York University School of Medicine. Exhibit II: Schedule of cost allocation I and II

	Total costs $	I Operation and main- and mainte- nance of plant $	II General and admin- istrative services $	Subtotal $
Operation and maintenance of plant	163,275	163,275		
General and administrative services	270,822	43,954	314,776	
Ancillary departments				
laundry	14,560	1,210	11,993	27,763
dietary	18,258	2,856	15,039	36,153
transportation	10,329		8,508	18,837
equipment development	14,087	2,508	11,603	28,198
blood storage	7,498	5,410	6,176	19,084
pharmacy	7,248	420	5,970	13,638
Animal husbandry				
chimpanzee compound				
general care cages	28,435	12,446	23,422	64,303
breeding cages	6,998	1,891	5,764	14,653
intensive care cages	7,069	1,891	5,823	14,783
isolation cages	29,315	9,887	24,147	63,349
monkey compound				
general care cages	52,376	29,574	43,142	125,092
breeding cages	12,549	9,100	10,337	31,986
intensive care cages	2,573	2,136	2,119	6,828
isolation cages	17,105	9,341	14,126	40,572
Research services				
intramural				
chimpanzee	12,653	3,882	10,422	26,957
monkey	41,743	7,270	34,384	83,397
extramural[1]				
chimpanzee	4,279	204	3,525	8,008
monkey	12,749	848	10,501	24,098
Clinical laboratories				
hematology	7,989	1,304	6,581	15,874
pathology and histology	24,481	860	20,165	45,506
microbiology	11,946	1,163	9,840	22,949
biochemistry	10,327	9,831	8,506	28,664
Surgery	23,035	3,441	18,973	45,449
X-ray	4,504	1,848	3,710	10,062
Total	816,203	163,275	314,776	816,203

1 Special projects.

Table IV (continued)

Exhibit II continued: Schedule of cost allocation III–V

	III Laundry $	IV Dietary $	V Transportation $	Subtotal $
Operation and maintenance of plant				
General and administrative services				
Ancillary departments				
laundry	27,763			
dietary	533	36,686		
transportation			18,837	
equipment development	697			28,895
blood storage	566			19,650
pharmacy				13,638
Animal husbandry				
chimpanzee compound				
general care cages	1,099	1,471		66,873
breeding cages	122	184	319	15,278
intensive care cages	241	184		15,208
isolation cages	2,857	3,126	5,428	74,760
monkey compound				
general care cages	10,943	23,721		159,756
breeding cages	2,053	4,229	7,343	45,611
intensive care cages	312	461		7,601
isolation cages	1,735	3,310	5,747	51,364
Research services				
intramural				
chimpanzee	977			27,934
monkey	3,741			87,138
extramural[1]				
chimpanzee	36			8,044
monkey	155			24,253
Clinical laboratories				
hematology	122			15,996
pathology and histology	122			45,628
microbiology	122			23,071
biochemistry	122			28,786
Surgery	1,208			46,657
X-ray				10,062
Total	27,763	36,686	18,837	816,203

1 Special projects.

Table IV (continued)
Exhibit II contined: Schedule of cost allocation VI–VIII

	Equipment Development VI husbandry $	VII research $	VIII Blood storage $	Subtotal $
Operation and maintenance of plant				
General and administration services				
Ancillary departments				
laundry				
dietary				
transportation				
equipment development	26,005	2,890		
blood storage			19,650	
pharmacy				13,638
Animal husbandry				
chimpanzee compound				
general care cages	1,043			67,916
breeding cages	130			15,408
intensive care cages	130			15,338
isolation cages	2,215			76,975
monkey compound				
general care cages	16,818			176,574
breeding cages	2,998			48,609
intensive care cages	325			7,926
isolation cages	2,346			53,710
Research services				
intramural				
chimpanzee		199	1,324	29,457
monkey		2,691	17,861	107,690
extramural[1]				
chimpanzee			41	8,085
monkey			424	24,677
Clinical laboratories				
hematology				15,996
pathology and histology				45,628
microbiology				23,071
biochemistry				28,786
Surgery				46,657
X-ray				10,062
Total	26,005	2,890	19,650	816,203

1 Special projects.

Table IV (contined)

Exhibit II continued: Schedule of cost allocation IX and total developed costs

	IX Pharmacy	Total developed costs		
		Animal husbandry	Research services	Clinical laboratories, surgery and X-ray
	$	$	$	$
Operation and maintenance of plant				
General and administrative services				
Ancillary departments				
laundry				
dietary				
transportation				
equipment development				
blood storage				
pharmacy	13,638			
Animal husbandry				
chimpanzee compound				
General care cages	549	68,465		
breeding cages	68	15,476		
intensive care cages	68	15,406		
isolation cages	1,162	78,137		
monkey compound				
general care cages	8,818	185,392		
breeding cages	1,572	50,181		
intensive care cages	171	8,097		
isolation cages	1,230	54,940		
Research services				
intramural				
chimpanzee			29,457	
monkey			107,690	
extramural[1]				
chimpanzee			8,058	
monkey			24,677	
Clinical laboratories				
hematology				15,996
pathology and histology				45,628
microbiology				23,071
biochemistry				28,786
Surgery				46,657
X-ray				10,062
Total	13,638	476,094[2]	169,909[2]	170,200[2]

1 Special projects.

2 Total developed costs: $816,203.

ever, this would require that the agencies underwriting their research become aware of the need for such a provision, and that the full costs be carefully defined and determined. In the United States, funding agencies have so far been reluctant to utilize research funds for providing an investigator's share of the core operation support. The problem, in part, appears to be the lack of reliable cost accounting data to justify the need for core support participation. A cost study was, therefore, carried out by us at LEMSIP to define and determine all the cost factors involved. Because of the similarity between services rendered to LEMSIP investigators (see table II) and services provided to patients in a hospital, we were able to draw on the extensive experience of the American hospital industry in defining cost and setting reimbursement rates [1]. The objectives of our cost finding study were: (1) To provide full cost information as a basis of establishing rates for services. (2) To provide information for use in negotiating reimbursements and in determining the amount of reimbursable cost. (3) To provide information for reports to hospital associations, government agencies and other external groups. (4) To provide information for use in managerial decision making in areas other than rate setting. Prerequisites for a cost finding study are: (1) Sound organizational structure. (2) Appropriate accounting system. (3) Adequate statistical information.

Examination of rates charged in connection with animal husbandry services at several educational and research institutions in the United States disclosed that published rate schedules are not necessarily based on the expectation of full recovery of cost. When rates are not based on full cost recovery, the investigator pays less than real cost for the services provided and is at a definite advantage when competing for research support with investigators required to provide for the full cost of their animal experimentation.

Equitable evaluation on the national level of costs of animal experimentation would require all institutions to disclose full cost, establish rates based on full cost and describe the sources for financing established rates. Our study attempts to satisfy these conditions for LEMSIP.

Table III (exhibit I) details in functional sequence costs associated with the operation of LEMSIP for the 12-month period ended August 31, 1971. Table IV (exhibit II) reflects the allocation of cost from nonrevenue to revenue-producing functions. For our purposes, exhibit I is complete. However, it should be noted that costs related to depreciation of buildings and of fixed and major movable equipment are not included in the computation because, in general, accounting practices of universities in the United States do not recognize depreciation accounting in the same way that commercial industry does [2]. Another cost factor not reflected in our study is the purchase price of primate animals. At present, the facility provides services and the principal[2] investigator purchases his primate animals. (Cost finding pertinent to the concept of reproduction of nonhuman primates as a method of supply should be the subject of another study.)

Methodology utilized in the allocation of cost from nonrevenue to revenue-producing functions is characterized as the 'step-down procedure'. Prior to 'stepping-down' (allocating) cost, fundamental decisions have to be made regarding the rate structure by which the facility expects to recover the cost of services provided. For our purposes, we decided that all costs associated with animal husbandry services would be developed as a part of an overall *per diem* rate. The same decision was made for animal research services.[3] To recover costs incurred in the operation of clinical laboratories, operating rooms and X-ray services, the fee-for-service approach was judged the most adequate method.

In applying step-down procedures for the allocation of cost, it is important to recognize that any nonrevenue-producing function may service any of the other nonrevenue-producing functions. The laundry department for example, a nonrevenue-producing function, occupies space within the total physical plant. The cost of operating and maintaining this space should be added, therefore, to other direct costs of the department prior to allocating the department's total developed cost to revenue-producing functions.

There are a variety of statistical bases which may be applied in implementing the step-down procedure. For example, square and cubic footage may be used as the statistical base for the allocation of cost associated with the operation and maintenance of the physical plant. Application of accurate statistical data for determining cost has a significant effect on the level of income required to meet financial needs. Therefore, the methods of accumulation of such data must receive considerable attention.

It is important to note that, in the development of our model, the cost of nonrevenue-producing functions was allocated to revenue-producing

2 He is encouraged to share their use with other investigators who may provide a partial contribution to his costs.

3 Animal research services are those required from the resource to assist the investigators, i.e. blood drawing, plasmapheresis, biopsies, 24-hour urine collections, etc. (animal husbandry, surgery, X-ray and diagnostic laboratory services are *not* included in animal research services).

Table V. Laboratory for Experimental Medicine and Surgery in Primates, New York University School of Medicine. Exhibit III: Rate setting

	Statistical data				Total	Cost	Less core support		Adjusted
	Cage complement	Average daily occupancy	Occupancy %	Total animal days	developed costs $	per day $	Total subsidy $	subsidy per day $	*per diem* rates $
Animal husbandry									
chimpanzee compound									
general care cages	16	16	100	5,840	68,465	11.72	58,364	10.00	1.72
breeding cages	4	2	50	730	15,476	21.20	12,302	16.85	4.35
intensive care cages	4	2	50	730	15,406	21.10	12,302	16.85	4.25
isolation cages	16	34	213[2]	12,410	78,137	6.30	70,667	5.69	0.61
total	40	54	135	19,710	177,484	9.00	153,635	7.80	1.20
monkey compound									
general care cages	248	258	104[3]	94,170	185,392	1.97	10,537	0.11	1.86
breeding cages	60	46	77	16,790	50,181	2.99	2,972	0.18	2.81
intensive care cages	12	5	42	1,825	8,097	4.44	540	0.30	4.14
isolation cages	80	36	45	13,140	54,940	4.18	2,972	0.23	3.95
total	400	345	86	125,925	298,610	2.37	17,021	0.14	2.23
Research services									
intranural									
chimpanzee				1,863	29,457	15.85	24,984	13.42	2.43
monkey				25,126	107,690	4.29	92,587	3.69	0.60
extramural[1]									
chimpanzee				58	8,085	139.40	7,348	126.69	12.71
Monkey				597	24,677	41.34	22,045	36.93	4.41
total				27,644	169,909	6.15	146,964	5.31	0.84
Core support funding	$317,620	49 %							
Charges to investigators	328,383	51 %							
	$646,003	100 %							

1 Special projects.
2 Doubling and tripling numbers of juveniles in adult cages.
3 Periodic overflow into breeding cage space.

Table VI. Laboratory for Experimental Medicine and Surgery in Primates, New York University School of Medicine. Exhibit III: Rate setting (continued)

	Statistical data		Total developed costs	Charge rates
	number of tests	other	$	$
Clinical laboratories				
hematology	2,388		15,836	6.63
pathology and histology		208[1]	45,628	219.37
microbiology	2,388		23,071	9.66
biochemistry	2,090		28,786	13.77
Surgery		468[2]	46,657	99.69
X-ray		792	10,062	12.71

1 Four complete pathology procedures per week × 52 weeks = 208.
2 18 h/week × 52 weeks = 936 h/year.
Use charge based on two-hour sessions.

functions in the following order: (1) operation and maintenance of physical plant; (2) general and administrative services; (3) laundry operation; (4) dietary services; (5) transportation; (6) blood storage (includes other biologicals); and (7) pharmacy operations.

Tables V and VI (exhibit III) reflect rates developed for animal husbandry services, research services, clinical laboratories, operating room and X-ray services. These rates are based on the total developed cost reflected in exhibit II, and do not necessarily agree with those actually applied by LEMSIP during the 1970/1971 fiscal year. In developing *per diem* rates for animal husbandry services and research services we divided total developed cost by the annual number of animal service days. Since in the accumulation of total cost we were able to identify and assign costs directly to four types of animal care,[4] as well as to the several types of research services, it was unnecessary to apply weighted factors in the rate setting process. Had it been impossible to identify and assign cost, it would have been necessary to develop averages for the setting of the various *per diem* rates. The average daily cost for the combined four types of animal care for monkeys is $2.37. When discounted by core support funding of $0.14, the balance to be

4 Four types of animal care (a) general, i.e. routine, (b) breeding, including normal pregnancies, (c) intensive, i.e. care calling for specialized and constant around the clock attention, and (d) isolation, i.e. care under quarantine to contain or exclude infectious diseases.

financed by the investigator is $2.23. The average daily cost for the combined four types of animal care for chimpanzees is $9.00. When discounted by core support funding of $7.80, the balance to be financed by the investigator is $1.20. The average developed rate for intramural research services and extramural (special projects) research services is also discounted by core support funding.

It was during the process of developing rates for clinical laboratory services, surgery and X-ray that we discovered that our available statistical data could not lend itself to a weighted application of factors in determining the relative value of different types of procedures performed. We, therefore, had to apply the total number of all procedures performed as the statistical basis for rate setting purposes.

The real value of a cost finding and rate setting study lies with the ability of management to make use of the data in planning for future financial needs. This model, therefore, may be applied to anticipated expense budgets in order to develop rates for the recovery of that expense.

Acknowledgements

We gratefully acknowledge the assistance of Mr. P. Hughes, Director, Health Services, Research and Planning, University Hospital, New York University Medical Center and Mr. D. E. Herman, Assistant Manager, Office of Research and Training Management, Controller's Division, New York University Medical Center.

References

1 American Hospital Association; Cost finding and rate setting for hospitals (American Hospital Association, Chicago, Ill. 1968).

2 American Council on Education: College and university business administration (American Council on Education, Washington, D.C. 1968).

Authors' adresses: Joseph H. Davis, Assistant Director, Laboratory for Experimental Medicine and Surgery in Primates, and Thomas A. Fitzgerald, Assistant Controller, New York University School of Medicine, 550 First Avenue, *New York, NY 10016* (USA)

Medical Primatology 1972. Proc. 3rd Conf. exp. Med. Surg. Primates, Lyon 1972, part I, pp. 50–60 (Karger, Basel 1972)

The Handling of Simians

S. S. KALTER and F. T. PERKINS

Division of Microbiology and Infectious Diseases, Southwest Foundation for Research and Education, San Antonio, Tex. and Hampstead Laboratories, National Institute for Medical Research, London

The time that elapses from the trapping of a simian to the time when it has terminated quarantine in a laboratory can be divided into three periods. Firstly the animal is caught and transported to a compound within the country of origin, secondly groups of animals are transported by air over long distances from the export to the import country and, finally, the animal is imported, transported and quarantined within the importing country.

Each of these periods that occur within a short time has unique problems but to each there are common factors:

(1) Each represents a different psychological stress.

(2) Each will present a change of diet.

(3) Within each there will be a different but certain risk of cross infection.

Our objective is to deliver the simians in the healthiest state to the importing country and, thereafter, overcome any inherent infections that they may have before they may be considered for use for biomedical purposes.

Conditions in the Exporting Countries

Trapping

There are several methods of trapping simians differing from country to country but in all there is need for improvement. It is difficult to generalize in this respect because of the diversity of methods dictated by the availability of local materials and trappers, but in all areas there are some fundamental principles that, if obeyed, would greatly improve the present situation.

(1) The design of the trap is most important and of greatest significance is the necessity to prevent injury during trapping. An injured animal is not wanted and is immediately placed at a disadvantage within the group. Furthermore, the traps should be clean and attempts should be made to sterilize them between trappings.

(2) The point of trapping is the moment of greatest psychological stress and the effect of this on the subsequent well-being of the animal cannot be overemphasized.

(3) From the time of trapping the animal has been prevented from foraging for food and an adequate and balanced diet with potable water must be provided.

(4) There is no doubt that an undernourished animal under psychological stress will succumb to disease more easily than a normal animal and, furthermore, inherent infections may be exacerbated.

(5) Attempts must be made to educate trappers and to develop better methods than those currently employed for the trapping of monkeys and apes. For example, the practice of picking up animals from natives must be eliminated. Killing of adult chimpanzees to obtain young must be discouraged.

Transportation

The modes of transport from trapping area to compounds within the various exporting countries, and perhaps between trapping areas, are as varied as the methods of trapping. These vary from hand carriage in a sack, taking many hours, to transport by lorry or boat, taking several days. The animals will indeed be fortunate if they escape injury during the journey and are adequately fed on the way. At all times the animal should be transported as quickly as possible after capture to the compound.

Holding Areas

Irrespective of the condition of arrival at the compound, it is at this point that the animals may come into contact with animals from different parts of the same forest or from different forests. This then is the first real chance of cross infection and it will occur most readily in undernourished animals held in overcrowded and inadequate quarters.

There are a number of principles that should be followed in holding compounds. These are:

(1) Animals should be held in buildings designed to exclude rodents, birds and insects.

(2) Animals should be housed in cages holding no more than one or, at the most, two animals per cage. Gang cages must be discouraged. This again is recommended in the interests of the prevention of cross infection and for this reason also the cages should be spaced well apart and they should be made of material that can be sterilized.

(3) The animals coming into the compound from a single trapper using a specific area should be kept in a separate room or well spaced from animals trapped from a different area.

(4) Since it must be expected that injured and sick animals may enter the compound, daily inspection of the animals is essential. If the necessary medicaments are not available for treatment, a ruthless culling policy must be adopted. Most animals should respond to treatment but if they fail to do so, they too must be culled.

(5) The animals may be held in a compound area for some days or weeks and, therefore, a nutritional diet with an adequate supply of potable water is essential. Local grain and fruits should be fed as part of the diet so that the transition to a manufactured food may not be so traumatic.

(6) Access to the compound should be strictly limited to those involved in feeding and handling and adequate records should be kept of intake, deaths and export. Initiation of a microbiologic and parasitologic surveillance on both animals and personnel should be encouraged.

It is appreciated that the ideal holding compound would be quite expensive but much can be done to improve the existing facilities in some areas without a great capital outlay. It cannot be overemphasized that each of these principles contributes towards the prevention of spread of disease and any improvement is better than nothing.

Transportation between Countries

Unlike the conditions in the country of origin, air transportation has changed markedly over the last ten years.

Animals are now carried in specially heated and pressurized compartments of the cargo holds of passenger jets. The time of travel is much shorter and the animals arrive in better condition. They have not been off-loaded or transshipped and they have not required feeding on the way. Nevertheless, there are principles to be followed in order to reduce stress and infection.

(1) Animals must be shipped in cages of an appropriate design. Several animals may still travel in the same cage. This leads to some fighting and cross infection; therefore shipping of animals in single cages is to be encouraged.

(2) The greatest stress that can occur by air transportation is extremes in temperature, especially if the animals are involved in a stopover in a country with an ambient temperature differing markedly from the export country. Sudden changes in temperature can exacerbate latent infections.

(3) In the event of transshipment there should be no contact with other animals. A problem still remaining in some countries, however, is the handling of animals by the customs authorities before the importer may take charge of them. It is not an infrequent occurrence for the animals to take as long to clear customs as they do to fly between countries and customs authorities should be made aware of the health hazards to which they are exposing their personnel by holding the simians on proximity to humans until formalities have been completed. The importance of spreading human diseases to these valuable animals also should be emphasized. The time between arrival in the importing country and housing in quarantine facilities must be reduced to a minimum.

Quarantine in the Importing Country

In spite of what improvements may have been made in the trapping areas and in transportation between countries, the importer has no alternative than to assume that each simian imported is potentially infected, probably with several infectious agents, and because the user requires animals that are healthy, the effort and expense in his quarantine quarters has far outweighed similar expenditure in the exporting countries. If any animal arrives with clinical signs of disease or damage immediate action must be taken to prevent further exacerbation of infection.

Many of the principles appropriate for the holding areas in the export countries are equally applicable to the quarantine in the importing country. There are three factors of importance in the prevention of spread of disease from simians to man in the quarantine and experimental areas: (1) the premises, food and immediate medications, (2) the transmissible diseases prevalent in simians, and (3) the code of conduct that may be suitable for the safest handling of simians.

(1) *The premises, food and immediate medications.* Any newly arrived group of animals must have separate quarters for their housing even if they are of the same species. It is even more unsatisfactory to house animals of different species in the same room even though they may have been quarantined. Quarantined animals are, because of their quarantine period, animals without overt disease but they may, and probably do, harbour many bacteria and viruses each of which may be transmissible to a susceptible host. A newly arrived group should remain together, therefore, and not be brought into contact with any group throughout their life within a laboratory. If, however, it is necessary to mix different groups of quarantined simians, a further quarantine period should be enforced upon the mixed group before any animals are used.

The premises awaiting the newly arrived group should be sterilized by appropriate methods. Cages and racking may be dismantled and transported to an autoclave or steamer but there is no point in assembling sterile cages and racking in an infected room. Simians are notorious for their ability to spread their infected excreta to all corners of the room, and therefore, the walls, ceiling and floor also must be sterilized. This may be achieved only by a chemical applied in the most effective manner in order to reach every nook and crevice.

Diet plays a most important part in the health of a monkey colony. It is surprising how many apparently intelligent persons responsible for the health of a simian colony believe that when once the animals have become accustomed to the diet that they cares to provide, then all will be well. Simians need a high protein diet, fruit, vegetables, roughage and water. A simian will eat food only if it is palatable and, as many have discovered, many would rather die than eat distasteful or uninteresting food. Pelleted diet as available in some countries should be regarded only as a background and supplementary food is necessary in order to provide fruit, vegetables and roughage (brown bread). The water supply should always be clean and fresh.

An animal that is sick should be isolated immediately, appropriate specimens taken to make an etiologic diagnosis, and if it does not respond to the appropriate treatment, it should be destroyed because it is a danger to the whole colony.

(2) *Transmissible diseases.* There are many bacterial and viral diseases known in monkeys that may be transmitted to man and a few diseases are worthy of special mention.

Bacterial Diseases

(1) Tuberculosis. By far the most worrying problem of the bacterial diseases is tuberculosis: how to detect the infected animal and prevent the spread of infection. The situation has been confused by the lack of appreciation that New World monkeys are much more resistant to infection than Old World monkeys and also by the lack of knowledge concerning the sensitivity of the different monkey species to tuberculins. (OT or PPD). In general, tuberculosis in the monkey is a disease developing rapidly. For this reason it is essential to carry out frequent tuberculin tests in order to detect the disease in the early stages before the animal becomes anergic. Furthermore, it is important also to use a dose of tuberculin known to be effective. Many have assumed that the monkey is as sensitive as man to tuberculin and have, therefore, used ten international tuberculin units (TU) as the diagnostic dose. It is now known that this is inadequate and it is necessary to use between 1,000 and 10,000 TU in order to detect the infected animals by reaction to tuberculin. For colonies of monkeys on long-term studies, tuberculin testing must be carried out every three to four weeks, and any animal giving a positive reaction must be culled and a *post mortem* done. A positive reaction also calls for careful observations on the remainder of the colony with frequent testing. It is known that frequent testing with OT or PPD over long periods will not in itself give rise to sensitive animals.

(2) Salmonellosis. There are no less than 1,100 different types of Salmonella and all of them may cause disease in humans and monkeys. There are some, of course, that show a host specificity, such as *S. pullorum* for chickens and *S. bublin* for cattle, but most strains cause disease in many species of animals. It follows, therefore, that diarrhœa in monkeys may be due to a salmonella that can infect man and must be treated as a hazard to the animal handler.

ROWE [1969] has reviewed the data on isolation of salmonella from simians in captivity and it is clearly a problem to be reckoned with.

(3) Shigellosis. This is also a disease communicable from animals to man and simians are no exception. BACH *et al.* [1931] described a simian to human outbreak of *S. flexneri* spreading from five guenon monkeys to 17 humans. Since that time there have been a number of outbreaks spreading from simians to man and each is included in the review by COOK [1969].

The recognition and treatment of infected animals and effective methods of hygiene by the staff handling the simians should be adequate to prevent cross infection.

(4) Other nonviral infections. There are other nonviral infections such as *Pasteurella pseudotuberculosis* and protozoal infections, though the reporting of their occurrence has been very infrequent. It is generally agreed that a good technique of hygiene is adequate to prevent the spread of these to man. In addition to these, pulmonary acariasis can be caused by the lung mite *Pneumonyssus simicola* and the pathological picture has sometimes been confused with tuberculosis. This was first reported by INNES *et al.* [1954] and although there has been no known transmission to man it is a disease that must be borne in mind.

Viral Diseases

HULL [1968] has described the isolation and characterization of no fewer than 30 RNA simian viruses and 28 DNA simian viruses, any one of which may be present in any simian on entry to the quarantine area. The incidence of viruses in the rhesus and cynomolgus monkey is well documented as a result of the use of their kidney tissues for the growth of poliovirus for vaccine. Indeed it has been reported that the vast majority (certainly more than 90%) of all monkey tissues are inherently infected with one or other virus.

KALTER and HEBERLING [1971] have recently shown by serologic studies that viruses of human and simian origin produce infection in most monkeys and apes as well as in man.

Some viruses that have been particularly associated over the years with transmission to humans and having a fatal outcome are worthy of special mention.

(1) Herpesvirus B. Since the first isolation of B virus by SABIN and WRIGHT [1934] no fewer than 18 deaths have occurred as a result of bites or scratches by infected monkeys. There have also been reports of infection and death in laboratory workers having no contact with the living animal but only with the tissue in culture. This herpesvirus which cross-reacts with herpes simplex virus is benign in its natural host, the rhesus monkey, and almost always fatal in man.

The only way of reducing the possibility of spread to man is by careful examination under anesthetic of the animals on intake and rigorous culling of any animal with a lesion in the mouth or on the tongue. Many investigators test their animals serologically and cull out those that are seropositive. PERKINS and HARTLEY [1966] have discussed the handling of simians to reduce the risk of infection with B virus.

(2) Rabies virus. Rabies virus has been reported in simians [BOULGER, 1966; KAUFMAN, in press; RICHARDSON, 1971]. Since there is an increase in sylvatic rabies throughout the world and since the animals used in Europe and the USA are being trapped in the feral state and come from regions where rabies is enzootic, it is surprising that so little has been seen in laboratories. This may be due to the incubation period of the disease being longer than the normal life of the monkey in some laboratories, but none has been seen in the sections of brains and cords of the thousands of monkeys used in the research and control of poliomyelitis vaccines.

Although this is likely to be a most infrequent hazard, nevertheless the technique of handling the animals should be such that the handler is not bitten, scratched or comes into contact on the bare skin with saliva.

(3) Marburg disease. Although the ultimate source of this infectious agent is not known, the transmission of the agent from vervet monkeys to man causing a haemorrhagic disease in 30 humans with seven deaths is well documented [HENNESSEN, 1968; MARTINI and SIEGERT, 1971]. It is important to appreciate, however, that had the animals been allowed to follow a normal six- to eight-week quarantine period and had the laboratory staff been adequately protected against the transmission of infections, this episode would not have been a hazard to man. There is some serologic evidence to suggest a focus of the Marburg disease among African monkeys [KALTER *et al.*, 1969].

(4) Hepatitis, simian haemorrhagic disease, poxviruses and measles. These viruses, while apparently not fatal for man, have been of importance in causing human illness as well as a problem in colony management. Several hundred cases of hepatitis among animal handlers and pet owners are associated with contact with newly imported young chimpanzees. Simian haemorrhagic disease is highly lethal for Asian simians resulting in mortality rates of 100% in colonies of macaques. Monkeypox is also highly infectious for various species of monkeys as well as for nonvaccinated (smallpox) humans. Measles virus which produces latent infections in a number of different species of monkeys and apes is highly lethal for the marmoset.

(5) Oncogenic viruses. There is a great interest these days in oncogenic viruses because of the pertinence to the virus etiology of human cancer. ALLISON [1969] has summarized the findings from the isolation of Yaba virus by ANDREWES *et al.* [1969] to the finding of SV40 [SWEET and HILLEMAN, 1960] and the effect of inoculating Rous sarcoma virus in monkeys. It is important to appreciate also that five of the 17 adenoviruses of the rhesus or cynomolgus origin induce tumours in newborn hamsters.

Many diseases have not been mentioned but it is believed that scientist and technicians well trained in the handling of simians will run few risks of becoming infected if they have a good technique to prevent cross infection.

Code of Conduct of Personnel

The two main objectives in the handling of simians are (1) to prevent the transmission of disease from the animals to man and (2) to prevent the spread of disease from room to room, batch to batch or species to species. How can these be reduced to a minimum? A code of conduct has been discussed [PERKINS and O'DONOGHUE, 1969] and the main features were:

(1) The animals must be housed in well-designed cages. It is important also to have a restraining system in order to immobilize the animal for examination or treatment.

(2) All personnel must be provided with protective clothing including gloves, footwear and masks or visors. Eating and smoking in the quarantine area should be prohibited.

(3) All bites, scratches or abrasions must be reported to the officer on duty and given the appropriate treatment.

(4) It is a wise precaution to take a blood sample of all personnel working with simians and to store the sera. In the event of a pyrexia of unknown origin diagnosis may depend on demonstration of the rise in titre to known pathogens. Continued surveillance of the animal colony by serologic and isolation procedures will do much to warn investigators of impending problems.

Much has been learned in the last 20 years during which simians have been handled in large numbers. It is now generally believed that modern methods of housing and handling simians go a long way towards markedly reducing, if not eliminating, the risks the scientist may run in handling these animals. Several reports are now available to the interested investigator emphasizing the need for concern and appropriate action in the handling of monkeys and apes. [WHO, 1971; National Academy of Sciences, 1968 and 1970; KALTER and HEBERLING, 1969; PERKINS and O'DONOGHUE, 1969].

References

ALLISON, A.C.: Oncogenic viruses in monkeys. Lab. Anim. Hb. *4:* 169–179 (1969).

ANDREWES, C.H.; ALLISON, A.C.; ARMSTRONG, J.A.; BEARCROFT, W.B.C.; NIVEN, J.S.F., and PEREIRA, H.G.: A virus disease of monkeys causing large superficial growths. Acta. Un. int. Cancr. *15:* 760–763 (1959).

BACH, F.W.; FULSCHER, J., und HARNACK, G.: Über eine durch Affen verursachte Ruhrepidemie. Veröff. MedVerw. *34:* 61–73 (1931).

BOULGER, L.R.: Natural rabies in a laboratory monkey. Lancet *i:* 941–943 (1966).

COOK, R.: Simian shigellosis. Lab. Anim. Hb. *4:* 83–91 (1969).

HENNESSEN, W.: A haemorrhagic disease transmitted from monkeys to man. Nat. Cancer Inst. Monogr. *29:* 161–171 (1968).

HULL, R.N.: The simian viruses. Virol. Monogr. *2:* 1–66 (1968).

INNES, J.R.M.; COLTON, M.W.; YEVICH, P.P., and SMITH, C.L.: Lung mites. Pulmonary acariasis as enzootic disease caused by *Pneumonyssus simicola* in imported monkeys. Amer. J. Path. *30:* 813–5 (1954).

KALTER, S.S. and HEBERLING, R.L.: The study of simian viruses. WHO Chron. *23:* 112–117 (1969).

KALTER, S.S. and HEBERLING, R.L.: Comparative virology of primates. Bact. Rev. *35:* 310–364 (1971).

KALTER, S.S.; RATNER, J.J., and HEBERLING, R.L.: Antibodies in primates to the Marburg virus Proc. Soc. exp. Biol. Med. *130:* 10–12 (1969).

KAUFMAN, A.K.: Primate zoonosis surveillance report; in GOLDSMITH and MOOR-JANKOWSKI Medical primatology 1972, part I, pp. 182–184 (Karger, Basel 1972).

MARTINI, G.A. and SIEGERT, R. (ed.): Marburg virus disease (Springer, New York 1971).

National Academy of Sciences: Nonhuman primates. Standards and guidelines for the breeding, care and management of laboratory animals. Publ. No. 1677 (National Academy of Sciences, Washington D.C. 1968).

National Academy of Sciences: Nonhuman primate procurement standards and guidelines for procuring, compounding, holding and transporting nonhuman primates (National Academy of Sciences, Washington, D.C. 1970).

PERKINS, F.T. and HARTLEY, E.G.: Precautions against B virus infection. Brit. med. J. *i:* 899–901 (1966).

PERKINS, F.T. and O'DONOGHUE, P.N. (ed.): Lab. Anim. Hb. *4:* (1969).

RICHARDSON, J.H.: Primate zoonoses surveillance. Center for Disease Control, Atlanta, Ga. Rep. No. 4 (1971).

ROWE, B.: Salmonellosis in simians and other nonhuman primates. Lab. Anim. Hb. *4:* 63–75 (1969).

SABIN, A.B. and WRIGHT, A.M.: Acute ascending myelitis following a monkey bite with the isolation of a virus capable of reproducing the disease. J. exp. Med. *59:* 115–7 (1934).

SWEET, B.H. and HILLEMAN, M.R.: Detection of a 'non-detectable' simian virus (vacuolating agent) present in rhesus and cynomolgus monkey kidney cell culture material. A preliminary report. 2nd Int. Conf. Live Polio-virus Vaccines p. 79 (Pan Amer. Hlth. Org., Washington 1960).

World Health Organisation: Health aspects of the supply and use of nonhuman primates for biomedical purposes. WHO Technical Report Series No. 470 (WHO, Geneva 1971).

Author's address: Dr. F.T. PERKINS, National Institute for Medical Research, Hampstead, *London NW3* (England)

Medical Primatology 1972. Proc. 3rd Conf. exp. Med. Surg. Primates, Lyon 1972, part I, pp. 61–72 (Karger, Basel 1972)

Husbandry and Breeding of *Macaca nemestrina*[1]

G. A. Blakley, W. R. Morton and O. A. Smith

Regional Primate Research Center Field Station, Medical Lake, Washington and Regional Primate Research Center, University of Washington, Seattle, Wash.

Husbandry of laboratory primates has become increasingly important to medical research. Because the sophistication of research protocols demands more specialized animal models than are available in nature, defined animals must be produced under controlled conditions. The Regional Primate Research Center at the University of Washington established a Field Station at Medical Lake, Washington for such a purpose.

Physical Facility

The building, originally designed as a maximum security prison, is constructed of reinforced concrete with a brick veneer. It has three stories with four wings perpendicular to a central core (fig. 1), and encompasses approximately 6,628 m^2 of floor space. Wings A, B and C (fig. 1) are avail-

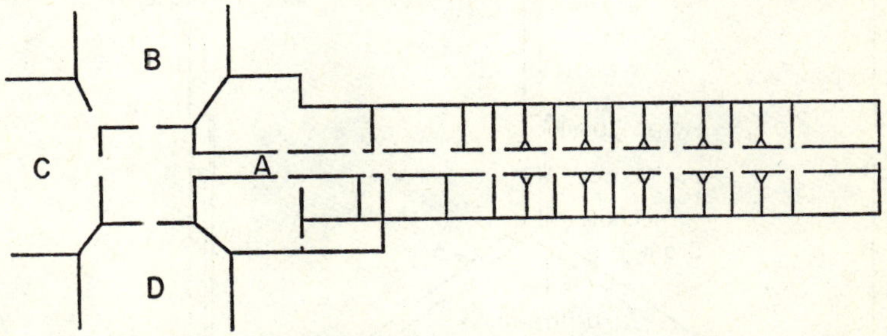

Fig. 1. Floor plan showing corridors surrounding central core.

1 Supported by grant RR 00166 from the National Institutes of Health, US Public Health Service.

able for animal housing and wing D is used as administrative space. An isolated basement area serves as a quarantine.

Typical Animal Holding Room

Animal holding rooms (fig. 2) measure approximately 2.2 × 3.4 m with a ceiling height of 2.8 m. Each floor also has two rooms at the end of each corridor measuring 3.4 × 7.75 m. Floors are covered with either quarry tile or trowelled seamless epoxy sloped to a 10-cm drain. Each room has one or more perch bars fixed to the wall approximately 30 cm from the floor. The entire surface of the room, except the floor, is painted with epoxy enamel. All rooms are individually lighted with a ceiling-mounted vapor-proof light fixture. This lighting is automatically controlled by a time clock which provides a light-dark ratio of 14:10. Dim illumination is provided during the dark period by lights located in the corridor and in the pipe chase adjacent to each room. A feed hopper is attached to each door and is filled from the corridor. Water is supplied *ad libitum* by an automatic valve (Lixit; Systems Engineering, Napa, Calif.). All animal rooms are interconnected by a pass-through door which is operated from the corridor.

Ventilation is accomplished by central intake from the corridor, exhausted via a collecting duct located at the angle of the ceiling and the out-

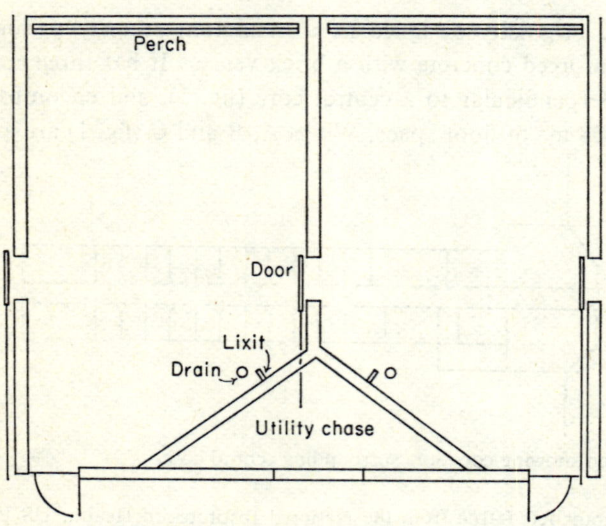

Fig. 2. Typical pair of free-housing animal rooms.

side wall. The system uses 100% fresh air and is designed to provide ten air changes per hour.

Support facilities consist of a surgical suite, clinical pathology laboratory, X-ray unit and a maintenance shop.

The currently unoccupied part of the building, when renovated, is projected to accommodate a colony population of approximately 3,000 primates.

Quarantine

The quarantine area (fig. 3) is located in the basement, physically isolated from the colony. There are four isolation rooms, each having 24 stainless steel cages. Six cages are mounted on racks in two tiers of three each. Connecting drop pans are located under each tier. A laminar flow water flush valve, located at the upper end of the top pan, is automatically timed to flush the pans periodically with a low velocity, high volume flow of water. This system materially reduces the labor required to care for the animals. The area also contains a cage washer room, feed storage, X-ray and treatment room.

Entry into the quarantine is restricted to animal technicians and veterinarians responsible for this area. Entry requires a complete change of clothing; exit requires a shower and another change of clean clothing.

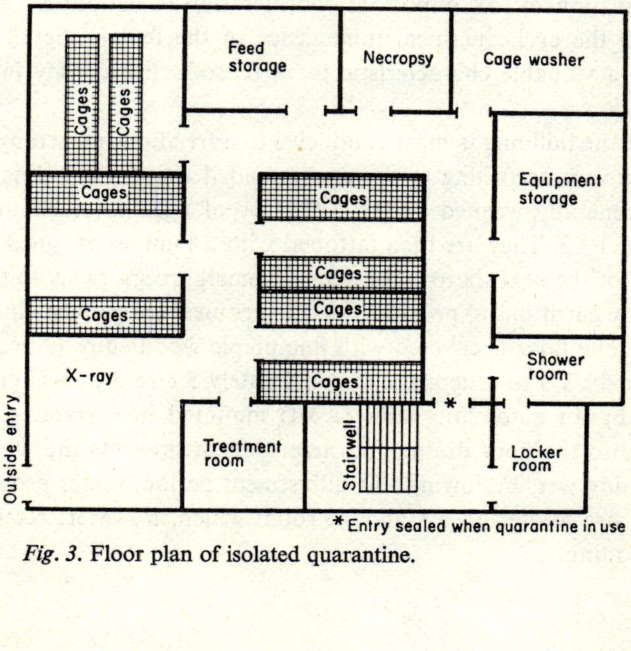

Fig. 3. Floor plan of isolated quarantine.

Newly acquired animals are admitted to the quarantine through an outside entrance. Animals remain in isolation for a minimum of 90 days, during which time routine screening procedures for tuberculosis, endo- and ectoparasites, and other diseases are conducted. Physiological data including body weight, hematology, and serum chemistries are also collected during this period.

Tuberculosis tests are conducted on a bi-weekly schedule during the first six weeks of quarantine. Mammalian tuberculin supplied by the United States Department of Agriculture is administered intradermally in the palpebra at a dose of 0.1 ml, containing approximately 25 mg of tuberculin protein [8]. Two additional tests are performed, the last one during the last week of the quarantine. All tuberculin-positive animals are euthanized and subjected to a complete necropsy.

Husbandry in the Stable Colony

The colony population of approximately 1,000 animals includes five species: *Macaca nemestrina, M. fascicularis, M. fuscata, M. mulatta,* and *Papio anubis*. Approximately 85% are *M. nemestrina* (pigtail macaque). The pigtail was selected as the basic research animal for this colony for two primary reasons: (1) the pigtail is generally less aggressive than the more widely used rhesus monkey, an important consideration in choosing a research animal; (2) the cyclic perineal tumescence of the female pigtail is easily recognized, a valuable characteristic for a reproductive colony in a free housing situation.

The design of the building is most conducive to a free housing arrangement. Group sizes vary according to age and intended use of the animals.

Infants are generally weaned at 100–120 days of age, provided that they weigh at least 1 kg. They are then tattooed with a number assigned at birth and housed for the next six to eight weeks in small groups of six to ten in rooms divided by partitions to provide two cubicles measuring approximately $1 \times 1.6 \times 2$ m. The floor is covered with fine maple wood chips (Ancare Corp., Manhasset, N.Y.) to a depth of approximately 5 cm. Water is provided *ad libitum* by an automatic valve (Lixit) mounted in a stand pipe directly connected to the floor drain. This arrangement prevents the wood chips from becoming wet. Following this adjustment period, larger groups of 15–20 animals are housed in a standard room which, however, retains the wood chip bedding.

Housing juvenile animals on wood chips until they are two years of age has materially reduced death losses in this group. This age group appears to be more active in rooms with wood chip litter and less inclined to huddle than they would be on bare masonry floors. The characteristic huddling on bare floors leads to continual exposure of the hair coat to urine which predisposes to skin problems and chilling stress. To further encourage activity, swinging 'trees' of galvanized pipe are suspended from the ceiling and a freely rolling cylinder made of metal rod is placed in the room.

Adult macaques are also housed as free groups in a room. Breeding groups are generally composed of one adult male and six to eight adult females. This group size was established by experience. Larger groups tend to be less stable and the incidence of wounds resulting from fighting is increased. An exception is made in larger rooms (3.4 × 7.75 m) where several adult males may be housed with 10–15 adult females. These groups are compatible and there have been few incidents of fighting among the males.

Groups of up to six adult males have on occasion been confined in the same room. The dominance hierarchy under these conditions is established almost exclusively by threats and only rarely does actual physical contact occur. By contrast, the dominance order in newly formed groups of females is invariably established by fighting, often resulting in severe wounds which require medical attention.

Once groups are established, the introduction of new members is avoided as much as possible. Again, an exception is made in large groups having several males and females. The females in such groups are generally subdominant animals taken from the smaller groups. Introducing new members does not seem to provoke fighting among these females as it does in the small harem groups. This may be due to the larger space and the less competitive nature of these subdominant females.

Cleaning Procedure

All animal holding rooms, except those having the wood chips litter, are cleaned daily. One room at each end of the corridor is free of animals. Prior to beginning the cleaning routine groups are shifted down one room by opening the pass-through doors sequentially, permitting free movement to the adjoining room. Cleaning is accomplished with a high pressure (500 lb/in^2 at the nozzle) washer which applies hot water containing a tuberculocidal detergent (LPH; Vestal Laboratories, Division of Chemed,

St. Louis, Mo.). After the room is rinsed with clear water, any excess is picked up with a wet vacuum and the animals are returned. Once each week four rooms on each floor are randomly selected for bacterial culture to monitor the effectiveness of the cleaning procedure. Rooms having wood chip litter are generally cleaned on a weekly basis and the litter is replaced.

Colony Health

Casual visitors are not allowed in the animal housing areas. Persons other than the permanent staff are required to wear face masks, clean laboratory coats and disposable footwear.

The present colony population of nearly 1,000 animals precludes mass tuberculosis testing. Therefore, monitoring is conducted on a random basis in the following manner. All animals are listed on a computer program and records are kept current on a daily basis. A bi-monthly print-out indicates any animal not tested within the last four months. Animals caught for pregnancy palpation, movement to other areas, or for any other reason, are checked against the print-out and tested accordingly. Females in late gestation and those nursing an infant are not tested until the infant is weaned. The advantages of this system are that it spreads the work load and provides a constant daily monitoring of the colony, thus avoiding the long lag periods inherent with periodic testing.

Body weight is recorded at the time of tuberculosis testing. Routine hematology checks are run twice yearly on all animals. Because accumulation of dental tartar is a chronic problem, all animal technicians carry dental scalers and routinely scale the teeth of any animal having tartar accumulation.

Diet

All animals in the colony are fed the same basic diet of Purina Monkey Chow (Ralston Purina Co., St. Louis, Mo.). Feed is distributed twice daily to all rooms. Efforts are made to insure that juvenile animals and lactating females have feed available at all times. Supplements such as fruit or vegetables are not fed in the rooms, but are given only to animals in cages when indicated for medical reasons. If fruit is introduced into a group, the most dominant animals eat all of it and the others receive none. In our experience, supplementing the standard diet appears to have no tangible value.

Breeding Program for Macaca nemestrina

The breeding program has several goals. It is designed to provide investigators with fetuses and infant monkeys, to supply eventual replacements to the colony, and to provide permanent subcolonies of genetically interrelated groups, gerontological groups or other special purpose animals.

Harem mating has proved to be a useful method for production in this colony. A harem generally is comprised of an adult male and six to eight adult females. Establishment of a new group is accomplished by introducing all animals into the room simultaneously [2]. New groups must be observed carefully for several weeks following assembly because dominance hierarchies are established at this time and often necessitate restructuring of the group. Wounds resulting from fighting often require that an animal be removed from the harem for medical treatment, and the animal is not then returned to the same group. Occasionally an animal must be removed from the group because of marginal nutritional status: even though she may not be physically attacked, she is denied access to the food supply.

The production of fetuses of known gestational age constitutes a significant part of the breeding program. For this purpose females are caged with proven sires for a period of 48 h in a specially designed breeding pen. This pen is actually a standard rack holding three cages, each 55 cm wide, 90 cm high and 70 cm deep. The cages are separated by acrylic panels. The male is housed in the center cage with a female on each side. When the acrylic panel is removed, a pen of adequate size ($100 \times 90 \times 70$ cm) for the breeding pair permits joining and separating without physical handling. Over the past three years a total of 92 timed conceptions have been produced with this approach.

Several methods for determining optimal mating time have been used, including perineal tumescence, vaginal cytology, counting from the first day of menses, and laparoscopy.

Perineal tumescence in the *M. nemestrina* is similar to that found in some other species of primates [4, 6]. This method is unsatisfactory if conception must be known with less than 72 h error. Vaginal cytology has been used extensively to predict optimal mating time. However, the phases of the vaginal epithelium of *M. nemestrina* are less well defined than those reported for *M. mulatta* [1,3] and the estrogenic effect is not so clearly defined in the immediate pre- and postovulatory phase. There is also a paucity of the mucus and leucocytes that proliferate in the vaginal tract of *M. mulatta*. A number of matings were based on the premise that ovulation occurs, on the average, near midcycle [9, 10]. Menstrual cycles were followed and 48-hour matings were arranged beginning on day 11 of the cycle (the first day of menses is designated as day 1). If conception did not occur, mating in the succeeding cycles was delayed two days, that is, to days 13, 15 and 17.

Each of these methods has one or more drawbacks. Perineal tumescence is too long and subject to extreme degrees of individual variation. The use of daily vaginal cytology is time-consuming and requires extensive handling of the animals. Matings based upon specific times of the menstrual cycle require the least amount of technician time and handling, but there are variations in individual cycle length.

Laparoscopy has recently been employed to gain more accurate information concerning cycle length and time of ovulation. Laparoscopic examination of the ovaries is performed on days 13, 15 and 17. Matings are discontinued on day 19, or when ovulation is observed. This procedure allows a long mating time and still meets the criterion of determining time of conception within 48 h.

The animal to be examined is sedated using ketamine hydrochloride (Bristol Laboratories, Syracuse, N.Y.). A light plane of anesthesia is achieved with halothane (Ayerst Laboratories, New York, N.Y.) administered by face mask. The combination of these agents has given satisfactory anesthesia and a rapid recovery allowing the female to be returned to the breeding pen, fully recovered within two hours. The number of conceptions achieved using this procedure indicates that it does not interfere with ovulation and subsequent conception.

Menstrual Cycle Length and Ovulation Time

The lengths of 75 menstrual cycles ranged from 19 to 45 days, for an average length of 30.4 days (fig. 4). This is in close agreement with the average reported by NAPIER and NAPIER [7]. Twenty-three of these were in quarantine, having been recently imported directly from Sumatra. The cycle lengths of this group ranged from 20 to 42 days with an average length of 30.4 days. It would appear that for this group the change from the wild to the laboratory had little influence upon menstrual cycle length.

Ovulation was determined by either vaginal cytology or laparoscopy in 27 *M. nemestrina* (fig. 5). Sixteen were observed via the laparoscope and 11 were determined by vaginal cytology and confirmed by timed mated conceptions. The results indicate that the optimal time for conception is day 14

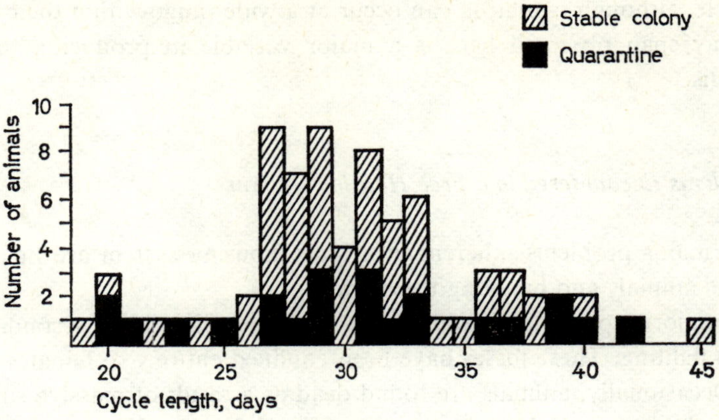

Fig. 4. Length of menstrual cycles in 75 *M. nemestrina;* average length is 30.4 days.

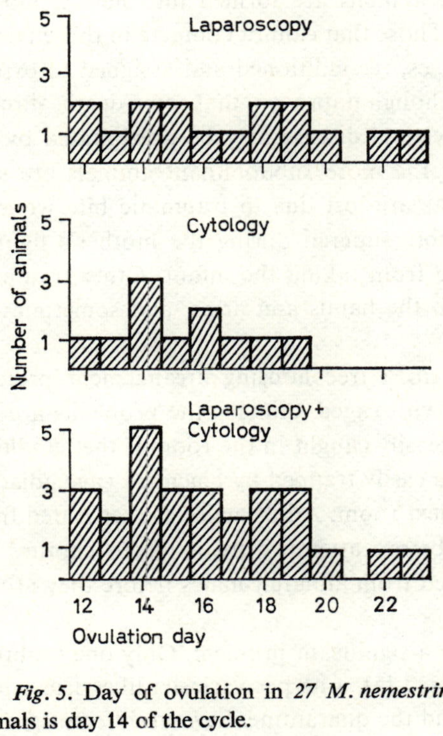

Fig. 5. Day of ovulation in *27 M. nemestrina;* optimal conception time for these animals is day 14 of the cycle.

of the cycle, although ovulation can occur at a wide range within the cycle. The 11-day span observed here is a major variable in producing timed conceptions.

Problems Encountered in a Free Housing System

Two major problems inherent in a loose housing system are fighting among the animals and handling the animals.

The majority of nonperinatal deaths are either directly or secondarily related to fighting. These losses have been confined entirely to females and infants. Occasionally, animals are found dead as a result of massive superficial wounds; however, they are more commonly found in a state of shock or depression. A high percentage of these animals die even though heroic efforts are made to combat the shock and associated septicemia. A closely related problem, previously mentioned, is that subdominant animals may be continually harrassed and denied access to food, so that they must be removed from the group. These animals are formed into new groups in which competition is less severe. Those that cannot compete in this environment are placed in individual cages, reconditioned and assigned to experimental projects. It is speculated, though unproven, that some deaths directly attributed to respiratory and intestinal disease may be precipitated by the chronic stress to which some of the more subdominant animals are subjected. A small number of infants are lost due to traumatic bite wounds. Some have crushed skulls, possibly suffered during the mother's fight to prevent a more dominant female from taking the infant. Other traumatic wounds are generally confined to the hands and arms, and sometimes require amputation.

Although handling animals in a free housing arrangement presents some problems not encountered with caged animals, the problems are not insurmountable. Females can be easily caught in the room if the handler is wearing heavy gloves. Groups are easily trapped by placing a cage adjacent to the pass-through door in the next room. Adult males are separated from the group by trap cage or net before attempts are made to capture the females. Canine teeth are extracted from all adult males before they are introduced into the colony.

Disease control has not been a significant problem. Only one outbreak of a contagious disease has occurred [5], a herpes varicella-like virus which infected both the stable colony and the quarantine. Several different groups of animals have shown clinical signs of this disease while in quarantine. Nearly all animals exhibited a titer rise to this virus, indicating an equal spread regardless of the type of housing.

Advantages of this Kind of Colony

The free housing system produces large numbers of infants with little necessity for close observation by technical personnel. The need to follow each individual and mate it at the appropriate time is eliminated, materially reducing the overhead cost per infant. Leaving the female in the group and weaning the infant at an early age reduces the birth-interval time, thus producing more infants from the same number of breeding females. Housing the animals in harem groups produces genetically defined infants of known parentage, a factor that is becoming increasingly important in biomedical research. The use of individual cages provides precisely timed pregnancies for those research projects having this requirement.

A major consideration in producing laboratory-raised primates is the expense. The free housing system incurs a lower cost per animal than caging systems, due primarily to lower labor costs. The initial capital investment is also greatly reduced since expensive caging is not required.

References

1 Allende, I. L. C. de; Shorr, E., and Hartman, C. G.: A comparative study of the vaginal smear cycle of the rhesus monkey and the human. Contrib. Embryol. Carneg. Inst. *31:* 1–26 (1945).
2 Bernstein, I. S.: Introductory techniques in the formation of pigtail monkey troops. Folia primat. *10:* 1–19 (1969).
3 Erikson, L. B.: Sampling of vaginal and cervical Muellerian duct derivatives in the adult rhesus monkey. I. Indirect assessment of ovarian function. Acta anat. *43:* 158–192 (1960).
4 Fooden, J.: Taxonomy and evolution of the monkeys of Celebes; in Fooden Bibl. primat., vol. 10, pp. 1–128 (Karger, Basel 1969).
5 Lourie, B.; Morton, W. G.; Blakley, G. A., and Kaufmann, A. F.: Epizootic vesicular disease in macaque monkeys. Lab. Anim. Sci. *21:* 1079–1080 (1971).
6 MacLennan, A. H. and Wynn, R. M.: Menstrual cycle of the baboon. I. Clinical features, vaginal cytology and endometrial histology. Obstet. Gynec. *38:* 350–358 (1971).
7 Napier, J. R. and Napier, P. A.: A handbook of living primates, p. 406 (Academic Press, New York 1967).

8 STUNKARD, J. A.; SZATALOWICZ, F. T., and SUDDUTH, H. C.: A review and evaluation of tuberculin testing procedures used for *Macaca* species. Amer. J. vet. Res. *32:* 1873–1878 (1971).

9 VALERIO, D. A.; MILLER, R. L.; INNES, J. R. M.; COURTNEY, K. D.; PALLOTTA, A. J., and GUTTMACHER, R. M.: Management of a laboratory breeding colony. *Macaca mulatta*, p. 60 (Academic Press, New York 1969).

10 WAGENEN, G. VAN: Optimal mating time for pregnancy in the monkey. Endocrinology *37:* 307–312 (1945).

Authors' addresses: Dr. GERALD A. BLAKLEY and Dr. WILLIAM R. MORTON, Primate Field Station, Medical Lake, *Washington;* Dr. ORVILLE A. SMITH, Regional Primate Research Center, University of Washington, *Seattle, WA* (USA)

Medical Primatology 1972. Proc. 3rd Conf. exp. Med. Surg. Primates, Lyon 1972, part I, pp. 73–80 (Karger, Basel 1972)

Rhesus Monkey Breeding Husbandry and Cost Accounting at the Primate Center TNO

G. A. Boorman, G. Fitzgerald, C. Goosen and H. Balner

Primate Center TNO, Rijswijk Z. H.

Introduction

There has been a marked increase in the utilization of nonhuman primates for medical research. The resulting literature attests to their usefulness suggesting that this trend will continue.

Similarly, the use of nonhuman primates at the Radiobiological Institute TNO (RBI) increased.

In 1967, with an average daily inventory of about 500 nonhuman primates at RBI, it was decided to place the management and care of these animals under a separate entity. The Primate Center TNO, as the new group was called, was charged with the responsibility of the care of all nonhuman primates for RBI and the Institute for Experimental Gerontology (IVEG) located in the immediate neighbourhood as well as developing its own research programs. Further it was expected to maintain contact and exchange information with other primate centers.

Since its conception the Primate Center has grown and now conditions, breeds, houses and rents monkeys, not only for RBI and IVEG, but also to other investigators both within and outside the Netherlands. Its developing research program includes nonhuman primate behavior and endocrinology, as well as many collaborative programs with the associated institutes. The average daily inventory is now about 700 animals, the majority of which are rhesus monkeys *(Macaca mulatta)*. Sixty-three chimpanzees *(Pan troglodytes)*, 140 stump-tailed macaques *(M. arctoides)* and 20 white-eared marmosets *(Callithrix jacchus)* constitute the remainder of the animals at the Primate Center.

In order to accurately know the complete expenses as well as plan for future growth, a detailed analysis of the complete costs has been undertaken. The purpose of this paper is to present our methods and costs of conditioning, breeding and housing rhesus monkeys. While some of the costs are unique to Holland (i.e. poldertax), the methods and relative costs should be of interest to other centers engaged in providing monkeys for medical research.

Since the structure of the Primate Center is determined by the needs of the major investigators, a brief description of the main research projects follows.

Research Programs

Bone marrow [VAN BEKKUM, 1971] and organ transplantation studies currently account for the utilization of most animals. These studies often involve long-term immunosuppression and, therefore, stringent conditioning procedures are required to assure a minimal incidence of infection. In connection with these projects and as an exercise in primate immunogenetics, the study of the leukocyte antigens of rhesus monkeys (RhL-A) and of chimpanzees (ChL-A) has become increasingly important [BALNER and VAN ROOD, 1972].

These studies require the availability of primate families and thus the breeding programs are designed for the production of siblings. Ethology, fundamental immunology, toxicology and providing aged animals for gerontological studies also have an influence on the structure of the primate center.

Methods and Cost Accounting

The cost of various functions and services have been determined in 1970[1] by an efficiency bureau (Bosboom and Hegener) and the results quoted herein are from that study.

The prices include land cost ($27.00/m^2), building cost ($40.00/m^3), building repairs (1.5% of original cost annually), installation repairs (5% of original cost annually), insurance, interest (7% of half of the total investment annually), salaries, and poldertax (a tax for polder drainage costs). The buildings are amortized over 30 years, quarantine trailers over five years, installations over ten years and animal cages over five years.

1 Mid-1972 costs are approximately 20% higher.

The present Primate Center evolved from RBI and thus currently many of the personnel and services are shared. Shared expenses are difficult to assess and so the study is based on the idea of the Primate Center as a self-sustaining unit. Therefore, buildings, salaries and expenses for a complete central facility for a primate center are calculated and these costs are added to the current expenses.

Central Facility

The total annual cost of a central facility for a primate center is shown in table I. This facility includes office and salary for director, secretary, academic personnel, administrator and chief animal caretaker. Rooms for coffeeshop, lectures, necropsies, operations, laboratories, showers, toilets, cage washing, storage, kitchen, boiler and hallways are included. Salaries, social benefits, equipment and buildings with amortization, interest, insurance repairs, expendables (gas, water, books, travel expenses, etc.), third party service (personnel, purchasing, etc.) and upkeep of buildings and lawns are also included.

Since in each area of the Primate Center labor costs are based on daily needs, the salaries of two reserve animal caretakers, two support personnel and a laboratory assistant are included in the cost of a central facility. Sharing of these personnel allows one to provide adequate animal care during weekends and vacations.

The total annual cost of the central facility is spread over five areas: quarantine 20%, breeding colony 50%, stock animals 10%, chimpanzees 10%, behavior research 10%. These percentages are not necessarily in proportion but do give an indication for calculation of present costs.

Table I. Annual cost of a central primate facility (US $)

Salaries	76,775.000
Installations and reserve cages	28,355.000
Interest, insurance and upkeep of installations	23,580.000
Expendables (gas, water, electricity, books, travel, etc.)	31,775.000
Third party services (personnel, purchasing, etc.)	11,450.000
Land and building	15,260.000
Total	187,195.000

Quarantine

The quarantine facilities were patterned after the existing facilities at LEMSIP, New York [DAVIS *et al.,* 1969] and consist of four low-cost trailers, three each capable of housing 48 monkeys (two rooms) and a central trailer housing heating and air-conditioning equipment, and providing space for showers and toilets and offices. These facilities are cheap, flexible and the 250-300 monkeys conditioned annually are presently sufficient for our needs, as well as our customers elsewhere.

The quarantine period lasts a minimum of 90 days. During this time the animals are tuberculin tested four times (10 mg OT, intrapalpebral), an anthelmenthic administered twice and the feces checked for the presence of enteric pathogens. Infectious diseases are treated with antibiotics as well as supportive care.

The cost of a conditioned monkey is shown in table II. The costs under housing and care are fixed costs and do not depend greatly on the number of animals conditioned. For the approximation of these costs, a maximum utilization of the quarantine facilities has been assumed.

Bacterial cultures and treatment are a significant portion of the total costs of a conditioned monkey. However, many studies here involve radiation or other forms of immunosuppression and, therefore, all animals have to be closely monitored for potential pathogens as well as antibiotic resistant flora.

Table II. Cost of a 'conditioned' rhesus monkey (US $)

Original cost (including 7% mortality)	52.00
Housing and care	
salaries (three caretakers)	11.00
trailer costs (amortized over five years)	18.00
interest, insurance and upkeep of installations	8.00
land (cost, interest, taxes, upkeep)	4.50
central facility support (20%)	52.00
Bacteriological cultures and treatment	30.00
Food	15.50
Cleaning supplies	2.00
Total	193.00

These fixed costs are based on maximum utilization of the quarantine facility.

Breeding

The breeding facilities are located in a separate building and have been previously described [GOOSEN, in press]. In brief, the breeding colony consists of six males and 60-80 females. Each male is assigned a panel of females, four to six of which are placed with him for a period of three weeks, after which they are removed and a new group introduced also for three weeks. The breeding period has been shortened since most conceptions occurred during the first three weeks. Six weeks following a mating period the females are palpated for pregnancy and those not pregnant are again available for introduction to the male. Approximately 40 offspring are born each year. Although this is insufficient to meet our needs, space and fund limitations make it impossible to increase the capacity at the present time.

Our breeding efficiency could be improved by replacing some of the breeding females. We are, however, reluctant to remove an animal that has already produced two or three young since the aim of the mentioned immunogenetic studies is the acquisition of families with maximum numbers of siblings. Currently we are investigating the use of clomiphene which has been reported to induce ovulation in infertile females [VALERIO and COURTNEY, 1968; BALIN and WAN, 1969].

The cost of producing a three-year-old laboratory-bred rhesus is shown in table III. An annual birth rate of 60 babies per 100 females is assumed, 40 of which reach three years of age.

Since the cost of the entire breeding facility must be recovered through the offspring, they are relatively expensive. We feel the 40-percent produc-

Table III. Cost of a three-year-old laboratory-bred animal (US $)

Breeding colony[1]	
salaries (eight caretakers)	255.00
installations and cages	227.00
interest, insurance and upkeep of installations	168.00
land, (cost, interest, taxes, upkeep)	90.00
building (cost, interest, taxes, upkeep)	94.50
food cost	315.00
cleaning supplies	8.50
Replacement of retired breeders (60 animals/year)	76.00
Central facility support (50%)	585.00
Total	1819.00

1 Includes 400 breeding females, 48 males and 180 babies under three years of age.

tion figure is conservative allowing for production of siblings and that production of 60 three-year-old rhesus per 100 females is possible. This would reduce cost of a laboratory-raised monkey by 50%.

The breeding colony is also supporting 50% of the cost of the central facility. This is a major portion of the expense of a laboratory-raised animal, however, the cost per individual animal again decreases as the primate center grows.

The cost of replacement of breeders as shown in table III is based on a reproductive life of seven years.

Stock Animals

Our stock animals consist of individuals being used for relatively short-term studies as well as the family stock which is being held for long-term periods. The average daily inventory is about 450 rhesus monkeys.

The *per diem* rates for rhesus monkeys are shown in table IV. Labor costs are based on 60-90 animals per caretaker. As mentioned earlier, weekend and vacation help is provided by reserve caretakers, the cost of which is included in the central facility.

Because of the variation in daily inventory the calculated rates for installations, cages, land and building assume that on the average only 67% of the cages are occupied. The cost of the central facility support is based on an average daily inventory of 360 animals.

The *per diem* costs in table IV include only routine handling and care. Any specialized services would have to be calculated as an extra expense.

Table IV. Per diem rate for rhesus monkeys (US $)

Salaries	0.22
Installations and cages	0.20
Interest, insurance and upkeep of installations	0.14
Cleaning supplies	0.01
Land (cost, interest, taxes, upkeep)	0.07
Building (cost, interest, taxes, upkeep)	0.04
Food	0.12
Central facility support (10%)	0.14
Total	0.94

Conclusion

Our calculated costs illustrate that monkeys are a relatively costly research animal. However, we feel our costs are complete and include buildings, land, tax, etc., many of which are not routinely included in calculating costs of research animals. When users of research animals are all located within the same institute, it is less important when some of costs are borne by the institute and *per diem* rates reflect only partial costs. However, since we provide services for investigators outside of our health research organization, an awareness of actual costs is essential.

While our costs occasionally seem high to individual customers, calculation of the complete costs at a less-specialized facility would probably reveal that it is not less expensive. In addition, better care and services can be provided at a larger central facility. In order to minimize costs, an investigator can rent monkeys for a number of innocuous experiments. Investigators with compatible research interest can often share animals, reducing both costs and the number of monkeys needed for research. Furthermore, following termination of an experiment, tissues and fluids are distributed to the various investigators. All of these procedures require planning and the investigators have to be aware of the necessity of maximum utilization of the animals.

The breeding of pedigree monkey families is providing invaluable material for biomedical research. Such animals cannot be obtained from the wild and are available in only a few breeding colonies. However, these animals are extremely expensive and not utilized for routine experiments. In other words, our breeding program is supplying material only for specialized investigations and a majority of our needs are still being filled with wild-caught animals. In order to preserve endangered species and to obtain more and better animals for research, an economical method of providing large numbers of monkeys, perhaps without the advantage of known parenthood, is urgently needed.

References

BALIN, H. and WAN, L. S.: A study of induction of ovulation in *Macaca mulatta.* J. repro. Med. *2:* 273–284 (1969).

BALNER, H. and ROOD, J. J. VAN (ed.): Transplantation genetics of primates (Grune & Stratton, New York 1972).

BEKKUM, D. W. VAN: Bone marrow transplantation. Transplant. Proc. *3:* 53–57 (1971).

DAVIS, J. H.; MCPHERSON, B. R., and MOOR-JANKOWSKI, J.: Maintenance and handling of primate animals for biomedical research. Ann. N.Y. Acad. Sci. *162:* 329–335 (1969).

GOOSEN, C.: Breeding macaques at the Primate Center TNO; in BEVERIDGE Primate breeding (Karger, Basel, in press).

VALERIO, D. A. and COURTNEY, K. D.: Treatment of infertility in *Macaca mulatta* with clomiphene citrate. Lab. anim. Care *18:* 339–345 (1968).

Authors' address: Dr. G. A. BOORMAN, G. FITZGERALD, Dr. H. BALNER and Dr. C. GOOSEN, Primate Center TNO, 151 Lange Kleiweg, *Rijswijk Z. H.* (The Netherlands)

Medical Primatology 1972. Proc. 3rd Conf. exp. Med. Surg. Primates, Lyon 1972, part I, pp. 81–87 (Karger, Basel 1972)

Chimpanzee Breeding and Management at the Primate Center TNO

G. A. BOORMAN, T. SPELTIE, C. GOOSEN and H. BALNER
Primate Center TNO, Rijswijk Z.H.

Introduction

The chimpanzee *(Pan troglodytes)*, as one of man's closest phylogenetic relatives, might be expected in some cases to yield information that more readily extrapolates to the human situation. For this reason and because of their availability and size, as compared to the other great apes, chimpanzees were obtained by our primate center in 1965.

The first project was studying their tissue antigens in relation to those of man. It turned out that chimpanzees could also be used for the production of immunoreagents to study human tissue antigens [BALNER *et al.*, 1967]. In addition, chimpanzees proved to serve as reliable surrogate hosts to test *in vivo* the efficacy and toxicity of anti-human lymphocyte sera [BALNER *et al.*, 1968].

As interests became more diverse (chimpanzee tissue antigens, blood groups, fetal immunology, behavior, endocrinology), a breeding program seemed to be a valuable extension of the chimpanzee program. This idea became feasible with the completion of new housing facilities in 1970. While our colony consists of 63 animals, most are immature and only 17 females and three males are included in the breeding program.

Because of the mentioned research interests, all breeding is directed toward the production of siblings. While no siblings have been produced as yet, three births have occurred and four other chimpanzees are pregnant, some for the second time.

It is the purpose of this paper to present the cost accounting of our chimpanzee colony as well as the methods and results of our breeding program.

Cost Accounting

The costs are based in part on a study undertaken by a Dutch efficiency bureau (Bosboom and Hegener) and the details have been reported elsewhere [BOORMAN *et al.*, 1972]. In brief, the costs are from 1970[1] and include buildings, installations, land cost, repairs, upkeep, taxes, interest, salaries and social services. The buildings were amortized over 30 years and cages and installations over ten years.

The cost of a planned central support facility with offices and personnel for the Rijswijk Primate Center has been detailed elsewhere [BOORMAN *et al.*, 1972]. At the present time, the personnel and facilities serving in this capacity are spread throughout the Primate Center, the Radiobiological Institute and the Institute for Experimental Gerontology, all located in Rijswijk. Ten percent of the total cost of this planned central facility is supported by the chimpanzee in the calculation of *per diem* rates.

Labor

Labor costs, usually a major share of animal care costs [WATKINS and COHEN, 1964], appear especially expensive with the chimpanzee (table I). The animal's size, special care, time required for handling and the need for highly trained personnel are all contributing factors. We find that the experience and skill are valuable attributes for a caretaker especially in a chimpanzee colony. The caretakers must also be aware of the behavior and temperament of each individual animal. We are fortunate in having a very low turnover in personnel; many individuals have been working with the colony since its conception and thus are familiar with each chimpanzee.

Table I. Per diem costs for chimpanzees (in US $)

Salaries (3.5 caretakers)	1.14
Installations and cages	0.69
Interest, insurance and upkeep of installations	0.61
Food	0.68
Cleaning supplies	0.03
Land (cost, interest, taxes, upkeep)	0.08
Buildings (cost, interest, taxes, upkeep)	0.28
Central facility support (10%)	0.77
Total	4.28

1 For mid-1972, all costs would be approximately 20% higher.

Another factor increasing labor costs is having the caretakers provide services such as obtaining blood, urine, and fecal samples from the chimpanzees. Minor surgical operations and anesthesia are also carried out by the animal caretakers. While these services increase daily labor costs, less time and effort from the various investigators are required. In addition, having these services performed by personnel familiar with the animal seems to facilitate the procedure and also results in less disruption of everybody's daily routine. Furthermore, having the animal care staff in charge of these procedures has resulted in excellent co-ordination and synchronization of the workload. Thus, when an animal is tranquilized, a maximum number of procedures are carried out simultaneously; for example, a routine tuberculin test, palpation for pregnancy, obtaining blood or bone marrow samples for several investigators, etc. Finally, this increased responsibility has resulted in a more varied and interesting day for the animal care personnel and may explain in part the stability of our work force.

Installations

A second major cost illustrated in table I is cages and installations. These installations include outdoor runs as well as complete enclosure of the outside area with heavy fencing material to contain any animal that may escape.

In order to evaluate caging systems for chimpanzees, our facility includes both galvanized and stainless steel cages of various designs. Advantages and disadvantages of the various cages are becoming apparent and the experience with a variety of systems will allow better planning for cage replacements when this becomes necessary.

The cost of the cages and installations were amortized over 10 years. Additional annual expense for installations and cages listed separately in table I are upkeep (5% of original cost), insurance (0.3% of original cost) and interest on investment (7% of half of the original cost).

Food

The food costs (table I) again illustrate that the chimpanzee is more expensive than smaller nonhuman primates. This is mainly due to the amount and variety of fresh fruit and vegetables in the diet. Our daily diet for an average-sized chimpanzee is 1 kg each of apples and carrots, an orange and a banana and 500 g of commercial monkey pellets. This is supplemented three times a week by 500 g of cooked rice containing vitamins, minerals and protein supplements. The chimpanzee seems to grow well on this diet. Similar chimpanzee diets utilizing fresh fruits and vegetables for a significant portion of the diet have been reported by others [DERWELIS *et al.*, 1969; BOURNE, 1971].

Breeding Program

Future inavailability, interest in genetic studies and fetal immunology are some of the factors which played a role in the decision to initiate a breeding program. As already mentioned, the interest in genetics directed all breeding towards the production of siblings. Currently one of three adult males is alternately housed with a group of females. The father can be determined with some degree of certainty by leukocyte typing of the infant and a female which has given birth will be mated only with the father of the first infant.

We are currently screening urine from all breeding females twice a month with a commercial human pregnancy test (Pregnosticon®, Organon, Oss, Holland). In nine animals, 132 negative results have been recorded with this test and no animal was later found to be pregnant. Thus it appears that false positive results are not common, suggesting that the levels of pituitary gonadotropin present in the urine during normal menstrual cycles of the chimpanzees do not interfere with this test. In eight animals positive results have been followed by two live births and two abortions. Four animals are currently pregnant (fig. 1). Two of the pregnant animals had nine and ten weekly consecutive positive urine samples before reverting to a negative status. The other two pregnant animals have only recently had positive urine samples. This test is helpful in the detection of pregnancy but also in detecting pregnancy loss which has been common in our facility.

Stillbirths and Abortions

Since its conception, our breeding program has experienced five confirmed abortions or stillbirths. This appears to be a high incidence as compared to other chimpanzee colonies [GUILLOUD, 1969; BUTLER, 1972]. Four were abortions occurring in the first trimester of pregnancy and one stillbirth, in which the fetus weighed 600 g, was estimated to be in the last trimester.

In this last case (Wanda), both fetus and placenta were available for histological examination. No lesions other than recent hemorrhage were seen in the fetus. There was reason to believe that this was related to trauma following birth. The placenta, however, contained large areas of necrosis, numerous acute inflammatory cells and acute vasculitis of the larger vessels. Special stains failed to reveal any infectious etiology and bacterial culture unfortunately, had not been done. The reason for this abortion (or premature birth), therefore, remains obscure. In three other cases, only pieces of decidua or embryos have been available for study. Bland necrosis has been a prominent feature in the decidua examined, which may be causative, although some necrosis has been reported in normal pregnancies [LUDWIG and BAUR, 1971]; whether the extent of necrosis in our cases was greater

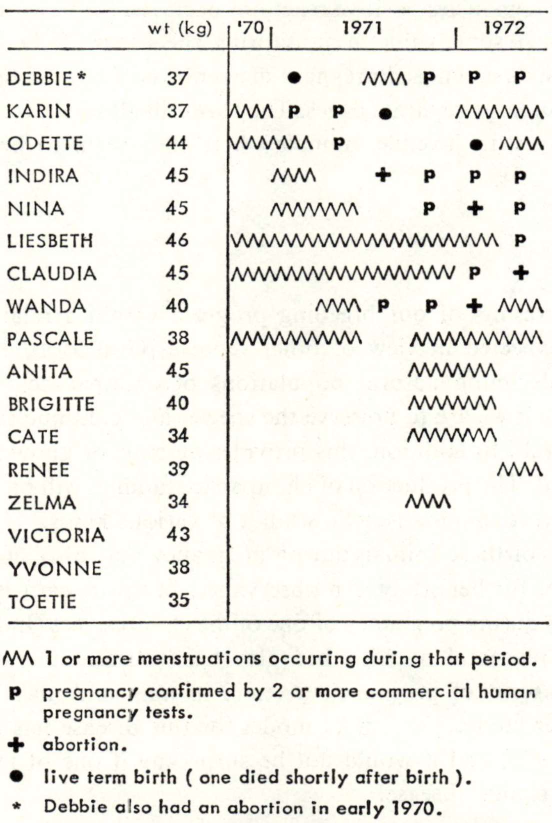

Fig. 1. Breeding record of chimpanzee females.

than normal is not known. The etiology of the early abortions is also unknown; however, it may be partially related to the age of the animal. Three of the animals who had aborted are again pregnant and the pregnancies appear to be proceeding normally.

Vaginal bleeding preceded the abortion by four to six weeks in Nina, Indira and Claudia (fig. 1). With Claudia, treatment with 15 mg of allysternol (Gestanon®, Organon, Oss, Holland) orally divided into three daily doses for seven days stopped vaginal bleeding. However, abortion occurred three weeks following cessation of treatment.

In two cases (Claudia and Nina), weekly screening of urine by Pregnosticon® revealed only five and six consecutive positive samples. Abortion occurred eight and 13 weeks, respectively, following the last positive test. With Indira only one positive urine sample was seen six weeks prior to abortion. Since in normal pregnancies positive reactions occurred for ten or more weeks, this test may have some value in monitoring the status of the pregnancy. Currently, animals diagnosed pregnant are removed from the breeding group and placed in large separate cages. They are still allowed access to outdoor runs but only with juvenile animals where they seem to be less subject to fighting.

Discussion

The encouraging results of our breeding program within a relatively short time is not unexpected in view of other reports [GUILLOUD, 1969; BUTLER 1972]. With declining natural populations of chimpanzees, such programs are necessary if we are to preserve the species and continue to use these animals in research. In addition, this provides animals of known age and genetic background. The production of chimpanzee families will provide unique material for future immunogenetic studies of various kinds.

Close observations of these animals during pregnancy may also suggest other possible areas for further study. An observation in this respect is the severe proteinurea seen during pregnancy of one of the chimpanzees (Indira). Widespread ischemic placental necrosis has also been occasionally seen. This is vaguely reminiscent of pregnancy toxemia in man and may be a fruitful area for further study. No animal model for this disease has been found so far [CHEZ, 1972] and it would not be surprising if one of man's closest relatives had a similar disease.

Our cost analysis revealed, not unexpectedly, that a chimpanzee is a relatively expensive research animal. They are, however, unique and will continue to be utilized in biomedical research. Of course, nobody would utilize chimpanzees for studies that can be performed in less expensive models. In addition, one should strive to obtain maximum utilization through 'sharing'; for example, immunological behavior and endocrine studies can all be performed on the same animals. Even the abortions have provided useful material: fetal cells for tissue culture, serum and blood for antibody characterization as well as material for histological study.

References

Balner, H.; Eijsvoogel, V. P., and Cleton, F. T.: Testing of anti-human lymphocyte sera in chimpanzees and lower primates. Lancet *i:* 19–22 (1968).

Balner, H.; Leeuwen, A. van; Dersjant, H., and Rood, J. J. van: Defined leukocyte antigens of chimpanzees. Use of chimpanzee isoantisera for leukocyte typing in man. Transplantation *5:* 624–642 (1967).

Boorman, G. A.; Fitzgerald, G. H.; Balner, H., and Goosen, C.: Rhesus monkey husbandry and cost accounting at the Primate Center TNO, The Netherlands; in Goldsmith and Moor-Jankowski: Medical primatology 1972, part I, pp. 73–80 (Karger, Basel 1972).

Bourne, G. H.: Nutrition and diet of chimpanzees; in Bourne: The chimpanzee, vol. 4, pp. 373–399 (Karger, Basel 1971).

Butler, T. M.: Reproductive history of the Holloman chimpanzee colony. J. med. Prim. *1:* 51–57 (1972).

Chez, R. A.: Monkeys, models, and midwives. J. med. Prim. *1:* 73–74 (1972).

Derwelis, S. K.; Douglas, J. D.; Fineg, J., and Butler, T. M.: Supply, maintenance and handling of chimpanzees. Ann. N.Y. Acad. Sci. *162:* 311–324 (1969).

Guilloud, N. B.: The breeding of apes. Experience at the Yerkes Regional Primate Research Center and a brief review of the literature. Ann. N.Y. Acad. Sci. *162:* 297–299 (1969).

Ludwig, K. S. and Baur, R.: The chimpanzee placenta; in Bourne: The chimpanzee, vol. 4, pp. 349–371 (Karger, Basel 1971).

Watkins, R. and Cohen, B. J.: A method of analyzing *per diem* rates for laboratory animal care. Lab. anim. Care *14:* 398–403 (1964).

Authors' address: Dr. G. A. Boorman, T. Speltie, Dr. H. Balner and Dr. C. Goosen, Primate Center TNO, 151 Lange Kleiweg, *Rijswijk Z. H.* (The Netherlands)

Medical Primatology 1972. Proc. 3rd Conf. exp. Med. Surg. Primates, Lyon 1972, part I, pp. 88–92 (Karger, Basel 1972)

The Separation of Baby Monkeys from Their Mothers

H. CHORAZYNA

Brain Research Unit, University of Brussels, Brussels

The separation of baby monkeys from their mothers is a problem which arises in nearly every facility where monkeys are reared. However, different approaches must be considered depending on whether the babies are separated from the mother immediately after birth, or later in life.

The effects on the newborns of a complete separation from their mothers has been studied by HARLOW and ZIMMERMANN [1958, 1959], HARLOW [1960], HARLOW and HARLOW [1965, 1966]. According to these authors, the mother deprivation influences the affectivity of the monkey for its whole life. Using inanimate surrogate mothers or rearing two baby monkeys together one can reduce, but not suppress, these affective consequences.

Our observations were carried out on five macaques and seven baboons which were separated from their mothers at varying ages, from birth to 14 months. In our rearing unit, the baby monkeys as a rule remain with their mothers throughout their first months. This constitutes the most economical method. We nurse by bottle only when for some reason the baby cannot be nursed by its mother. We have thus reared from the feeding bottle three baboon babies: one of them was born at term but was not accepted by its mother and two of them were born by caesarian section about three weeks before term.

Each of the three babies was placed in a rearing box of plexiglas with an electrical heating pillow inside. They were given milk every three hours, day and night during the first two months which provided plenty of possibilities for touch contacts with me. They were carried to my home for some months, every night and on weekends. This means that, even when remaining in their box, they saw me and remained aware of my presence. They often were allowed to run about freely outside their box. When they were afraid of some new stimuli they used to run either towards their accustomed box or towards me and remained clinging to my leg. Two of these three baboons were removed at three to four months of age from the plexiglas boxes and put in the iron cages used for adult monkeys. This transfer represented an important psychological shock for the babies. In fact the transfer has to be carried out gradually in about ten days, thus allowing the babies often to escape from the iron cage to my arms or to cling to my leg. At last they got accustomed to their new home cages provided I remained in view. As soon as I went out of the room they protested fiercely which they did only rarely in their plexiglas box. From the age of five months on, the young monkeys did not see me more than a few times a day, but that change was carried out very gradually. In one instance only, one of these baby monkeys was abruptly deprived of my presence for some months. He developed a nervous depression much like what appears in young monkeys separated from their mothers. After my return when he was seeing me again several times a day he gradually recovered apparently to a normal behaviour.

These young baboons were so well accustomed to live in the presence of people that they showed averse reaction to the presence of other young baboons. It was a long process, which took the second year of their lives, to accept and get used to living in the company of other baboons.

The youngest of the babies is presently six months old. In his room the plexiglas box is kept as a place of security. This allowed him to be deprived earlier (at 3 months) of my constant presence. By now I only pay a visit to him some times each day and in these conditions the young animal looks well adapted.

Two of these baboons are females and one is male. This male baboon has displayed since 2 years old an apparently normal heterosexual play. It will be of interest to follow up these monkeys when they reach sexual maturity for their ability for reproduction and maternal affection.

It now remains to consider the babies which were reared under the care of their monkey mothers. The first macaque born in the Unit remained caged with his mother up to the age of 14 months. As a consequence it was underfed and to a certain extent rachitic, his mother not allowing him to eat fruit and vegetables which she preferred herself. This food was available in limited amount, only Purina chow was provided without limit. This confirms previous observations [CARPENTER, 1942] according to which female monkeys usually claim food priority over their own infants. At the age of 14 months the young macaque was abruptly deprived of his mother. He started then to show marked stereotyped traits of behaviour such as turning endlessly inside the cage.

The other four baboons and four macaques were separated from their mothers at two and a half to three months of age, as soon as they started to eat fruit and vegetables. It was thus possible to provide them with supplementary food without it being eaten by the mother. The separation from the mother was always a shock to the baby. It ran about the cage crying or cowered in a corner.

The effects of separation from the mother, which could have resulted in a severe neurosis were attenuated as follows. For the first day after separation the young monkey found some comfort in my arms and invested its affection in me. It was only gradually that it got used to staying alone in its cage but I continued to visit it sometimes. The next baby monkey was separated from its mother progressively. During about ten days the baby monkey was taken from its mother for some hours a day only and gradually it got used to being deprived of the presence and the milk of its mother. This method seems to be advantageous for the baby and it saves time for the persons in charge. However, this method is possible only if it is not dangerous to take repeatedly the baby from its mother, since drugs cannot be used to soften her resistance because of nursing.

The next four babies after separation from the mother were caged in pairs. At that time there were two male baboon babies and one male and one female macaque. In each cage we placed a pair of young monkeys of different species according to their own preference. They remained most of the time in a ventral-ventral clasp which is named by HARLOW 'together-together' pattern. This behaviour is typical for infants raised in pairs without mothering for the first reflex stage. The babies which with their mothers had already displayed some exploratory behaviour, when put with an age-mate, returned to the pattern of the first age. This regression of the behaviour may be due to the sudden separation from the mother.

The baby monkeys were caged in pairs for one month; later on, to improve their conditions they were all put together in a playroom with one adult female baboon. Not every female baboon is a good adoptive mother. We once tried to give a newborn baby to a baboon female which had been earlier separated from her own child. She remained indifferent towards the the newborn. In another case we gave to a female baboon a three-month-old baby which was already able to feed himself. The female held the baby very tight on her chest and did not allow him to eat.

It is not easy to find a female which will be a good 'aunt' for the group of babies, but we were successful with an old female baboon. We do not know much about her previous history, but while in our colony she did not get pregnant. She was kind towards other animals, easy to handle by people and above all she disclosed a protective behaviour towards young animals. We could observe her while she was holding the young monkeys by their tails to prevent them from falling from the top of the cage and to protect them from all other dangers. We found her often holding one or another baby in her arms with no apparent species preference. From the moment this group was formed, the behaviour of the baby monkeys was changed significantly. It was characterized above all by interactive plays which correspond to the normal social relations among peers of these ages. The abnormal pattern of behaviour 'two together-together monkeys' disappeared completely. We have never observed the 'choo-choo' formation which is a chain of infants each clinging to the back of its companion in front. It is also a pattern characteristic for young monkeys reared in a group without the mother.

This raises the question whether in this group the affection of an adult female monkey did replace effectively for the babies the affection of their mothers. These data were gathered occasionally and one should try experiments where the mother would be replaced from birth on by another female. Some observations of HARLOW and HARLOW [1965] in this matter are very promising.

As a *conclusion,* I would like to stress that the affective links between animals are delicate matters. Each change introduced in the affectional relations should be carried out gradually. The peer affectional ties cannot adequately replace the mother-infant ties. However, the mother can apparently be replaced by another adult female.

Acknowledgements

This research was supported by the Fonds de la Recherche scientifique médicale to the Brain Research Unit and its Primatology Section.

I am very grateful to Prof. J.E. DESMEDT and Dr. J. DEBECKER for encouragement and advice.

References

CARPENTER, C. R.: Societies of monkeys and apes. Biol. Symp. *8:* 177–204 (1942).

HARLOW, H. F.: Primary affectional patterns in primates. Amer. J. Ortopsychiat. *30:* 676–684 (1960).

HARLOW, H. F. and HARLOW, R. K.: The affectional systems; in SCHRIER, HARLOW and STOLLNIK Behavior of human primates, vol. 2, pp. 287–334 (Academic Press, New York 1965).

HARLOW, H. F. and HARLOW, R. K.: Learning to love. Amer. Scient. *54:* 244–272 (1966).
HARLOW, H. F. and ZIMMERMANN, R. R.: The development of affectional responses in infant monkeys. Proc. amer. phil. Soc. *102:* 501–509 (1958).
HARLOW, H. F. and ZIMMERMANN, R. R.: Affectional responses in the infant monkey. Science *130:* 421–432 (1959).

Author's address: Dr. H. CHORAZYNA, Brain Research Unit, 115, bd de Waterloo, *B-1000 Brussels* (Belgium)

Medical Primatology 1972. Proc. 3rd Conf. exp. Med. Surg. Primates, Lyon 1972, part I, pp. 93–94 (Karger, Basel 1972)

The Owl Monkey *(Aotus trivirgatus)* in Research

R. N. T-W-Fiennes

Nuffield Institute of Comparative Medicine, Zoological Society, London

The value of the owl monkey in research first became apparent from the discovery that – hominoidea apart – it was the only primate which could be infected with human malignant subtertian malaria, caused by *Plasmodium falciparum*.

Interest was enhanced when these monkeys were found to develop malignant lymphoma or generalised sarcomatosis in response to infection with *Herpesvirus saimiri* or spider monkey herpesvirus.

Abnormal responses of these animals to infection or the presence of unusual pathological conditions consequently calls for report. This short paper draws attention to two such conditions.

Malaria Complicated by Herpes

Three adult owl monkeys, used by my colleague Dr. A. Voller in experiments with *Plasmodium brasiliense,* succumbed unexpectedly when suffering from malaria.

In two, histological examination revealed the presence of herpes type intranuclear inclusions and cell necroses typical of herpes, in the lymphatic glands of the neck, adrenals, kidneys, and spleen. There was pyelonephritis, probably due to herpes, and changes in the livers together with pigment deposition characteristic of malaria.

The third monkey suffered primarily from lobar pneumonia with giant cell involvement, indicating viral pneumonia. A few small necrotic foci were found in the adrenals associated with scanty intranuclear inclusions. The only other site of inclusions to be found was the oesophagus. Here, numerous inclusions were present in the deeper layers of the epithelium associated with characteristic necrotic foci.

None of these monkeys had been exposed to the risk of infection with exogenous herpesvirus, except possibly *H. simplex* from a human source from which they usually die. The infection was more likely due to exacerbation of one of the three recognised owl monkey H-viruses. It would appear that the combination of malaria and herpes accounted for their deaths, in one case complicated by viral pneumonia.

Myocardial Infarction

A fourth of Dr. VOLLER's monkeys died suddenly without experimental interference. In this animal, an infarct was discovered in the cardiac muscle. There was no thrombus in the major coronary arteries and the infarct was only found on histological examination. A number of thrombi were, however, present in some small arterioles adjacent to the lesion, and others showed typical small focal atheromatous changes affecting chiefly the intima, the internal elastic lamina and, to a lesser extent, the media. No material was available for fat staining. The infarct was somewhat diffuse and ill-defined, as would be expected from one arising from small vessel involvement.

Comment

Owl monkeys may well prove to be the 'hamsters' of the Primate world, having atypical responses to infectious and oncogenic influences. A more general study of these responses would appear to be justified.

It is also indicated that the small arteries, especially of the myocardium, should be routinely examined at autopsy, because naturally occurring atheroma is so rare in other primates.

Author's address: Dr. R. N. T-W-FIENNES, Nuffield Institute of Comparative Medicine, Zoological Society, Regents Park, *London NW1* (England)

Medical Primatology 1972. Proc. 3rd Conf. exp. Med. Surg. Primates, Lyon 1972, part I, pp. 95–99 (Karger, Basel 1972)

Drug Safety Evaluation in Primates

R. Heywood

Huntingdon Research Centre, Huntingdon

Introduction

Toxicological studies are now required before a new compound is introduced into man's immediate environment. These animal studies should be approached objectively and not as speculative evaluations. It is perhaps unfortunate that government regulatory agencies have produced recommendations – for to many it means that toxicological tests are conducted simply to fulfil these regulations.

Choice of Species

A fundamental consideration is the choice of species. At the end of each study one is left with the problem as to how the animal data can be extrapolated to man. The usual argument for the use of monkeys in toxicological studies is the phylogenetic one. It is, however, much more important to establish the similarities of metabolic pathways than the anatomical and physiological similarities between man and simians.

Metabolic Studies

Metabolic studies are carried out in our laboratories using compounds labelled with ^{14}C. The label is put into the metabolically stable part of the molecule. Two studies are carried out, each involving six primate animals, to establish excretion and plasma level patterns.

The excretion of radioactivity is measured in the urine and feces for five days. The animals are maintained in the standard metabolism cages, but some modifications are necessary, such as fixing transparent plastic sheets onto the front walls of the cages in order to prevent the loss of urine from the front of the cage. It is also advisable to have a very steeply sloping urine collection tray.

Plasma concentrations are measured in the same animals but at different times. Blood samples are taken pre-dose, and then at eight intervals post-dose for a period of 24 h. The nature of the plasma radioactivity of unchanged drug and the total metabolites are determined chromatographically, and the amount of unchanged drug or its measured metabolite in plasma is recorded at each time. The plasma protein binding of the drug is measured at the peak plasma concentration. The studies provide information on absorption characteristics of the drug and the elimination half-life of the radioactivity of unchanged drug.

The monkey is also a convenient animal in which to study bioavailability, that is, comparing absorption of a new product against a similar marketed product.

General Toxicity Studies

When carrying out any toxicological study, the drug should be administered by the intended clinical route. With an experienced team of primate handlers, there are few problems in dosing untranquillized monkeys by any of the normal clinical routes. One of the main arguments against using simians in general toxicological research is the difficulty one experiences in clinically defining which system is being affected. Well-designed toxicity studies always endeavour to induce toxic symptoms in the high dose level group, and it is more satisfactory to define the target organ in life as well as at *postmortem*.

Problems exist in the interpretation of the haematological and biochemical parameters. Table I presents normal haematological and biochemical values, and compares rhesus monkeys with baboons and cynomolgus monkeys. The haematological values are within similar ranges, and there is little difficulty in their interpretation. The biochemistry, however, shows much wider variations so that its diagnostic value is more limited. Simple tests for liver function, such as determination of serum alkaline phosphatase and serum glutamic-pyruvic transaminase, are very valuable in

Table I.

Parameter	Units	Rhesus (n = 28)		Baboon (n = 60)		Cynomolgus (n = 44)	
		mean	95-percent range	mean	95-percent range	mean	95-percent range
Haematology							
Packed cell volume	%	42	35 – 49	46	41 – 52	44	33 – 54
Haemoglobin	g %	12.4	10.5– 14.3	14.2	12.5 – 15.8	11.8	8.2 – 15.3
Red cell count	$\times 10^6/mm^3$	5.2	4.4– 6.0	4.98	4.25– 5.71	6.04	4.36– 7.72
Mean corpuscular haemoglobin concentration	%	30	28 – 32	31	29 – 32	27	23 – 30
Mean corpuscular volume	μm^3	80	75 – 86	93	81 –105	73	62 – 83
Total white cell count	$\times 10^3/mm^3$	11.4	4.4– 18.4	11.0	4.5 – 17.6	15.7	8.7 – 22.7
neutrophils	$\times 10^3/mm^3$	6.1	1.7– 10.5	4.8	0 – 11.4	9.3	4.0 – 14.5
lymphocytes	$\times 10^3/mm^3$	5.2	1.3– 9.1	6.0	2.2 – 9.8	6.0	2.6 – 9.4
Platelet count	$\times 10^3/mm^3$	235	146 –324	238	127 –349	294	156 –432
Prothrombin Index	%	99	92 –106	101	95 –106	98	88 –107
Biochemistry							
Urea	mg %	37	24 – 51	29	15 – 44	35	16 – 54
Glucose	mg %	82	49 –116	95	70 –120	72	47 – 96
Total serum protein	g %	7.8	7.0– 8.6	7.5	6.7 – 8.3	7.5	6.5 – 8.5
albumin	g %	4.0	2.8– 5.2	4.5	3.7 – 5.4	3.1	2.4 – 3.8
α_1-globulin	g %	0.5	0.1– 0.9	0.20	0.04– 0.35	0.5	0.3 – 0.7
α_2-globulin	g %	1.1	0.4– 1.7	0.8	0.4 – 1.2	1.2	0.7 – 1.7
β-globulin	g %	1.3	0.7– 1.9	1.3	0.7 – 1.9	1.8	1.1 – 2.4
γ-globulin	g %	0.9	0 – 2.0	0.7	0.14– 1.22	0.9	0.5 – 1.2
SAP	KA units	86	39 –132	129	76 –183	39	9 – 69
Serum glutamic pyruvic transaminase	SF units	44	6 – 82	40	31 – 49	50	0 –117
Serum glutamic oxylacetic transaminase	SF units	40	18 – 63	46	30 – 62	47	21 – 73
Leucine amino-peptidase	GR units	248	51 –446	140	101 –180	330	0 –754
Electrolytes							
Sodium (Na^+)	mEq/1	153	140 –166	154	145 –163	144	133 –155
Potassium (K^+)	mEq/1	4.6	3.3– 5.9	4.9	3.8 – 6.2	5.2	3.9 – 6.5
Urinalysis							
pH		8.2	7.6– 8.8	7.9	6.9 – 9.1	8.4	7.9 – 8.8
Volume	ml	35	4 – 65	52	0 –121	39	1 – 88
Specific gravitiy		1.028	1.012–1.044	1.031	1.016 –1.047	1.025	1.016 –1.033

the diagnosis of liver disease in man and dog, but more unreliable in monkeys. Serum glutamic oxaloacetic transaminase is routinely measured in most simian studies. However, this parameter is greatly affected by non-specific circumstances such as fright, handling, clinical procedures and bleeding. We have evaluated the serum leucine amino-peptidase test as an additional determination of the liver function but we found it somewhat unreliable.

Inhalation Toxicity Studies

In recent years the direct administration of drugs into the respiratory system has become a new method for the administration of compounds direct to the bronchi and lung. Novel systems have been devised for this purpose. The whole body chamber may be quite satisfactory for some industrial chemicals, but for aerosols and smokes it is too crude a method. A system has been devised whereby a tube can be fitted over the larynx and the aerosol administered directly into the trachea.

Lung function measurements can be made on unanaesthetised animals seated quietly restrained on chairs, fitted with face masks. The dead space in the mask is minimised by using individual rubber liners moulded to fit each monkey, then an air-tight seal is achieved by using a rubber dam. A saline filled nylon catheter of 0.6 mm diameter is inserted into the 4th intercostal space, approximately 3 cm lateral to the mid-sternal line, and this is used for intrapleural pressure measurements. A wide variety of measurements can be made, and these include tidal volume, respiratory minute volume, duration of expiratory phase as percentage of duration of the complete respiratory cycle, pressure change between end tidal volume at inspiration and expiration, dynamic lung compliance, pulmonary resistance during inspiration, pulmonary resistance during expiration and average pulmonary resistance. Blood gases and blood pH can be measured on 200-μl samples taken from the femoral artery of restrained but fully conscious supine monkeys. Initially it was our practice to isolate the artery and place it under a skin flap for ease of puncture; now with experience, arterial samples can be obtained from 90% of monkeys by direct arterial puncture.

The mechanical characteristic of respiration in the rhesus monkey, cynomolgus monkey and the baboon are basically similar for animals of the same size [BINNS *et al.*, 1972].

We are concerned with evaluating the effect of cigarette smokes on the respiratory system of simians. An exposure system has been designed incorporating a mask with a valve/timer. The type of apparatus used is critical, for the smoke must penetrate deeply into the lungs. The face mask must fit closely, and the smoking device be placed so that the butt end of the cigarette is as close to the animal as possible. The animals must not be able to avoid smoking by breath holding; therefore it has proven necessary that smoking be linked to the active part of inhalation. The smoke inhaled should be controlled.

The monkeys are restrained in chairs fitted with masks. Each monkey has an individually moulded face mask. Attached to the top part of the mask is a valve system through which the animal can breathe room air. There is a load head valve mounted against a notched cam which revolves clockwise once per minute. When the notch in the cam moves a quarter of a turn from the vehicle, the slide valve is pushed into the notch, the cam stops rotating and the animal is no longer able to breathe room air. At this stage the animal is connected directly to a lighted cigarette fitted into a holder. The cigarette holder contains an adjustable pressure sensing device and, when the animal takes a breath from the cigarette, a signal from the pressure senser reactivates the notch cam, the slide valve opens and the animal is then exposed again to the room air. The depth of inhalation is controlled by this pressure senser. The animals inhale from the cigarette at minute intervals. This system has proven satisfactory in use with cigarettes containing the radioactive component (^{133}I).

Toxicology is an inexact and at times an empirical science. It is apparent that the use of the subhuman primate in these studies will only produce valuable data if the species is defined as the animal of choice on as many parameters as possible.

References

BINNS, R.; CLARK, G. C., and SIMPSON, C. R.: Lung function and blood gas characteristics in the rhesus monkey. Lab. Anim. *6:* 189–198 (1972).

Author's address: Dr. R. HEYWOOD, Huntingdon Research Centre, *Huntingdon* (England)

Medical Primatology 1972. Proc. 3rd Conf. exp. Med. Surg. Primates, Lyon 1972, part I, pp. 100–104 (Karger, Basel 1972)

Husbandry Procedures in an Indoor Breeding Squirrel Monkey Colony

E. W. HUPP

Biology Department, Texas Woman's University, Denton, Tex.

Squirrel monkeys *(Saimiri sciureus)* have gained increased acceptance as laboratory primates because of their relatively small size, ease of housing, low cost of purchase and maintenance, certain desirable anatomic and physiologic characteristics, and because some of their diseases are of comparative medical interest. Techniques for housing, feeding and maintaining adult monkeys are well established [National Research Council, 1968; ROSENBLUM and COOPER, 1968; HARRIS, 1970; and others]. However, establishing successful breeding colonies has proved to be more difficult. Some workers have reported relatively good results over limited periods of time [CLARKSON *et al.*, 1968], but the observation of COOPER [1968a], 'It has been our experience that only about one-third of all pregnancies result in viable offspring, another one-third to one-half terminate with premature, stillborn, or nonsurviving infants, and the balance end in abortion', still seems to describe the general degree of success. Since data on procedures used in successful colonies is still limited, the following description of techniques employed and results obtained in the Texas Woman's University colony is presented.

The monkeys which made up the breeding colony were obtained in three separate shipments. All were apparently of the Peruvian type [COOPER, 1968b]. The initial group which consisted of two males and three females was obtained in August, 1966. Two additional groups were added: ten pregnant females were obtained in mid-November, 1967 and seven females and five males were obtained in October, 1967. The latter males were subsequently used as sires for colony-bred offspring produced by the ten females obtained while pregnant.

The breeding animals and their offspring were maintained in four rooms in two different buildings: one was a modern laboratory facility, but without separate temperature and humidity control, and the other an old frame building. The mean animal room temperature in all rooms was 25 °C, but varied between 21 and 30 °C, with higher temperatures in summer. Humidity was not controlled; however, it was increased during periods of very low humidity in winter by evaporation of water from wet towels. The relative humidity varied between 10 and 90 %. All rooms were well shielded from outside light with an automatically-maintained constant 12-hour light-dark cycle using fluorescent lighting.

Most monkeys were kept in 90 × 60 × 75 cm cages containing one male and three or four females; the first group obtained was caged in a 75 × 75 × 185 cm cage. Females remained in the cage with the male and other females during pregnancy, delivery and lactation. Infants were removed at four to six months of age. The adult monkeys caged in the larger cage appeared to be more content and may have been more healthy, since no deaths occurred in this group during five and a half years in the colony; no difference in reproductivity between animals caged in the two different size cages was noted.

The animals were fed three times each day at approximately 8 a.m., noon, and 4 p.m. Numerous modifications were made in the diet in an attempt to encourage the monkeys to eat enough to maintain them in good condition. The basic diet consisted of Purina Monkey Chow (Ralston Purina Co., Checkerboard Square, St. Louis, Mo.) containing approximately 15 % protein. This chow was fed at two or three meals each day. Purina Monkey Chow 25, containing 25 % protein was also fed, usually at one meal per day, but in some cases, made up a part of a second meal each day. The amount of the 25-percent protein chow was frequently reduced to combat diarrhea which appeared to be associated with feeding increased quantities. The chow was soaked before feeding. Reconstituted powdered milk was used for the morning feeding and reconstituted frozen orange juice or Tang (General Foods Corp., White Plains, N.Y.) was used for evening meals. Warm water was used for soaking the pellets at the other meals. Fresh fruits – apples, bananas and occasionally grapes or oranges were fed at noon along with three unsalted peanuts in the shells, soda crackers, and few raisins. Water was *ad libitum* from ball watering tubes. The water bottles were changed twice daily to provide fresh water.

Careful observation for symptoms of illness, prophylactic treatment where indicated, and prompt treatment of illness minimized health problems. Shortly after arrival in the colony, the monkeys were treated for internal parasites with Thibenzate (Merck, Sharp & Dohme, West Point, Pa.) and at intervals for up to two years as indicated by examination of the feces. Treatment for intestinal parasites was not necessary after the animals had remained in the colony for more than two years. Diarrhea was the most common health problem, but appeared to be associated with diet and was successfully treated by altering the diet. No animals died due to this cause. Only one instance of a severe respiratory inflammation in the colony was observed following a prolonged period of low humidity. This was successfully treated with Sulmet (American Cyanamid Co., Princeton, N.J.).

Bite wounds were a frequent problem. During the early years of the operation of the colony, several males were severely injured in fights. Females have been bitten by males or other females. Minor bite wounds were treated with topical administration of antibiotic (usually Panalog ointment [E. R. Squibb & Sons, Inc., N.Y.]). More severe wounds were treated with antibiotics i.m. and oral Terramycin (Charles Pfizer & Company, Inc., N.Y.). One male died following a reaction to Combiotic (Charles Pfizer & Company, Inc., N.Y.) and one female died from an abscess at the base of the skull probably caused by a bite wound. One three-month-old infant was killed by an adult male after it was returned to its cage, the cause of this attack was unknown since infants had been previously returned to this cage numerous times by the same handler.

Tail-chewing was a frequent problem among young monkeys. In some cases monkeys sucked or chewed their own tails until they became sore or infected, but more frequently they were chewed or bitten by other occupants of the same cage. This problem has been combatted by providing objects for the young monkeys to play with and by rearranging groups of monkeys. Infected tails have been treated with antibiotics by various routes, usually topically. One animal was lost resulting from infection of the tail.

Some difficulties with dystocia were encountered. These occurred only in animals that were in the colony for two years or longer. One dead infant was delivered by cesarean section; the mother made an uneventful recovery. Two other females were sacrificed after it was determined that the infants were dead and it was not economically advantageous to perform cesarean-sections in order to deliver the young. All three females were multiparous and had previously experienced uneventful delivery and lactations. The cause of the difficulties were not determined; all infants appeared to be at term, but in no case were the fetuses larger than had previously been delivered by the female concerned.

Table I. Summary of reproductive performance of squirrel monkeys, by years of birth

Year	Number of females	Number of conceptions observed	Number of aborted and stillborn	Number born alive	Number weaned
1967[1]	3	2	1	1	1
1967[2]	10	10	3	6	6
1968	20	2	0	2	2
1969	20	13	2	11	9
1970	19	15	4	11	7
Total		41	10	31	25

1 Young adults received in August, 1966.
2 Pregnant monkeys received in November, 1967.

The results obtained during four breeding seasons are presented in table I. During this period, 31 monkeys were born alive, an additional 10 observed conceptions resulted in abortions or still-born infants. Four of the infants died within 72 h, an additional two died before weaning; thus a total of 25 offspring were weaned during this period. Considering the length of time the various monkeys were in the colony, there were 0.54 live births and 0.42 offspring weaned per monkey per year.

The low number of young born in 1968 apparently is due to seasonality of breeding. The jungle-conceived animals were all born in November and December, 1967, except for one born in January, 1968. The dams of these infants did not produce offspring again until 1969, an average of 17 months after the birth of the previous infant. This change in season of birth in jungle captured animals following transport to the northern hemisphere has been described by other workers [ROSENBLUM and COOPER, 1968]. Apparently the young females received in October were received too late in the season to conceive for the following spring and thus were in the colony an average of 20 months before giving birth to young.

Table II. Number of live births per female squirrel monkey during four years

Group	Number of females	Number of live born				Total young produced
		0	1	2	3	
1	3	0	2	0	1	5
2	7	2	1	4	0	9
3	10	0	5	3	2	17
Total	20	2	8	7	3	31

Group 1: Young adults received in August, 1966.
Group 2: Young adults received in October, 1967.
Group 3: Pregnant monkeys received in November, 1967.

The number of live young produced per female in the various groups of monkeys is presented in table II. Group 2 which contained ten females initially obtained during the second half of pregnancy was the most productive. Seven offspring were born alive and successfully weaned as a result of these pregnancies. An additional ten young were born to these females in the succeeding breeding seasons. Ten of the 20 females produced two or more offspring during this period.

In this study nine females failed to produce live young in 1969; one of these died before the 1970 breeding season and only one of the surviving eight females weaned a baby in the 1970 season. Thus, it is probable that a rigorous selection program with retention of only those animals that remained healthy and those that produced an offspring each year, or at least two years of three, would greatly increase the reproductive efficiency of the colony.

References

CLARKSON, T. B.; BULLOCK, B. C.; LEHNER, N. D. M., and FELDNER, M. A.: The squirrel monkey as a laboratory animal; in Animal models for biomedical research, pp. 64–87 (National Academy of Sciences, Washington 1968).

COOPER, R. W.: Small species of primates in biomedical research. Lab. anim. Care *18:* 267–279 (1968a).

COOPER, R. W.: Squirrel monkey taxonomy and supply; in ROSENBLUM and COOPER The squirrel monkey, pp. 1–29 (Academic Press, New York 1968b).

HARRIS, R. S. (ed.): Feeding and nutrition of nonhuman primates (Academic Press, New York 1970).

National Research Council, Subcommittee on Primate Standards, Institute of Laboratory Animal Resources: Nonhuman primates, standards and guidelines for the breeding, care and management of laboratory animals (National Academy of Sciences, Washington 1968).

ROSENBLUM, L. A. and COOPER, R. W.: The squirrel monkey (Academic Press, New York 1968).

Author's address: Dr. EUGENE W. HUPP, Department of Biology, Box 22923, TWU Station, *Denton, TX 76204* (USA)

Medical Primatology 1972. Proc. 3rd Conf. exp. Med. Surg. Primates, Lyon 1972, part I, pp. 105–114 (Karger, Basel 1972)

Baboon and Chimpanzee Husbandry at Southwest Foundation for Research and Education[1]

S. S. KALTER and G. T. MOORE

Southwest Foundation for Research and Education, San Antonio, Tex.

It is apparent that most biomedical investigators still do not consider husbandry to be of major importance. While there is voluminous literature concerning the use of nonhuman primates, little is available on the correct, detailed procedures for handling and maintaining them, nor are the personnel at most research facilities familiar with what constitutes good husbandry. Recently the World Health Organization suggested better methods for the supply and use, as well as development and maintenance of simian colonies [1]. This information becomes important when one realizes that most experimental nonhuman primates are substandard in terms of their projected use and that the depletion of wild stocks will eventually restrict the supply of animals.

This presentation provides information on proven husbandry methods for baboons and chimpanzees. It is based on the practices used at Southwest Foundation for obtaining, monitoring and maintaining high-grade, standardized animals for experimental use.

The same standard management and preventive medicine procedures are applicable to both species. In addition, there are certain precautions that are particularly applicable to chimpanzees. These will be noted where appropriate in the following sections.

Acquisition

Most of our baboons *(Papio cynocephalus)* and approximately half of our chimpanzees *(Pan troglodytes)* were obtained directly from their

1 WHO Regional Reference Center for Simian Viruses.

native habitat, not through commercial sources. Although it is certainly not practical for most laboratories to maintain a capture and holding facility, an understanding of the problems inherent in obtaining animals commercially may eventually help to alleviate the present unsatisfactory conditions. For example, direct capture allows unwanted animals to be rejected before shipment and a well-designed holding facility eliminates or minimizes such problems as the mixing of species, substandard caging, crowding, poor nutrition, inadequate water and excessive contact with people in exporting countries who maintain poor health standards. All of these factors contribute to the physical and psychic stress of the newly captured animals, which, in turn, results in animal loss and in generally poor animal quality.

The 18 animals in our original chimpanzee colony were obtained directly from the Congo. They were maintained for a short stabilization period in Africa and then transported to San Antonio where they were routinely quarantined, examined and tested. Another group of 20 chimpanzees were acquired from the 6571st Aeromedical Research Laboratory at Holloman Air Force Base, N. Mex. The chimpanzee colony presently totals 44 animals: 22 adult females, six adult males, three juvenile females, two juvenile males, five infant females and six infant males. This colony was established to obtain baseline biological data on the chimpanzee and to study their breeding habits in captivity. The breeding program has progressed quite well, since 24 conceptions have been recorded for 23 females in the past two years. Twenty conceptions resulted in live births, four were stillbirths.

Our baboon holding facility in Kenya provides trained personnel, clean individual cages and facilities to separate species. New arrivals are also segregated from other animals temporarily. The animals are fed with fresh fruit which is gradually replaced with a milled preparation of the following approximate composition: crude protein, 20.0%; fat, 4.5%; fiber, 6.0%; NFE, 60.5%; Ca, 1.2%; P, 0.7%; vitamins and minerals, 7.0%; and FE, 100 ppm.

After a quarantine period of at least two weeks, the undesirable animals are rejected and the remaining animals shipped to San Antonio. During this quarantine, the animals are tuberculin tested, vaccinated against yellow fever [5], and treated for parasites with six doses of thibenzole (25 mg/lb.). The animal's general physical condition is noted and blood, throat and rectal specimens are collected. This information forms the initial entries on the animal's biological profile record on which all pertinent information is subsequently recorded.

Our animals are shipped from Africa by the most direct air route possible; however, stops in London and Chicago are generally unavoidable. Arrangements are made with the staff of the animal hostels in the stopover cities to minimize contact with people. The animals usually arrive in San Antonio about 40 h after leaving Nairobi. With the exception of mothers with infants, animals are individually crated in wooden cages made especially for shipment. The cages are not normally reused because of contamination. Fresh food and water are provided in the cages.

Quarantine

Upon arrival in San Antonio, the animals are quarantined in a separate building in individual cages. Animals from different shipments are never housed together, even though they are of the same species.

After a two- to three-day stabilization period, testing of the animals is begun. Tuberculin tests (0.1 ml OT intrapalpebrally) are given every two weeks until three consecutive negative tests are recorded; additional blood, throat and rectal specimens are collected two or three more times. These specimens are examined with a battery of tests for evidence of bacterial, mycotic, parasitic and viral infections. Hemograms are obtained and aliquotes of serum from each animal are stored for future comparisons. Each animal is tatooed with a permanent identification number, which also appears on a disc fastened around the neck. The animals are carefully observed during the entire quarantine period by qualified primate veterinarians. All information gathered on an animal is entered in the profile which was begun at time of capture. The minimum length of quarantine is four weeks, but it may be extended if any disease conditions or other problems arise in animals from a single shipment. No animal is released for experimental or breeding purposes until its clinical tests have been carefully evaluated.

Maintenance

Once the animals are released from quarantine, they are under the care of the Department of Animal Resources. The type of caging selected depends upon how the animals will be used.

Baboons are maintained in gang or colony cages, individual indoor/outdoor cages, or individual indoor cages. Each outdoor colony cage houses a troop of approximately 20 baboons, including a dominant male and two to four juvenile males. This type of caging provides the troop with a living situation approximating their natural environment. It permits maximum freedom of movement, shelter from inclement weather in 'caves' built into rock piles, radiant heated floors in the shelters, fresh running water and troughs where food is always available. Fortunately, San Antonio weather is very similar to that of Kenya and, therefore, maximum use of the relatively inexpensive outdoor facilities can be made. The ratio of males to females permits natural social interrelationships to be formed, and the structure and size of the colony cage provides places for an animal to escape an attacker whenever necessary.

All of the colony cages are connected with tunnel-like chutes which lead to separating cages and a weighing unit. This arrangement permits the separation and removal of any animal within 10 min, and provides a run for the animals while the cages are being cleaned and the feed troughs filled.

The individual indoor/outdoor cages are used for animals that are on special experiments that require frequent monitoring, for sick or injured animals, and for controlled breeding experiments. The inside half of these cages also connects with separating chutes and a weighing unit.

Individual cages are used for animals on infectious studies or other closely-controlled experiments. For newborns, infants, or young juveniles, commercial rack-type cage units with six or eight cages to a unit are quite satisfactory. For adults, cages must be large enough to permit adequate movement. Our larger animal cages are constructed to our specifications, but similar cages are also available commercially. All cages should have squeeze backs, which are essential for handling large primates, and containers for food and water. It is important that individual cages be constructed so that debris cannot collect, since such debris provides an ideal medium for developing infectious agents. For this reason, mesh cages are unsuitable and we have converted to bar-type cages whenever possible. Metals that are resistant to corrosion by urine, excreta and strong disinfectants should be used for cage construction and in any other areas with which the animals are likely to come in frequent contact. The rooms that house individual cages must be equipped with temperature and humidity control. Those used specifically for infectious studies should have individual air intake and exhaust mechanisms; otherwise, all animals in the same room become infected in a short period of time. Ample space must be left between cages to prevent all direct contact and to avoid spattering of feces, urine, discarded food and saliva. It is important to remember that each animal, even one considered to be 'normal' carries organisms that are potentially dangerous to its neighbor.

In contrast to the baboon breeding colonies, our chimpanzee facility is composed of indoor/outdoor units, each of which is large enough to house three adult chimpanzees; they can be enlarged by opening sliding doors between the cages. Our first facility, which has since been enlarged, is illustrated in figure 1 [3]; basic construction is concrete block and No. 6 galvanized wire. A two-way locking, remotely controlled, guillotine door separates the indoor and outdoor halves so that both areas can be cleaned. The indoor sections have self-feeders, water 'lixits' (Alco Manufacturing, Napa, Calif.) and concrete sleeping pads with radiant heat for cold weather. Two females and one male are generally housed in each unit, but pregnant females and juveniles must be caged separately. An infant nursery is maintained for baby chimpanzees rejected by their mothers and for infants separated for experimental purposes.

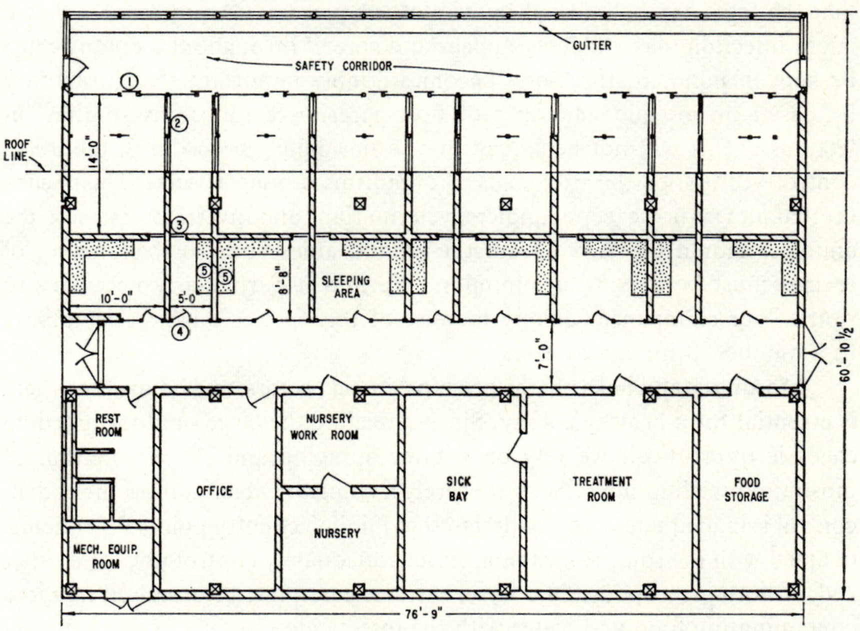

Fig. 1. Floor plan of building housing chimpanzees in the Southwest Foundation for Research and Education. 1 = Sliding gate (woven wire); 2 = sliding solid metal door (1'–8" × 2'–0"); 3 = guillotine solid metal door (1'–7¼" × 2'–3"); 4 = hinged gate (woven wire); 5 = concrete sleeping pallet with radiant heat (7' × 1'–8" × 1'–8"). Scale: ⅛" = 1'–0".

The general health and welfare of animals is grounded in a rigid system of sanitation. Any variation or shortcut in the established routine may cause widespread dissemination of a microorganism. Sufficient data are now available demonstrating that animals in a single herd and animals that have been in contact with humans or other animals rapidly exchange organisms [4]. Normally, the spread of organisms is minimized by the immune system of the invaded animal, but if this system breaks down or if other factors such as diet, medication, etc., become involved, overt disease occurs. How organisms are transmitted to a single animal or are spread throughout a population is fairly well understood, but these infectious pathways are too frequently ignored or overlooked by the maintenance staff [3, 4].

The existence of inapparent or latent infections is a new concept in epidemiology. An infectious agent may exist harmlessly within the tissue of an animal for unknown periods of time, only to appear suddenly in response to some other factor or factors. These infections can even be transmitted vertically, i.e. to the offspring via the ovum. It has also been demonstrated that the fetus can be infected by way of the placenta. An awareness that a latent infection may suddenly appear and spread throughout a colony must be kept in mind. It, therefore, becomes doubly important that husbandry be based on an understanding of host-parasite relationships so that infectious agents will not be spread due to unsanitary procedures, failure to control vectors, or the exchange of organisms among species. This means (a) that all visitors, except handlers, veterinarians and investigators using the animals, should be barred from the animal areas and (b) that mixing of species must be kept to a minimum. We find it particularly necessary to restrict our chimpanzee colony because of this species' high susceptability to a number of human diseases.

Mosquitos and ticks are the principal insect vectors, and controlling them is essential for a healthy colony. Since screening the large outdoor breeding cages is impractical, we rely on routine spraying and the elimination of mosquito breeding sites. Ticks are rarely a colony problem unless the rodent control is inadequate. Facility design that minimizes entry points for rodents, coupled with the routine spraying, should adequately control ticks and other rodent-born arthropods. Of course, extreme care must be exercised to avoid contaminating food and water with the insecticides.

Pressurized hot water not only cleans, but the steam is germicidal. Detergents are helpful, but must not be considered 'complete' disinfectants, since they may be inactivated by reacting chemically with organic material in animal waste products, body fluid, hair and feed. Portable, high pressure sprayers are relatively inexpensive and effective. Though somewhat slower, high-use areas such as sleeping pads, walls near sleeping and eating areas, water 'lixits' and feeders may need to be scrubbed with brushes because of the heavy build-up of body oils and feces. We use high pressure hoses and chemical detergents twice daily to clean the colony cages and the indoor/outdoor baboon and chimpanzee cages. Disinfectants are applied on alternate days. The smaller, individual cages are also cleaned once or twice a day with chemical detergents and disinfectants. Tatooing equipment, syringes, needles, handlers' gloves, etc., must be handled with extra care in order to avoid transmitting infectious agents. Routine culturing of material from different parts of the cages is an effective check on the adequacy of cleaning methods, but care must be taken to see that all disinfectants are removed before obtaining samples.

The baboons can be trained to cooperate in the cleaning routine by using both food rewards and their dislike of pressurized water to make them enter the chutes. These must be washed down after the animals have been returned to their cages.

Because a complete and satisfactory sanitation program is not easy to maintain, it should be supervised regularly by trained people who have a clear understanding of public health problems, sanitation and epidemiology. Even though a person has an advanced degree, he may not be trained in these subjects and so may not be qualified to supervise this function. Also, animal handlers have not usually had advanced education and may not be familiar with the problems of general hygiene and public health. If a well-designed, complete training program is not available to familiarize all personnel with these problems, they often resort to short cuts that result in a breakdown of the routine and in the development of serious disease problems.

Nutrition

One of the major husbandry problems in any primate colony is the transition from natural food to a laboratory diet that is acceptable and nutritious. In the bush, baboons (and chimpanzees) are primarily vegetarians, although males may become carnivorous, feeding on rodents or young game. With our baboons, the transition is accomplished by feeding fruits which are gradually replaced by a milled diet which is, in turn, replaced by a biscuit ration. This ration was originally developed by KOEHN [6] and subsequently modified [2]; it is mixed to our specifications (table I) by the Ralston-Purina Corporation, St. Louis, Mo. Our chimpanzees have been adapted to the baboon ration supplemented with fresh fruit (apples, oranges or bananas). The animals are fed *ad libitum,* and it is important to provide adequate food in the colony cages so that even the subordinate animals will be able to get enough to eat.

Newborn baboons that are removed from their mothers are bottle fed on a formula of equal parts SMA (Wyeth Laboratories, Philadelphia, Pa.) and water. For the first two weeks they are fed approximately every 4 h. During this time, the infant gradually learns to feed itself with a bottle fastened to the side of the cage by a special adapter.

Chimpanzee infants are first fed formula of one part SMA to three parts water, supplemented with vitamins. After four weeks, some bulk is provided in the form of mixed baby cereals. By 12–15 months, the baby chimpanzees begin to eat the standard biscuit and fruit diet.

Handling

Animals that are 4–5 lb. or larger usually require sedation for safe handling. While routine, visual examinations can be done with unanesthetized

Table I. SFRE baboon ration, %

Corn, yellow ground	35
Standard wheat middlings	20
Meat and bone meal	10
Soybean oil meal	10
Skim milk, dried	10
Alfalfa meal, dehydrated	2
Bone meal, steamed	2
Corn oil	6
Molasses, cane	2
Salt, iodized	1
Vitamin B, premix	1
Vitamin A, stabilized concentrate, sufficient to furnish 10, 100 IU/lb. ration Irradiated yeast, sufficient to furnish 70 IU vitamin D/lb. ration Ascorbic acid, 30 g/100-lb. ration Isonicotinic acid hydrozide, 0.015 %	1

This ration provides the following approximate analysis: protein (40 % animal protein): 20.85 %; fat: 9.35 %; carbohydrate: 46.70 %.

animals, collection of samples or other experimental procedures should only be performed when the animals are sedated. For baboons, Sernylan (Bioceutics Laboratories, Inc., St. Joseph, Mo.) gives excellent results when administered at 1.0 mg/kg of body weight. Within 5–10 min after i.m. inoculation, anesthesia that is sufficiently intense to perform minor surgery is induced. Sernylan has a wide nontoxic range in baboons, since 10 mg/kg doses have been given with no apparent ill effects. It does not appear to be addicting nor to produce side effects; however, continued use can result in increased resistance to the drug.

With the chimpanzees, Sernylan may also be used at 1.0 mg/kg of body weight. It produces cataleptic anesthesia in 10–15 min and allows approximately 45 min for examination. Ketamine (Parke-Davis Co., Detroit, Mich.) given i.m. at 12.0 mg/kg of body weight produces similar anesthesia in 4 min or less and persists for about 20 min.

The use of squeeze cages is essential for the quick, safe inoculation of animals. All individual cages should be equipped with this mechanism. Animals in the colony cages are run through the chutes to separating cages where they can also be weighed. The chimpanzees are separated by attaching a capture box to the guillotine door and opening it. A squeeze cage can then be attached to the capture box. The animals are trained to cooperate with this procedure using food rewards.

For both animal species, these capture periods can be used for observation of changes in the appearance and turgesence of the females' external genitalia. The estrous cycle of most animals can be calculated by these changes. If timed pregnancies are desired, they are best done in individual cages after suitable time for familiarization has been allowed.

Whenever an animal is sampled, treated with any preparation, sedated, mated, etc., the fact is noted in its profile. This continuing record provides for complete characterization of the animal's background and documents any factor which may subsequently effect the outcome of an experiment.

Acknowledgements

This study was funded in part by USPHS grants RRO5519, RROO278, RROO361, RROO451 and AIO5374 and contract NIH 69-2011; National Multiple Sclerosis Society grant 682-B-2; and World Health Organization grant Z2/181/27. This laboratory serves as the WHO Regional Reference Centre for Simian Viruses.

References

1 Health aspects of the supply and use of non-human primates for biomedical purposes. WHO Technical Report Series, No. 470 (World Health Organization, Geneva 1971).
2 HUMMER, R. L.: Observations on the feeding of baboons; in HARRIS Feeding and nutrition of nonhuman primates, pp. 183–203 (Academic Press, New York 1970).
3 HUMMER, R. L.; MAY, H. C., and KNIGHT, W. F.: Observations during the first year of operation of a chimpanzee breeding colony. Z. Versuchstierk. *11:* 207–221 (1969).
4 KALTER, S. S. and HEBERLING, R. L.: Comparative virology of primates. Bact. Rev. *35:* 310–364 (1971).
5 KALTER, S. S. and JEFFRIES-KLITCH, H.: Yellow fever vaccination of primates. Amer. J. trop. Med. Hyg. *18:* 466–469 (1969).
6 KOEHN, C. J.: The feeding of baboons *(Papio cynocephalus* and *P. doguera)*. Lab. anim. Care *16:* 178–184 (1966).

Authors' address: Dr. S. S. KALTER and Dr. G. T. MOORE, Southwest Foundation for Research and Education, *San Antonio, TX* (USA)

Medical Primatology 1972. Proc. 3rd Conf. exp. Med. Surg. Primates, Lyon 1972, part I, pp. 115–119 (Karger, Basel 1972)

Oncogenic Virus Studies in Nonhuman Primates under the Special Virus Cancer Program

R. F. KINARD and H. RABIN

Viral Leukemia and Lymphoma Branch, National Cancer Institute, National Institutes of Health, Bethesda, Md. and Department of Virology and Cell Biology, Litton Bionetics, Inc., Kensington, Md.

In 1962, the National Cancer Institute (NCI) initiated viral oncology programs in nonhuman primates. The history and composition of this overall program along with a statement of its principal goals have been reviewed recently [3, 4]. Programs for production and inoculation of simians were begun with an aim to induce leukemia of human origin in an experimental primate host and to recover the causative virus, establish laboratory strains of it, confirm its pathogenicity for humans through seroepidemiological studies, and develop an effective control measure for the disease. These goals basically remain unchanged. The program was originally conceived by Dr. G. ZUBROD and Dr. W. R. BRYAN. In 1964, the Special Virus Leukemia Program (SVLP) directed by Dr. F. J. RAUSCHER, assumed responsibility for the primate studies. In 1968, the SVLP was broadened to include other neoplasms and was renamed the Special Virus Cancer Program (SVCP). The SVCP, currently headed by Dr. J. B. MOLONEY, continues its support of primate production and inoculation programs.

As this is a rather dynamic program its projects have varied extensively since its inception. The purpose of this report is to present the composition of the current program. In doing so, we will discuss the project, the principal primate species used, the major viruses under study, and the general work scope. We are including in the discussion only those projects in which breeding colonies are an essential part of the individual program.

The Litton Bionetics, Inc. Program, Kensington, Md.

This contractor has an established breeding and inoculation program of macaques, mainly *Macaca mulatta* but also including *M. fascicularis* and

M. radiata. These animals are all hand reared and are maintained in isolators when under study. Current production of rhesus monkeys is set at 225 newborns per year. Attempts are being made to establish similar although somewhat smaller colonies of several New World species, principally *Saguinus fuscicollis, Aotus trivirgatus, Saimiri sciureus,* and *Cebus albifrons*. Adults of these New World species, obtained commercially, are currently available for study. A small colony of the prosimian species *Galago crassicaudatus* is located at this facility, along with small numbers of inoculated individuals of several other species being held for observation. The work scope and viruses under study include: investigations of the natural history and possible oncogenicity of the Mason-Pfizer monkey virus, an agent associated with a breast tumor of a rhesus monkey [1]; studies of oncogenicity and immunology of *Herpesvirus saimiri,* a herpesvirus of squirrel monkey origin [7]; attempts to infect and transform rhesus monkey cells *in vitro* and produce tumors in rhesus monkeys with RD-114 virus, an agent associated with a human rhabdomyosarcoma [5]; and attempts to transmit lymphosarcoma in rhesus monkeys from material originating from spontaneously-occurring rhesus monkey lymphosarcomas [8]. In addition to these programs, inoculations with other selected virus and tumor materials are also performed. The program at Litton Bionetics, Inc. is presently under the direction of Dr. D. A. VALERIO and Dr. H. RABIN.

The Rush-Presbyterian-St. Luke's Medical Center Program, Chicago, Ill.

This program is under the direction of Drs. L. G. WOLFE and F. DEINHARDT. The principal species bred are *S. fuscicollis* and *S. nigricollis*. Other species in use include *S. oedipus* and *Saimiri sciureus*. The number of newborn marmosets produced per year is greater than 100. The program consists of studies concerning bioassay, antigenicity, oncogenicity, and enzymology of simian sarcoma virus type 1 and simian sarcoma associated virus type 1 [9, 10]; attempted oncogenesis with RD-114 virus; and studies on the natural history and oncogenicity of HVS and of the characterization of HVS-infected lymphoid cell lines. Other studies include investigations of feline sarcoma virus and the Epstein-Barr virus.

The Lincoln Park Zoo Program, Chicago, Ill.

This program is solely concerned with the breeding of marmosets *(S. fuscicollis* and *S. nigricollis)* under the direction of Dr. L. FISHER. There are approximately 107 adult monkeys in this colony, and the over 100 viable young produced per year are used either as breeding stock or are made available for experimental programs within the SVCP but principally for the Rush-Presbyterian-St. Luke's Medical Center Program.

The Southwest Foundation for Research and Education Program, San Antonio, Tex.

A program for the breeding of chimpanzees, *Pan troglodytes,* is supported with Dr. S. S. KALTER as director. The goal of this program is to provide the SVCP with newborn chimpanzees with two of three newborn animals designated for experimental study as determined by the NCI. A total of 15 animals is presently in the colony, including two breeding age males, six breeding age females, one juvenile male, two juvenile females, three infant males, and one infant female. Approximately five newborn chimpanzees are produced annually.

The Yerkes Regional Primate Research Center, Emory University Program, Atlanta, Ga.

This program is directed toward documenting and characterizing tumors which arise in a colony of irradiated and/or aging rhesus monkeys and in breeding selected animals of this group to evaluate possible genetic damage and occurrence of neoplasia in their progeny. At present, the colony includes 65 adult monkeys ranging in age from 13 to 19 years, of which 48 were exposed to various radiation regimens in 1956–1958. 16 are nonirradiated, and one was added to the study due to the presence of an abdominal adenocarcinoma. There are also eight offspring ranging in age up to four years. Of the total number of deaths in this colony, 33% were related to tumors. Of the last four monkeys which died, three had malignant neoplasms. At present, three animals in the colony have tumors including an adenocarcinoma, a seminoma, and an uncharacterized tumor adjacent to the ileum. Specimens from the tumors and from the monkeys bearing them are made available for study within the SVCP. The program is directed by Dr. H. M. MCCLURE.

New England Regional Primate Research Center, Harvard University Program, Southboro, Mass.

A program is being initiated with Dr. L. V. MELENDEZ as director. The work scope will involve investigations of herpes-type viruses isolated from various species of New World monkeys. Principally, investigations will be made of the biological characterization and of the pathogenesis for several species of New World and Old World primates of *Herpesvirus ateles,* a recently-described virus of black spider monkey *(Ateles geoffroyi)* origin [6], and of the effectiveness of attenuated HVS in preventing disease inducible by parent virus. Additionally, several other herpes-type virus isolates will be investigated for oncogenic and oncolytic properties in nonhuman primates. The primate species under study will include five New World species: *A. trivirgatus, S. oedipus, C. albifrons, Ateles* sp., and *Saimiri sciureus;* and three Old World species: *M. mulatta, M. fascicularis,* and *Cercopithecus aethiops.*

The Comparative Oncology Laboratory, University of California, Davis Program, Davis, Calif.

A program for the study of simian sarcoma virus type 1 and an oncornavirus associated with spontaneous gibbon lymphosarcoma [2] is directed by Drs. L. K. Bustad and T. G. Kawakami. Investigations of these viruses are in areas of biological and physicochemical characterization, pathogenicity, and immunology. The laboratory maintains a small breeding colony of marmosets *(S. fuscicollis* and *S. oedipus)* and is attempting to establish a breeding colony of gibbons *(Hylobates lar).*

Former Programs

The breeding program to produce and inoculate newborn baboons *(Papio anubis* and *P. cynocephalus)* operated by Baylor College of Medicine, Houston, Tex. and the Southwest Foundation for Research and Education under the direction of Dr. J. L. Melnick has been discontinued. However, a number of baboons previously inoculated at Baylor are still being maintained under observation.

The program at the Institute for Comparative Biology of the Zoological Society of San Diego, San Diego, Calif. has also been terminated. This program emphasized breeding of small species of primates (adult about 1 kg body weight or less) and had been successful in breeding several species of primates including *Cercopithecus (Miopithecus) talapoin, Cercopithecus pygerythrus, Samiiri sciureus, Callithrix jacchus, Saguinus mystax* and *Galago crassicaudatus*. Tumor virus studies performed on this program included those with Rous sarcoma virus, feline sarcoma virus, simian sarcoma virus type 1, and HVS. Dr. R. Cooper, formerly director of this program, is currently serving with the Peace Corps in Colombia, South America where he is participating in population studies of squirrel monkeys.

References

1 Chopra, H. C. and Mason, M. M.: A new virus in spontaneous mammary tumor of a rhesus monkey. Cancer Res. *30:* 2081–2086 (1970).

2 Kawakami, T. G.; Huff, S. D.; Buckley, P. M.; Dungworth, D. L., and Snyder, S. P.: C-type virus associated with gibbon lymphosarcoma. Nature New Biol. *235:* 170–171 (1972).

3 Kinard, R.: A program for inoculation of primates with potentially oncogenic viruses; in Goldsmith and Moor-Jankowski Medical primatology 1970, pp. 895–902 (Karger, Basel 1971).

4 Kinard, R.: Cancer viruses in primates. Science *169:* 828–831 (1970).

5 McAllister, R. M.; Nicolson, M.; Gardner, M. B.; Rongey, R. W.; Rasheed, S.; Sarma, P. S.; Huebner, R. J.; Hatanaka, M.; Oroszlan, S.; Gilden, R. V.; Kabigting. A., and Vernon, L.: C-type virus released from cultured human rhabdomyosarcoma cells. Nature New Biol. *235:* 3–6 (1972).

6 Melendez, L. V.; Daniel, M. D.; Barahona, H. H.; Fraser, C. E. O.; Hunt, R. D., and Garcia, F. G.: New herpesviruses from South American monkeys. Preliminary report. Lab. anim. Sci. *21:* 1050–1054 (1971).

7 Melendez, L. V.; Hunt, R. D.; Daniel, M. D.; Garcia, F. G., and Fraser, C. E. O.: *Herpesvirus saimiri*. II. Experimentally induced malignant lymphoma in primates. Lab. anim. Care *19:* 378–386 (1969).

8 Stowell, R. E.; Smith, E. K.; Espana, C., and Nelson, V. G.: Outbreak of malignant lymphoma in rhesus monkeys. Lab. Invest. *25:* 476–479 (1971).

9 Theilen, G. H.; Gould, D.; Fowler, M., and Dungworth, D. L.: C-type virus in tumor tissue of a woolly monkey (*Lagothrix* spp). with fibrosarcoma. J. nat. Cancer Inst. *47:* 881–889 (1971).

10 Wolfe, L. G.; Deinhardt, F.; Theilen, G. H.; Rabin, H.; Kawakami, T. G., and Bustad, L. K.: Induction of tumors in marmoset monkeys by simian sarcoma virus, type 1 (*Lagothrix*). A preliminary report. J. nat. Cancer Inst. *47:* 1115–1120 (1971).

Authors' addresses: Roy Kinard, DVM, Viral Leukemia and Lymphoma Branch, National Cancer Institute, National Institutes of Health, *Bethesda, MD 20014;* Harvey Rabin, PhD, Director, Department of Virology and Cell Biology, Litton Bionetics, Inc., 5510 Nicholson Lane, *Kesington, MD 20795* (USA)

Medical Primatology 1972. Proc. 3rd Conf. exp. Med. Surg. Primates, Lyon 1972, part I, pp. 120–129 (Karger, Basel 1972)

Preventive Medicine in a Primate Colony

E. C. MELBY, jr.

Johns Hopkins University School of Medicine, Baltimore, Md.

Introduction

There is growing recognition of the importance of nonhuman primates in biomedical research and testing as well as an increasing concern over problems in primate conservation. Over the past 25 years a large number of primate centers and colonies of various sizes have been established throughout the world in order to support the research and testing demands of science and technology. Much has been learned, largely through trial and error, concerning the trapping, compounding, transportation, nutritional and other demands of these animals in order to make them available for use in laboratories. The variety of primate species now required to meet these demands speaks of the unique potential of these animals as models for the study of disease processes in man. No single species nor group of animals possess all of the unique characteristics required and undoubtedly the number of species will expand as more information becomes available. We simply cannot predict which primate will become an invaluable resource for the study of a given problem at some future date. Serious problems in supply brought about through political uncertainties, destruction of natural environments and related factors make it imperative to conserve these valuable resources [1]. The interrelationships of various disciplines such as husbandry, behavior and clinical medicine must be understood if we are to be successful in conserving our primate resources and develop programs which can address the legitimate demands of the biomedical researcher as well as the concerns of the naturalist who views nonhuman primates as important components of their natural habitats.

The development, maintenance and operation of a primate colony requires the knowledge and skills of many different areas if it is to be successful. Depending upon the size of the program, this expertise may be represented through the assembly of permanent staff covering specialties such as animal husbandry, animal behavior, clinical medicine, surgery, pathology, diagnostic laboratory support, physiology, pharmacology, engineering and laboratory animal technology. Smaller programs, whether they recognize it, must touch on these areas in one way or another, perhaps relying on the use of consultants, a good library or experience gained through trial and error. The important thing to recognize, however, is that a successful program encompasses all of these and, therefore, the development and operation of a primate colony program is not to be taken lightly. It is the manner in which interrelationships of many disciplines and areas of interest effectively develop which will determine ultimate success, or failure, and scientific productivity. The primary emphasis of this paper, however, will be that of the development of an adequate program in preventive medicine as it relates to successful colony management.

A common definition of the term 'disease' states that it implies a deviation of the body, or an organ of it, from health or normality. Taken in its broadest sense, 'disease' is any state of change and strict adherence to this definition would imply that most animals and man are in a state of continuous 'disease' at any given time. A more acceptable definition would state that 'disease' occurs whenever the physical or functional state is altered in such a way that it is at least measurable whether by clinical observation, diagnostic procedures or other means. Since a major concern of any primate colony operation is to develop and maintain animals in good health or free from 'disease', it follows that preventive medicine is of extreme importance if one is to achieve this goal. To this end, it is necessary to consider all factors affecting the animals which result in conditions delterious to their health. Much has been written concerning the problems of capture and compounding of primates in their respective countries of origin. Transportation still remains a major problem although the advent of high-speed jet aircraft in permitting us to rapidly bring the animals from their place of origin to the laboratories has proven of immeasurable value in lessening mortality and morbidity. Problems still remain and every effort must be made to lessen the amount of stress, improve nutrition, prevent contamination by pathogens and, in other ways, move the animals as carefully and expeditiously as possible from their native habitat to the controlled environment of the laboratory. An excellent review of this problem has recently been presented by HARRISON [1]. Since it is generally impossible for the staff of a primate colony to control most events prior to arrival at the laboratory, the implementation of a program in preventive medicine essentially begins upon the arrival of the animals. A general understanding of events, which have occurred prior to arrival, must be taken into consideration since these factors will play a major role in the health status of animals, especially during the first few weeks in the laboratory [4].

In establishing a program in preventive medicine one must consider the health of the animals, the primate colony staff, investigators or others who will come in contact with the animals, and the type of research or other use for which the animals are destined. It is most important that the administrative staff of the colony clearly understand what is at stake, and firmly support concepts of preventive medicine for this must serve as the basis for any successful colony operation. The total environment of the animal must be carefully considered. Is the housing and caging adequate to meet the needs of the species of animal being used? The subject of adequate space for the housing of various species of animals, especially nonhuman primates, is widely debated and there is no unanimity of opinion. Professional judgement must be exercised in all cases, taking into consideration the species to be housed, the nature of the work for which the animal is destined, length of confinement and other similar factors. The Institute of Laboratory Animal Resources (ILAR) of the National Research Council, National Academy of Sciences in the United States has developed guidelines covering the minimum space requirements for various species of animals and these documents will be helpful in critically evaluating these requirements [2, 3].

Quarantine

It is most important that adequate quarantine facilities be provided in which individual shipments of animals can be isolated. In general, a series of rooms should be devoted to this purpose which are isolated from the main colonies so as to prevent any possibility of cross infection [4]. Within the past few years new information concerning subclinical or latent viral infections in certain species of primates which may be fatal when transmitted to other species has become available, thus making it mandatory to further separate or segregate primates by species. For example, *Herpesvirus tamarinus* may cause mild stomatitis in squirrel monkeys *(Saimiri sciureus)* but fatal epizootics have followed natural transmission of the virus from squirrel monkeys to owl monkeys *(Aotus trivigatus)* and some marmosetts *(Saguinus* sp.); thus, squirrel monkeys should not be housed in the same room with these latter species [5]. Also, it is well known that *Herpesvirus simiae* (B virus) is seen in Asian monkeys, especially in the rhesus *(Macaca mulatta)* and cynomolgus monkeys *(Macaca fascicularis)*. It is, therefore, best to house them separate from other primate species. Primates of African origin should possibly be housed apart in order to protect against the spread of viruses such as Yaba or Yaba-like and Marburg. Undoubtedly, as more information becomes available, it will be found necessary to further isolate primates by species both in the quarantine as well as in the holding colonies [3]. Although there is no unanimity of opinion as to length of quarantine, experience has shown that an average of 90 days is adequate to identify and control major disease problems prior to introducing new animals into the main primate colony [2].

During the quarantine or isolation period, various clinical and laboratory diagnostic procedures should be carried out. Tuberculosis remains one of the major problems in primate colony management and strict adherance to a well-defined tuberculosis control program should be followed [6]. The recommendations promulgated by the ILAR have proven helpful in establishing a good tuberculosis surveillance and control program [2].

Nutrition

The nutrition of simians must play an important part in any program of preventive medicine. A knowledge of individual species requirements is, of course, necessary although it has been found that the specialized primate diets prepared by several of the laboratory animal feed manufacturers can serve as a basis for the feeding of many different species of primates. Peculiar requirements such as the need for vitamin D_3 in South American primates must be recognized [7]. It is most important that the food be presented in a palatable manner which prevents fecal contamination and that it be provided in suffiucient quantities to permit normal growth in immature animals and maintenance in adults. The feeding of primates should be carried out by personnel who will do more than just mechanically provide food to the animals. Feeding time should be a period in which observations as to general behavior and health status can be made and recorded. The frequent weighing of animals, perhaps in conjunction with testing for tuberculosis, should be carried out and recorded on individual animal records. Examination of these weight records may be the first indication of disease, under or over feeding and various other problems.

Diagnostic Facilities

The provision of adequate diagnostic laboratory support must receive high priority whether this be provided within the colony or by access to other laboratories which can provide prompt back-up support. Several primate colonies have attempted to identify carriers of salmonella or shigella through scheduled stool samplings [8]. Although such procedures are of value in confirming diagnoses of clinical bacillary dysentery, thus assisting in treatment, it is doubtful that methods currently available for the treatment of either salmonellosis or shigellosis can satisfactorily eliminate the carrier state of individual animals. For these reasons we do not recommend the routine screening of colonies for either of these organisms assuming that a relatively high percentage of individual animals will shed bacilli at one time or another. Retrospective studies of these problems as they relate to incidence of disease, receipt of newly arrived animals and the development of multiple antibiotic drug resistance have been published [9, 10]. Control of dysentery in a primate colony is dependent upon the interrelationship of adequate housing, sanitation, nutrition, the minimizing of stress, prompt clinical diagnosis, laboratory confirmation and treatment.

Depending upon the use for which the animals are destined, the size of the colony and other factors, serologic monitoring for virus diseases may be indicated. A recent review article [11] has provided insight as to problems which may be encountered and identified through serologic sampling. At the same time it may be indicated to check for the presence of malaria and other parasitic organisms. Information has been published concerning the parasites of various primate species, and most colonies do administer appropriate anthelminthics, especially during the quarantine period. Here again, proper sanitation and procedures which minimize contact with excreta, thus interrupting parasitic life cycles, will do much to lessen parasitic disease in the colony. Some parasites, such as the *Acanthocephalids,* do not respond to currently available methods of treatment and, therefore, reliance must be placed on the interruption of life cycles of the parasite. Strict control of vermin, especially cockroaches, is of importance in the colony.

Effect of Stress

In the management of disease, especially as related to preventive medicine, the need to minimize stress has gained increasing recognition. All primates are undoubtedly subject to stress in one degree or another during their entire lifetime in the wild. Since confinement in a primate colony undoubtedly increases the stress, it is necessary to recognize this as major factor in disease control. The provision of controlled environments, adequate space and nutrition has already been mentioned. A major source of stress is often that of the animal technician and other personnel who must deal directly with the animals. Personnel must be carefully selected and trained in such a way as to minimize the amount of additional stress they will cause. Not all people are capable of working with primates and it is often only through close observation by supervisory staff that those who can work well with animals can be identified. No program in preventive medicine will be successful unless it takes under consideration the manner in which animals are handled and otherwise cared for. The best of medical attention, diagnostic laboratory support, physical facilities and nutrition will not overcome basic inadequacies in technical services which result in either physical or psychological trauma to the animals.

Reproduction

The problem of primate reproduction is receiving increasing attention in colonies and centers throughout the world. Although much of our basic knowledge of human reproduction was gained through the study of primates, largely *Macaca mulatta,* there is still much to be learned about reproduction in most of the other primate species. Many breeding colonies have been developed in order to provide biological models for the study of growth and development or support studies in areas such as reproductive physiology, immunology and surgery, so that it has been possible to gain both experience and knowledge in certain aspects of primate reproduction. There is growing concern, however, that the supply of primates from their natural habitats is not inexhaustible [1]. Vast areas of South America, especially in Brazil, are being deforested for agricultural purposes as well as to provide room for the growth of new cities. Concern for potential changes in the policies of governments of the major countries supplying primates and the increasing emphasis on conservation of resources can result in the sudden closing off of established supply lines for various species of primates urgently required in biomedical programs. It would appear that we must now seriously consider establishing centers for the production of various species of nonhuman primates outside of their normal countries of origin in order to assure a continuous and orderly supply. If such centers are to come about and be successful, it will require the close co-operation of the many disciplines required to implement sound programs in preventive medicine. Involved will be the provision of facilities specifically developed to meet the different requirements of primate species to be housed therein, close attention to aspects of nutrition, handling, clinical medicine, disease diagnosis and control, primate behavior and laboratory animal technology. We simply do not have sufficient information available at the present time to satisfactorily launch major breeding programs for many of the important primate species now used in research.

Whenever attempting to undertake such endeavors, however, well spelled out programs in preventive medicine should be developed in advance of proceeding with the construction of facilities or attainment of breeding stocks of primates.

As an example, efforts are currently underway to develop a new commercial breeding colony of *Macaca mulatta* on an island in the Bahamas. Since the island has not previously been inhabited by nonhuman primates, or for that matter by man, it should offer excellent possibilities to clearly define the disease profiles of the animals to be introduced and thus enhance the value of the subsequent offspring which will be available for research. It is hoped that before any breeding stock are introduced a well-defined disease control program will have been established which will carefully select those animals to be introduced. The selection process should screen for known pathogens as well as for physical vigor, size and desirable behavioral characteristics. All too often these factors have not been considered until a program is well along and experiencing serious problems.

Personnel Health and Safety

In speaking of preventive medicine in a primate colony, one must not fail to include the development of a well thought out program in personal hygiene and personnel health. An occupational health program is mandatory for all personnel with significant animal contact as well as for personnel working in the laboratory animal facilities. Such a program should include both pre-employment and periodic physical examinations for the occupational hazards that may exist should be recognized. For example, it is generally recognized that it is important to immunize all animal care personnel against tetanus. It is generally accepted that all personnel should be tested for tuberculosis at least annually and those with direct animal contact at six-month intervals. For personnel working with newly imported chimpanzees the hazard of the transmission of infectious hepatitis must be considered. In addition to carefully instructing such personnel as to proper means of handling the animals and personal hygiene, some programs attempt to provide protection through administering γ-globulin to all persons considered to be at high risk. Protective clothing such as full length gowns, gloves, masks and face shields should be available for use when handling primates although it is recognized that experienced primate handlers will often shun the use of some of these protective devices. The administrator of the colony should weigh all factors, set reasonable standards for personnel and then establish strict guidelines which must be enforced.

A retrospective study of disease incidence and personnel health within the central primate colony of the Johns Hopkins University may serve to illustrate the importance of implementing an adequate personnel health program [12]. During the course of a 30-year period, one technician, working in the same colony, experienced separate episodes of laboratory confirmed disease including severe colitis caused by entamoeba histolytica, shigella paradysenterica, strongyloides stercoralis as well as a positive tuberculin conversion with right hilar lobe adenopathy and increased density in the right lower lobe. No progression of the disease has occurred. Numerous, severe bite wounds and scratches have also been recorded for the same individual and at one point he suffered from a severe herpetic infection of the mandibular branch of the 5th cranial nerves. No serologic testing for viral antibodies has thus far been conducted. Two other technicians developed infectious hepatitis while caring for newly imported chimpanzees, and a fourth converted to positive on routine tuberculin testing. Periodic stool samplings have also been carried out on primate colony personnel, comparing incidence of specific enteric organisms in this group with a matched group of individuals without simian contacts. Continuation of this study should provide information as to the potential for cross-species transmission of enteric organisms.

We must also recognize that we are probably dealing with many unknown problems which only time and research will unfold. Only within the past few years have we learned about the transmission of hepatitis from simians to man, the fact that Yaba-like virus can cause infection in man, or the disease potential of herpesviruses of South American monkeys. It is important, therefore, that an adequate surveillance program be developed as an essential part of the occupational health program. The program should include a permanent case record of individual work assignments so that it is possible to identify specific animal contact, should subsequent experience require this information. Such records should include the occurrence of bite wounds or other injuries and any illnesses. It is also strongly suggested that both preplacement and postemployment sera be obtained and stored for future diagnostic purposes. All personnel should be instructed to notify their supervisors of any suspected health hazards or signs of illness. The zoonoses and surveillance program should be looked upon as a means to not only provide as much protection as possible to personnel but also make it possible to perform retrospective studies which would lead to further insight into this increasingly important area of investigation.

References

1 HARRISON, B. (ed.): Conservation of nonhuman primates in 1970. Primates medicine, vol.5 (Karger, Basel 1971).

2 RIOPELLE, A. F.; COOPER, R. W.; DELANNOY, C. W.; HUMMER, R. L., and MIDDLETON, C. C.: Nonhuman primates. Standards and guidelines for the breeding, care and management of laboratory animals. Publ. No. 1677 (Institute of Laboratory Animal Resources, National Academy of Sciences, Washington, D. C. 1972).

3 COHEN, B. J.; CLARKSON, T. B.; FLYNN, R. J.; HIGHTOWER, D.; LANG, C. M.: MELBY, E. C.; RABSTEIN, M. M.; SOAVE, O. A.; TRUM, B. F., and USENIK, E. A.; Guide for the care and use of laboratory animals. Publ. No. 1024 (Institute of Laboratory Animal Resources, National Academy of Sciences, Washington, D. C. 1972).

4 MELENDEZ, L. V.; DANIEL, M. D. H.; BARAHONA, C. E. O.; FRASER, R. D.; HUNT, and GARCIAS, F. G.: Problems associated with the use of nonhuman primates. Lab. anim. Sci. *21:* 997–1001 (1971).

5 MELENDEZ, L. V. *et al.*: New herpesviruses from South American monkeys. Preliminary report. Lab. anim. Sci. *21:* 1050–1054 (1971).

6 STUNKARD, J. A.; SZATALOWICZ, F. T., and SUDDUTH, H. C.: A review and evaluation of tuberculin testing procedures used for macaca species. Amer. J. vet. Res. *32:* 1873–1978 (1971).

7 NAVIA, J. M.: The preparation, use and nutrient stability of nonhuman primate diets; in HARRIS Feeding and nutrition of nonhuman primates (Academic Press, New York 1970).

8 WEIL, J. D.; WARD, M. K., and SPERTZEL, R. O.: Incidence of shigella in conditioned rhesus monkeys *(Macaca mulatta)*. Lab. anim. Sci. *21:* 434–437 (1971).

9 LINDSEY, J. R.; HARDY, P. H.; BAKER, H. J., and MELBY, E. C.: Observations on shigelloses and development of multiple resistant shigellas in *Macaca mulatta*. Lab. anim. Sci. *21* (1971).

10 Hardy, P. H.; Lindsey, J. R., and Melby, E. C.: Transferable multiple drug resistance in enterobacteria from nonhuman primates. I. Development and characteristics of resistance in *Shigella flexneri*. Johns Hopk. Med. J. *123:* 29–37 (1968).
11 Kalter, S. S. and Heberling, R. L.: Comparative virology of primates. Bact. Rev. *35:* 310–364 (1971).
12 Melby, E. C. and Brayton, J. B.: Interrelationship of nonhuman primate disease and personnel health and safety in a primate colony (in preparation).

Author's address: Dr. Edward C. Melby, jr., Johns Hopkins University School of Medicine, *Baltimore, MD 21205* (USA)

Medical Primatology 1972. Proc. 3rd Conf. exp. Med. Surg. Primates, Lyon 1972, part I, pp. 130–137 (Karger, Basel 1972)

Design and Management of a Primate Research Facility for an Academic Institution Located in an Area of Mild Climate[1]

A. F. MORELAND

University of Florida, Division of Comparative Medicine, Gainesville, Fla.

Introduction

Gainesville, Florida, is the home of the University of Florida and the J. Hillis Miller Health Center. It is located at latitude 29° 41' North and longitude 82° 16' West. This location places it approximately equidistant from the waters of the Gulf of Mexico and the Atlantic Ocean. The Gulf Stream flows along the Atlantic coast of Florida approximately 8 mi off shore at this latitude. The average mean temperature is 69.9 °F and extreme mean temperatures are 91.9 °F in August and 45 °F in December. Thirty-year record high is 104 °F and low is 12 °F. During an average winter approximately 14–21 days of temperatures below freezing point are encountered, for the most part no lower than 25 °F and occurring for brief periods between midnight and 8:00 a.m. These low temperatures normally occur only two or three times per winter and only for very brief periods. Day-time hours are normally sunny and even though freezing temperatures may have been experienced briefly at night the afternoon temperature usually reaches 60 °F or more.

The first phase construction of the medical center was completed in 1955. Unfortunately, facilities for subhuman primate research were not in-

1 Supported in part by grants RR00421, RR05003 and RR00685 from the Animal Resources Branch, NIH, USPHS.

cluded. Faculty members desiring to do research with primates had to travel to other institutions.

Planning the Primate Research Facility

Realizing the uniqueness of subhuman primates compared with other laboratory animal species, the special attention required in their handling and manipulation, their attractiveness as curiosities to the public, and the potential for problems regarding both animal and human disease when unlimited human contact is allowed, the decision was made to locate the primate research resource on the periphery of the campus in a secluded area where human traffic could be minimized and controlled, and yet near enough to the Health Center to require only 5 min transit time to reach it. Another advantage of this site lay in the fact that the medical center had already begun to develop it as a conditioning station for dogs, cats, and larger animal species so that certain physical resources were already available. For example, a small building originally constructed for another purpose was of a design easily converted to a quarantine facility for subhuman primates. Additionally, cage sanitation, food storage, necropsy and incineration facilities already existed at this site and did not have to be duplicated. Electricity, well water, and septic tank sewage disposal was available. There was also ample room for expansion should the building, as originally planned, prove inadequate in size.

Discussions with our faculty revealed that reproductive biology and neurophysiology-psychology would be the main users of the primate facilities. Financial resources were limited; therefore, the decision was made to provide only minimal laboratory facilities at the site and to concentrate on providing sufficient animal holding areas, a place for sterile surgery, and plan to transport animals to medical center laboratories in those rare instances required by research protocols.

Because of the dichotomous research effort, the building was designed with an administrative and service area at the center of its long axis and laboratories and animal holding areas at each end. The research laboratories and the surgical facilities were located on one side of the central corridor and on the opposite side animal housing spaces were provided. The long axis of the building was located in an east-west direction with the animal housing spaces on the south side of the corridor allowing the exterior pens maximal exposure to sunlight.

Construction

The building was constructed of concrete block and wood ceiling and roof. The exterior pens were constructed of chain link wire over a concrete slab and divider panels. Laboratories, surgery, and indoor animal housing spaces were heated and air conditioned on a zone control basis.

Because of our belief that housing of simians in small groups reduces the stresses which lead to disease outbreaks and minimizes the potential losses should spread of disease occur, the animal rooms were designed to accommodate per room a maximum of 16 macaque-size animals or 25–30 smaller New World species. Near the west end of the building a series of six rooms were planned for indoor-outdoor housing of groups of smaller animals or one or two chimpanzees per room. Guillotine doors in the wall which separates the room from the exterior pens are operated by counter weighted cables from the corridor, thus facilitating management of the animals during the daily sanitation and handling processes. Twenty-nine other exterior pens were provided which included hutches constructed of plywood for animal shelter from wind, rain and temperature extremes. The floors of these pens were constructed of concrete, sloped to a drainage gutter; galvanized steel pipe perches (or shelves) were provided. A single panel of green translucent fiberglass sun screening was fixed by tie wires over the chain link wire above the perches. The hutches were mounted flush with the top of the pen. Their floors were made of 1-inch mesh stainless steel grids, to allow excreta to fall to the pen floor preventing soiling of the hutch interior. During periods of extremely cold weather a solid panel of plywood can be inserted to form a floor beneath the grid thus effectively preventing direct exposure to chilling winds. Old World species have not required temperature regulation of their hutches; however, New World primate hutches are equipped with infrared lamps, protected from tampering by wire grids, and controlled by thermostatic sensors set to tuın on when the ambient temperature drops below 50 °F.

We deliberately chose wood as a construction material for the hutches because of its insulating properties and its low thermal conductivity. Metal and plastic pens are extremely cold in winter and extremely hot when exposed to direct sun in summer. By selecting an exterior grade of plywood and sealing and painting it well, we have been able to sanitize these hutches on a daily basis by rinsing with cold water and steam sterilizing once weekly for extended periods. The average life of the plywood hutches has proven in service to be approximately six years. All exposed edges of the original hutches were covered with galvanized sheet metal flashings, however, it was necessary to replace this material with stainless steel because it quickly rusted and became a source of injury to the animals. Use of the steel flashings has minimized the chewing of the wood by the animals.

Commercially available stainless steel portable caging, some of which allowed for a dry litter system and some of which utilized the water rinse cleaning system, was selected for the indoor animal housing. Daily changing of litter or rinsing of drop pans is accomplished within the facility; all cages are rotated and sanitized in a mechanical cage wash machine in an adjacent building once weekly.

The southern exposure of the animal pens is extremely advantageous in that it warms the concrete in winter, hastens drying of the floor after the daily cleaning and prevents the growth of algae which invariably occurs in this area on damp concrete surfaces protected from direct exposure to sunlight. As originally completed in 1963 the facility consisted of approximately 8,000 ft^2 of floor space including the exterior pens. Additional pens were added later bringing the total area to approximately 10,000 ft^2. In 1970 the building was expanded by the addition of a 5,500-ft^2 wing which provides additional research laboratory space so that the total size is about 15,000 ft^2.

The quarantine building is located approximately 75 yd from the primate research facility and consists of a series of nine rooms varying in size from 10 × 16 to 16 × 16 ft, thus allowing for separately housing simians in groups until properly conditoned. These rooms are air-conditioned with 100 % fresh air. The building is constructed of concrete blocks. Rooms are equipped with fluorescent lighting, epoxy painted walls, smooth troweled concrete floors, floor drains and sinks.

The Quarantine and Care Program

Duties of technicians servicing the quarantine building are restricted to simians in the quarantine station, whereas duties of technicians servicing the primate facility are restricted to that facility.

Simians are acquired from importers, either unconditioned or conditioned, depending on the wishes of the investigator. In either case on arrival they are given a general physical examination by a veterinarian experienced in primate medicine and are routinely dewormed with thiabendazole. Intradermal tuberculin tests are administered either immediately or within three days after arrival in two sites, namely, the skin of the palpebrum and the skin of the midabdomen. Our experience has been that it is not uncommon for the abdominal site to react when there may be no reaction at the palpebral site. This may possibly be due to the difficulty of ensuring an *intradermal* inoculation in the palpebral skin. All simians are held in quarantine for 90 days and receive three such tuberculin skin tests during this period. Veterinarians supervise daily the examination of all animals and technicians report abnormalities regarding stool consistency, appetite, coughing, etc. Immediate attention is given to animals showing signs of abnormalities. If tuberculosis is diagnosed in any group, the affected animal is immediately removed and destroyed, and the remainder of the group is requarantined for a period of 90 days from the time the tuberculous animal was removed from the group; during that period the entire group must pass three additional tuberculin tests. At the successful completion of the 90-day quarantine each animal is radiographed in both anterioposterior and lateral positions. These radiographs are carefully examined for lesions of tuberculosis. If the radiographic examination is negative the animals are tatooed and transferred into the primate research facility for permanent housing. Once within the primate research facility they are under daily surveillance for abnormalities, are tuberculin skin tested at six-week intervals and undergo quarterly radiographic examinations of the chest. Sick animals receive immediate attention by the veterinary staff and are treated as indicated. All simians that die within either the quarantine facility or the research facility from spontaneous causes or experimental procedures are necropsied and management decisions made depending on the outcome.

Personnel who come into contact with the animals wear face masks and gloves to reduce possibility of disease transmission. Females and small immature males of the Old World species are handled by technicians equipped with special leather primate handling gloves or modified fish dip nets. New World species are also handled in this manner. Adult males of the Old World species and adult chimpanzees are handled only after tranquillization administered via a dart from a CO_2 gas operated pistol. Leather gloves and the fish dip netting is sterilized at frequent intervals in an ethylene oxide gas sterilizer.

Animal Nutrition

Old World simians on arrival at our quarantine station are placed on the standard pelleted food (Ralston Purina Monkey Chow [15% protein] Ralston-Purina Company, St. Louis, Mo.). It is normally fed dry, but occasionally sick or debilitated animals will only partake after it is moistened with water or milk. Animals who refuse to eat are force-fed via a stomach tube with diets consisting of a thin slurry made of the standard diet pellets and milk or a milk substitute such as Similac (Ross Laboratories, Columbus, Ohio) or Mulsoy (The Borden Co., Pharmaceutical Div., New York, N.Y.). *Cebus* species also normally readily accept this diet. If not they are treated similarly to *Saimiri* as discussed below.

Saimiri frequently are reluctant to take the standard diet or are too weakened to bite and chew the hard pellets. Therefore, during the initial days while adapting to the new environment and recovering from shipping stress, we offer the diet previously fed by the importer if we can obtain this information. Otherwise we feed a milk and cereal formula similar to that used by Tarpon Zoo, Inc., Tarpon Springs, Fla. [HARRIS, 1970]. Our standard laboratory diet for New World species (Ralston Purina Co. Monkey Chow 25 [25% protein]) is offered from the outset. Gradually the milk and cereal formula is withdrawn and the animals accept the laboratory diet. Weak and severely-debilitated animals are fed via a stomach tube with a diet similar to that given Old World species in similar condition.

Aotus are offered the milk and cereal formula *and,* initially a small bit of ground beef and lettuce leaves. Gradually these are reduced in amount and replaced by the standard laboratory diet. Weakened animals are treated as previously described for Old World species.

All animals are offered portions of orange and apple twice weekly during quarantine and once weekly in the research facility. Otherwise they receive the standard laboratory diet.

In 1969 we experienced an outbreak of tuberculosis in the macaques housed in exterior pens of the reproductive biology portion of the primate research facility. Rigorous testing and isolation procedures were begun but after a year we were still discovering an occasional infected animal (a total of 21 animals were destroyed) and all macaques in this portion of the facility were placed on the standard diet fortified with izoniazid and pyridoxine. The drugs are added to the diet by the manufacturer at the level of 182.5 g of izoniazid per ton of food and 6.81 g of pyridoxine per ton of food. We have experienced no further difficulty with tuberculosis in any of this group nor with animals in other parts of the research facility. As far as can be determined the medication has had no effect on breeding performance or other measurable experimental parameters.

The Research Program

The research program in reproductive biology has dealt primarily with placental physiology, specifically with the gaseous and chemical exchanges across the placenta. The breeding program which has supported this research effort has involved both individual and harem breeding systems. In carrying out the individual breeding program animals are trained to present the perineum for vaginal washings which are obtained daily and checked for onset of the menses by tests for occult blood. Females are placed with the male on day 11 following onset of menses and removed on day 13. Although numerous animals require repeated breeding before conception, the annual conception rate has been approximately 80%.

Under the harem system animals are housed in exterior pens in groups of either five females and one male or ten females and one male. During the breeding season these animals are examined at biweekly periods utilizing rectoabdominal palpation to determine pregnancy; estimations of conception date are based on the examination intervals and the uterine size. The harem system has resulted in annual conception percentages varying from 60 to 85%. *M. mulatta* and *M. arctoides* have been utilized in the reproductive physiology program. We routinely extract the canine teeth from adult males and have observed no reduction in the animals' breeding performance or libido.

Neurophysiology-psychology has utilized *M. mulatta, M. arctoides, Cercocebus torquatus atys, Cebus albifrons* and *Saimiri sciureus*. The experiments have varied, including discrimination reversal tasks and the effects of lesions to the central nervous system, spinal somesthetic pathways and the mechanisms of pain, and the development of visual and auditory prosthetic devices.

Since the initiation of the primate resource a number of projects have been begun, which have not fallen directly within the two major subject areas. Considerable work has been done with *Aotus trivirgatus* regarding herpes infections and resultant ulcerations of the cornea, their pathogenesis and the chemotherapeutic means for treatment. Also testing of contact lenses, corneal transplant procedures, storage of corneal tissue, cryosurgery, and treatment of occular lesions with the laser beam have been accomplished. Studies involving heavy metal toxicity, alcoholism, prosthetic bone devices, immunologic mechanisms, and Charcot joint syndrome have all been undertaken within this primate resource.

It is our feeling that the primate resource has been invaluable to the research program of our medical center. We have taken advantage of the area's mild climate to facilitate our care and management program. Exposure of the animals to normal seasonal and diurnal rhythms has not proven to be a handicap to our breeding or management programs. Pregnancies have occurred during all months of the year although most occur during the fall and early winter. This program of primate research has provided a useful service at a reasonable cost.

References

HARRIS, R.S.: Feeding and nutrition of nonhuman primates, p. 32 (Academic Press, New York 1970).

Author's address: Dr. ALVIN F. MORELAND, University of Florida, Division of Comparative Medicine, *Gainesville, FL* (USA)

Medical Primatology 1972. Proc. 3rd Conf. exp. Med. Surg. Primates, Lyon 1972, part I, pp. 138–142 (Karger, Basel 1972)

The Mountain Gorillas and Dwarf Chimpanzees at the Zoo of Antwerp

Housing, Husbandry and Breeding Problems

J. MORTELMANS

Royal Zoological Society of Antwerp, Antwerp

Mountain gorillas *(Gorilla gorilla berengei)* and dwarf chimpanzees *(Pan paniscus)* are very rare animals. Their natural habitat is situated in the Zaire republic: the first species in the volcanic mountains of the Kivu province, the latter species in the tropical forest on the left side of the Zaire (Congo) river.

At the Zoo of Antwerp, important efforts have been made to keep these animals in captivity under good conditions and to create the most suitable facilities for breeding and eventually for research.

A new ape house was built in 1957 on the premises of the zoo [VAN DEN BERGH, 1959]. This building harbors both lowland and mountain gorillas, orangutans, white-faced chimpanzees and dwarf chimpanzees. Monkeys live in other buildings and are served by other attendants.

Inside of their buildings the apes have spacious dayrooms and small nightrooms; outside they have three large playgrounds, exposed to sunshine and separated from the public by a deep ditch four or five meters wide. As apes are not able to swim, they can fall into the ditch accidentally and drown; during the hot season, mountain gorillas may enter the water to bathe. Therefore, at the side of the playground, the ditch is shallow and a wire mesh barrier is placed between the deep and the shallow water area to save the animals and to help them to get out.

During the cold season the animals remain indoors.

Inside the ape house the animals are separated from the public by 'polyglass' which consists of two glass sheets, 2 and 1 cm thick, separated by 1 cm of dry air. The sheets have been tested to withstand the charge of an adult gorilla and have proved successful. The ape units have radiant heat from the walls and paved floor; humidity, ventilation and supplementary heating is supplied by an air-conditioning system; this allows us to supply only filtered air. The cages used during the day are equipped with stone wall-fixed shelves and iron tubular equipment for exercise and play; one gorilla cage has an automatic balance. A small iron reinforced wooden door and a second independently locked grille separates the cages from the service corridor. The attendants can look into each cage through a transparent brick in the wall or through a magnifying glass in the wooden door. Another door leads to an antichamber from which the animals can go into their separate 'bedrooms'; sometimes the animals sleep alone, sometimes two or three occupy the same cage, but they sleep on separate wooden shelves. The bedroom doors shut vertically and can be manipulated from a distance by a lever which is provided with a lock.

The service corridor can only be entered from the outside; there are two entrances provided with a wooden door and each has a second iron key-locked grille.

The animals receive the following food: sweetened skimmed milk in the morning, followed at noon by a pasta containing meat, eggs, cooked rice, meal, vegetables, vitamins and minerals; at night they get fruits and vegetables in their bedrooms. A small quantity of straw is available in the dayrooms to be used for play, to make nests or to search for grains and peanuts.

The main problems arise when the animals arrive from Africa. They have to undergo a long quarantine period, to be checked and eventually treated for tuberculosis, salmonellosis and shigellosis. We vaccinate the animals against poliomyelitis (3 types) and they receive all their lives daily 5 mg/kg isoniazide as a prophylactic measure against tuberculosis. Generally they harbor a lot of intestinal nematodes which have to be treated by broad-spectrum anthelminthics. Occasionally we could find cestodes but these are not of any importance and can be easily treated. Amoebiasis is a real problem; it is difficult to cure. When lesions of the liver or large intestine abcesses occur, the animals are bound to die. Other intestinal protozoan parasites such as *Balantidium coli, Troglodytella abrassarti,* etc., may be found too, but they do not have the same significance as *Entamoeba histolytica* [MORTELMANS *et al.,* 1971].

Besides the problems of infections and parasitic diseases, there is a real need to ensure the young new animals arriving from the bush against psychological stress and to give them permanent maternal help and warmth. We are convinced that this latter is as important as the medical care, especially during the first months after arrival.

At this moment, on May 1, 1972, the collection of the Zoo of Antwerp harbors two adult male mountain gorillas about 17 to 18 years old, two (pregnant) females about 12 to 14 years old and a four-year-old female, which was born at Antwerp. We have three adult female dwarf chimpanzees, about 12 to 16 years old, and a four-year-old female born at the Zoo of San Diego; its father was one of our males sent to the San Diego Zoo several years ago.

We never noted pregnancy in our female dwarf chimpanzees because we never had the opportunity to put them together with an adult male. We have had several young male dwarf chimpanzees in our collection; some of them were sent to other zoos, the others died before they reached sexual maturity. The causes of death were amoebiasis, strongyloidiasis, salmonellosis.

Our mountain gorillas were very young when they arrived at Antwerp [GIJZEN and TIJSKENS, 1971]: the male 'Kaisi' arrived on September 25, 1957 (± two years old) and weighed 7.5 kg; its weight increased to 36 kg on January 1, 1961, to 63 kg on February 12, 1963, to 107 kg on January 31, 1965, to 130 kg on February 3, 1966, to 142 kg on February 1, 1967, to 163 kg on March 8, 1968, to 172 kg on January 16, 1970, and to 190 kg on May 1, 1972. The other male 'Kisubi' arrived on November 25, 1957 (± two and a half years old) and weighed 14 kg; its weight increased to 40 kg on January 30, 1961, to 63 kg on February 12, 1963, to 96 kg on January 31, 1966, to 122 kg on February 1, 1967, to 132 kg on March 8, 1968, to 133 kg on January 17, 1970 and to 150 kg on May 1, 1972. The difference in weight gain is due to a chronic intermittent colitis which started about 1964; since that time the animal has been receiving at regular intervals and with very satisfactory results an aspecific antidiarrheal treatment.

The female mountain gorilla 'Quivu' arrived on March 24, 1963 when she was about four years old. The animal had been kept in captivity in different African countries and showed symptoms of pulmonary tuberculosis which was confirmed by radiography. Bacteriological cultures of stomach lavage demonstrate *Mycobacterium tuberculosis* (human type). The animal has been treated with streptomycine (six weeks) and isoniazide (two years) and the tuberculosis appears arrested. Its weight increased from 42 kg on October 1964 to 43 kg on January 31, 1965, to 61 kg on February 3, 1966, to 67 kg on February 1, 1967, to 84 kg on January 8, 1968, to 90 kg on January 16, 1970, and finally to about 110 kg on May 1, 1972. The other female 'Pega' arrived at Antwerp on July 12, 1962 when she was three to four years old and weighed 28 kg on October 17, 1962. Its weight increased to 46 kg on February 6, 1964, to 54 kg on January 31, 1965, to 65 kg on

February 3, 1966, to 73 kg on February 1, 1967, to 96 kg on March 8, 1968 and, finally, to about 110 kg on May 1, 1972.

A female 'Victoria' was born on June 9, 1968 from Quivu and Kisubi after an estimated 254- to 274-day pregnancy; the animal was raised by its mother until September 13, 1968; at that time she weighed 3,550 g. The separation occurred without any psychological troubles for mother and child. The infant grew up very well and weighs about 20 kg at this time.

On August 20, 1970, abortion occurred and the immature fetus (1,075 g) died 2 h later; the parents were Quivu and Kisubi. The same parents mated again and at present the urine tests for pregnancy are positive and we hope for a second birth from this pair.

The other female 'Pega' aborted twice; on September 7, 1967 we found an estimated six-month-old fetus and on March 10, 1970, another seven- to eight-month-old (790 g) female fetus. Pega is pregnant for the third time.

Kisubi was the father in all the pregnancies in both females, as the other male Kaisi refused to mate; we do not know why, but it is not the first time that we observed such a phenomenon in apes as well as in other animal species.

The average length of the oestrous cycle of the two mountain gorillas observed by TIJSKENS [1971] at the Antwerp Zoo was about 33 days (17 cycles observed). When a male and a female live together we regularly examine the urine at two-week intervals; when two positive pregnancy reactions have been observed, the female is kept alone in a separate room to avoid any physical disturbance which could provoke abortion. We hope that this method will give us excellent results in the future.

It is the real wish of the Zoo of Antwerp that these rare valuable animals may reproduce not only for the show but also for scientific purposes.

Acknowledgements

I am greatly indebted to Mr. VAN DEN BERGH, Director of the Zoological Society of Antwerp, and to my colleagues Mrs. A. GIJZEN and Mr. J. TIJSKENS of the same zoo. I also wish to thank Mr. A. SAMSOM, headkeeper, and his colleagues of the Anthropoid house for their valuable observations and help.

References

GIJZEN, A. and TIJSKENS, J.: Growth in weight of the lowland gorilla *(Gorilla g. gorilla)* and of the mountain gorilla *(Gorilla g. berengei)* Int. Zoo. Yb. *11:* 183–193 (1971).

MORTELMANS, J.; VERCRUYSSE, J., and KAGERUKA, P.: Three pathogenic intestinal protozoa of anthropoid apes. *Entamoeba histolytica, Balantidium coli* and *Troglodytella abrassarti;* in BIEGERT and KUMMER Proc. 3rd Int. Congr. Primatology, Zurich 1970, vol. 2, pp. 187–191 (Karger, Basel 1971).

TIJSKENS, J.: The oestrus cycle and gestation period of the mountain gorilla. Int. Zoo Yb. *11:* 181–183 (1971).

BERGH, W. VAN DEN: The new ape house at Antwerp Zoo. Int. Zoo. Yb. *1:* 7–11 (1959).

Author's address: Dr. J. MORTELMANS, Royal Zoological Society of Antwerp, *B-2000 Antwerp* (Belgium)

Medical Primatology 1972. Proc. 3rd Conf. exp. Med. Surg. Primates, Lyon 1972, part I, pp. 143–147 (Karger, Basel 1972)

Treatment of Acute Gastric Dilatation Syndrome[1]

E. MUCHMORE

Laboratory for Experimental Medicine and Surgery in Primates (LEMSIP) of New York University Medical Center, New York, N.Y.

Death from acute gastric dilatation has been a partially unsolved mystery in monkey colonies. The Primate Zoonoses Surveillance Report for 1970[2] shows gastric bloat as cause of death in six animals: one vervet, two *Macaca mulatta,* two *Macaca arctoides,* and one unspecified Old World monkey. These were all reported to be mature animals; all but two were held for longer than one year. There are no cases of gastric dilatation cited in the nonfatal group. Individual laboratories have reported varying, somewhat higher, incidence of death from that cause. Most recently, TURNER and COWEY, in the July 1971, Laboratory Primate Newsletter,[3] reported the death of 12 of 64 *M. mulatta,* 10–38 months after arrival in their colony at Oxford. In 1968, SMITH *et al.* [10] reported the death of 24 of 500 rhesus monkeys during one year in a colony which housed an average of 800 animals. NEWTON *et al.* [7] reported the loss of four of 12 *M. arctoides* which had been in the colony for more than two years. CHAPMAN [1] cites eight cases: five *M. mulatta* and three *M. speciosa;* three cases were diagnosed prior to death, and one lived six hours after evacuation of stomach contents. SEIBOLD and WOLF[4] found among 700 consecutive primate necropsies that

1 Supported in part by USPHS, NIH grant RR00316.

2 Primate Zoonoses Surveillance Report No. 4, Annual Summary 1970, Center for Disease Control, Epidemiology Program, Bacterial Disease Branch, Atlanta, Ga. This report lists mortality and morbidity data from naturally occurring diseases in all species of nonhuman primates in participating primate facilities, whose average monthly population at risk was 7,069 animals.

3 Laboratory Primate Newsletter, vol. 10, No. 3, edited by A. M. SCHRIER, Brown University, Providence, R.I., July 1971.

4 Primate Zoonoses Surveillance Report No. 5. A classification by primary cause of death of 700 consecutive necropsies. January-March 1971.

acute gastric dilatation was the cause of death in three *M. mulatta,* three *M. arctoides,* two *Cercopithecus aethiops,* one *Papio anubis,* one *Papio papio,* five *Erythrocebus patas,* two *Saimiri sciurea,* and one *Lagothrix lagotricha.* These 18 deaths constituted almost two-thirds of the deaths attributed to noninfectious diseases in this group of autopsies done during the period from June, 1967, to May, 1971, at the Delta Regional Primate Center. VALERIO [11] reported the death of approximately 25 female *M. mulatta* and *M. fascicularis* from acute gastric dilatation in a five-year period. During a major outbreak in July, 1970, seven monkeys died and 25 others, found bloated in one 48-h period, were relieved by gastric intubation and survived.

The present report deals with experience in cases of gastric dilatation at the Laboratory for Experimental Medicine and Surgery in Primates (LEMSIP) including successful therapy for this condition. During the three-year period, from October, 1968, to October 1971, LEMSIP had 11 cases of acute gastric dilatation: one 1½-year-old female *M. fascicularis,* four adult female *Papio* sp., two adult female *M. mulatta,* one juvenile female *Cynopithecus niger,* one juvenile female *Theropithecus gelada,* and two *Papio hamadryas,* an adult female and a juvenile male. Time in our colony for these animals ranged from 6 months to 6½ years. Six of our 11 cases were diagnosed before death and nasogastric suction was initiated on four, of which two survived and recovered; these are the first reported cases of recovery from gastric dilatation in baboons. Five of the six cases diagnosed before death were found in varying stages of distress between 10 p.m. and midnight; the other at 10 a.m. In both cases that survived, abdominal distention and lethargy were noted before the animals collapsed; therefore, they were kept under close surveillance and brought to treatment within half-hour from the time that they were sufficiently depressed to be removed from the cage and put on the treatment table without need for sedation.

The surviving animals were one old adult female *P. hamadryas* weighing 9.6 kg and one juvenile, 10-kg female *T. gelada;* both were typically in good condition before this episode. After removal of 920 ml, of sweetish fermented, fruity-smelling liquid accompanied by considerable gas from the gelada, and 450 ml of similar material in the case of the hamadryas, the animals were so improved and active that it was not considered feasible to attempt fluid replacement therapy without anesthesia, which was considered not indicated. The animals were kept without food or water for eight hours or until the following morning when they were fed later in the day. The gelada was given no other treatment, but considering the possibility of *Clostridium* being involved [7], the hamadryas was given 600,000 units Bicillin (Wyeth

Laboratories, Philadelphia, Pa.) i.m. daily for four days. In both, recovery was immediate, uneventful, and with no sequellae during the subsequent years in our colony, and both are still alive.

The gross and microscopic pathological findings in our deceased animals were comparable with those cited in the literature [1, 7, 10]. At autopsy there were three cases of stomach rupture, three of diaphragm rupture and two of rectal prolapse. These changes were felt to have occurred *post mortem*, because there was no evidence of bleeding at the sites. The chyme found in all cases had a characteristic fermented odor and was actually effervescent in the case of one baboon from which 1,300 ml of such fluid was removed at necropsy performed within one hour of death. Culture of ingesta removed immediately *post mortem* from the stomach of another, has shown many colonies of streptococci and moderate number of colonies of yeast-like organisms which were not further identified.

Though nasogastric intubation has largely eliminated acute gastric dilatation as a postoperative complication in human patients, there is still interest in the pathophysiology of the mechanism of vascular collapse associated with the acute gastric dilatation syndrome as it is seen in patients. Three theories have been proposed: neurogenic shock, mediated mainly via splanchnic pathways as proposed by WILLIAMS [12]; hypovolemic shock due to loss of fluid and electrolytes into the stomach as considered by DRAGSTEDT [4] and shock due to decreased venous return caused by compression of the inferior vena cava by the distended stomach. In 1969, DOPPMAN [2] did a study showing that, in supine *M. mulatta*, acute gaseous distention of the stomach, to pressures not clinically unrealistic, resulted in complete occlusion of the upper abdominal cava which was reflected in a concurrent drop in blood pressure. He found the hypotension thus induced to be completely reversible by deflating the stomach.

The etiology of the gastric dilatation syndrome, however, has not been defined. Human cases, other than postoperative, have been attributed to the cast syndrome [3], the air sucker's syndrome [9], and acute distention following large meals in chronically-starved subjects [6]. The last seems most closely related to the situation seen in nonhuman primates. The 12 cases of bloat syndrome in *M. mulatta* reported by TURNER and COWEY in the Laboratory Primate Newsletter (op. cit.) were in animals on a behavioral study with monitored dietary intake five to six days per week. It was considered significant that ten of the 12 animals which died had been fed *ad libitum* the previous one to two days. D'VER [5] found a similar situation in his colony of rhesus and squirrel monkeys on stressful behavior and/or food deprivation when fed *ad libitum* after the study, but not when fed at a metered rate when taken off study as well.

NIEMANN [8] has seen cases in macaques related both to stressful studies involving some food deprivation followed by freedom to feed, and also related to changes in feeding routine such as may be experienced over weekend or holiday periods followed by normal routine. Ten of the 11 LEMSIP cases can be directly correlated with a weekend or holiday period when feeding schedules were modified, though two of the animals were on a study with rigidly dictated diet of minimal food intake and should have had no opportunity for change or *ad libitum* excesses. The majority of NIEMANN's cases were found dead on Tuesday morning, suggesting that the animals were underfed on the weekend and overate on the following day. The LEMSIP cases correlated with weekends or holidays were found both on the second day following the holiday or a holiday evening as reported by TURNER and COWEY (op. cit.). This would suggest that our weekend or holiday change in routine may result in underfeeding followed by overeating, if one accepts the gorging after deprivation theory.

Thus, the immediate cause of death in the acute gastric dilatation syndrome might be attributed to shock from vena cava compression; predisposing conditions can be traced to changes in dietary routine. As our experience has shown, nasogastric suction is the treatment of choice, provided it is initiated early enough. Successful therapy, therefore, depends on surveillance of the animal colony 24 h/day, 7 days/week. Some studies [7] point to *Clostridium perfringens* as the etiologic agent, and in this connection it should be noted that a recent analysis of commercially-prepared diet fed to some of the animals in our series, contained 72 organisms of *C. perfringens* per gram. Bacteriological examination of ingesta from primate bloat cases has not shown *C. perfringens* in many instances, however, so that the question of the initiating factors is still open.

References

1 CHAPMAN, W. L., jr.: Acute gastric dilatation in *Macaca mulatta* and *Macaca speciosa* monkeys. Lab. anim. Care *17:* 130–136 (1967).

2 DOPPMAN, J. L. and JOHNSON, R. H.: The mechanism of shock in acute gastric dilatation. An angiographic study in monkeys. Brit. J. Radiol. *42:* 613–618 (1969).

3 DORPH, M. H.: The Cast syndrome. New Engl. J. Med. *243:* 440–442 (1950).

4 DRAGSTEDT, L. R.; MONTGOMERY, M. L.; ELLIS, J. C., and MATTHEWS, W. G.: The pathogenesis of acute dilatation of the stomach. Surg. Gynec. Obstet. *52:* 1075–1086 (1931).

5 D'Ver, A. S.: Personal commun.
6 Markowski, B.: Acute dilatation of the stomach. Brit. med. J. *ii:* 128–130 (1947).
7 Newton, W. M.; Beamer, P. D., and Rhoades, H. E.: Acute bloat syndrome in stump tailed macaques *(Macaca arctoides)*. A report of four cases. Lab. anim. Sci. *21:* 193–196 (1971).
8 Niemann, W. H.: Personal commun.
9 Olsen, V. B. and Foley, T. J.: The Air-Sucker's gastric rupture syndrome. Amer. J. Surg. *102:* 602–606 (1961).
10 Smith, A. W.; Casey, H. W.; LaCroix, J. T., and Johnson, D. K.: Acute Bloat Syndrome (gastric dilatation) in *Macaca mulatta*. J. Amer. vet. med. Ass. *155:* 1241–1244 (1969).
11 Valerio, D. A.; Valerio, M. G.; Ulland, B. M., and Innes, J. R. M.: Clinical conditions and diseases encountered in a large simian colony; in Biegert and Kummer Proc. 3rd Int. Congr. Primatology, Zurich 1970, vol. 2, pp. 205–212 (Karger, Basel 1971).
12 Williams, J. S.: Hemodynamic alterations in acute gastric dilatation in the dog. Surg. Forum *16:* 335–336 (1965).

Author's address: Dr. E. Muchmore, Laboratory for Experimental Medicine and Surgery in Primates (LEMSIP) of New York University Medical Center, 550 First Avenue, *New York, NY 10016* (USA)

Medical Primatology 1972. Proc. 3rd Conf. exp. Med. Surg. Primates, Lyon 1972, part I, pp. 148–151 (Karger, Basel 1972)

'Normal' Values

A Critique and a Proposal

MARYEVA W. TERRY and ANNE C. MORROW

Primate Information Center, Regional Primate Research Center, University of Washington, Seattle, Wash.

A consideration in using wild-caught animals for biomedical research is: Are they truly healthy or have events in their unknown history induced some chronic disorder that is not readily apparent but that could confuse experimental results? By analogy with human medicine, it should be possible to answer this question by performing a battery of clinical tests. Interpretation of the results of such tests depends, however, on availability of well-defined ranges of 'normal'.

For six years, we have been extracting from the literature hematologic, biochemical and physiologic data on nonhuman primates in the hope of defining some of these ranges for at least the more commonly used species. Repeatedly, we find ourselves frustrated in this goal, not by paucity of animals or data but by lack of significant details on the conditions under which the data were obtained. Authors and editors apparently have no standards for what to include in the 'Methods' section of a paper reporting quantitative data. Consequently, they provide no basis for statistical comparison of one paper with another. The following items seem to be minimal documentation.

Taxonomic designation. Incredibly, despite all that has been written and said about the need for proper zoological description of laboratory primates, many authors are still publishing 'control' and 'normal' values for 'monkeys'. We sympathize with the difficulties of securing accurate identification of animals from the wild, but we sincerely question whether quantitative data from monkeys of unknown genus and species have sufficient value to justify the costs of collection and publication.

Source. Animal importers seem reluctant to provide information on the countries from which they obtain animals. In most cases, of course, these people do not know the exact locale at which specific animals were trapped, but a statement of the city from which they were exported can be useful, even though this information does not allow precise genetic identification. For example, in analyzing original data from the breeding colony at the Primate Field Station (Medical Lake, Wash.), which imports some animals directly, we found a distinct difference in cholesterol values, both initially and after the quarantine period, according to export point [BLAKLEY *et al.*, in press]. Data on the colony-bred generation suggest that this difference was not genetic.

Length of Time in Colony

Weight. Even if the ages of the animals are known, these data provide a better basis for comparison with animals of unknown age. If the group is composed of animals of different ages or both sexes, the weight ranges should be stated by age, sex and number of animals in each category, even though values are reported for the group as a whole.

In assessing age of imported postpubertal animals, many persons, more accustomed to dealing with rodents and carnivores, tend to describe as 'mature' or 'adult' monkeys clearly in the early adolescence. Like man, many species of nonhuman primates have a period of postpubertal growth extending over several years [NAPIER and NAPIER, 1967]. There is a need to study whether animals in this phase of their lives are physiologically different from juvenile and fully adult animals.

Number of Animals

Interval between repeated determinations. It has been demonstrated [KRISE, 1960; WISECUP *et al.*, 1968] that too frequent sampling of blood can affect the values obtained. Also, since determinations are usually made on growing animals, it should be possible to estimate whether age changes have affected results.

Diet

Housing and conditioning. Since the psychological state of the animal obviously affects many more labile variables, the question of whether it was chaired, caged individually or housed in groups is quite important.

Conditions of sampling. Was the animal under physical or chemical restraint? If the latter, what drug was used in what dosage? It is surprising the number of values reported for sedated animals in which neither the fact of sedation nor its type is mentioned. It is probably reasonable to develop ranges to be expected in sedated or anesthetized animals, since such restraint is safer for personnel, but this cannot be done if relevant values are not identified as such.

Method of determination. If the method has been 'previously described', give the exact citation. If the method has been 'modified', describe the modifications.

Unit of Measurement

Mean, range and standard deviation. Use of the above checklist would greatly improve comparability of data and might very well lead to establishment of reliable ranges for the more stable parameters. It would not, however, resolve the problem that, once the data on individual animals have been reduced to a mean, the 'code' cannot be broken to test for effects of individual factors on the more labile parameters. Our work with original data from such large colonies as those at the California, Oregon, Washington and Wisconsin Regional Primate Research Centers, and particularly our success in identifying subgroups within those colonies from value differences, has led us to believe that, if we could go back to original data underlying published means for small colonies, we might well be able to determine which factors have significant effects on which variables.

Assuming that the costs of publication will force journals to continue to publish only statistical evaluations, how can access to original data for comparative purposes be achieved? One possibility is creation of a multi-institutional, computerized data bank. Since such a program seems a logical extension of the Primate Information Center's normal values program, we have examined this approach in some detail. Our conclusion is that contributors would find it burdensome to feed the system, and thus would be reluctant to participate. Also, even though we have access to suitable computer programs, the overall costs of establishing and operating such a data bank could well be disproportionate to the value of the results.

Another possibility is more extensive use of the National Auxiliary Publications Service, sponsored by the American Society for Information

Science and operated by the CCM Information Corporation. There are costs to both the depositor and the withdrawer of data, but they are minimal compared to the value of unsnarling some of the current confusion.

Acknowledgement

This paper was supported by grant RR 001 66 from the National Institutes of Health U.S. Public Health Service.

References

BLAKLEY, G.A.; MORROW, A.C., and MORTON, W.R.: Intraspecies variation in serum cholesterol levels in imported *Macaca nemestrina*. Lab. anim. Sci. (in press).

KRISE, G.M.: Hematology of the normal monkey. Ann. N.Y. Acad. Sci. *85:* 803–810 (1960).

NAPIER, J.R. and NAPIER, P.H.: A handbook of living primates (Academic Press, New York 1967).

WISECUP, W.G.; HODSON, H.H., jr.; LOVETT, R.L.; PRINE, J.R., and HANLY, W.C.: Anemia in chimpanzees resulting from serial collection of blood. Amer. J. vet. Res. *29:* 1823–1830 (1968).

Authors' address: MARYEVA W. TERRY and ANNE C. MORROW, Primate Information Center, Regional Primate Research Center SJ-50, University of Washington, *Seattle, WA 98195* (USA)

Medical Primatology 1972. Proc. 3rd Conf. exp. Med. Surg. Primates, Lyon 1972, part I, pp. 152–155 (Karger, Basel 1972)

Primates Used in Biomedical Research at New England Regional Primate Research Center

B. F. Trum

New England Regional Primate Research Center of Harvard University, Southborough, Mass.

At the time of writing, the New England Regional Primate Research Center had exactly 1,100 simian primates, of which there were specimens of 26 different species. This is not a large number of monkeys nor have we used many since the Center was founded. The New England Regional Primate Research Center is a Center for biomedical research without facilities for the accumulation or breeding of any larger group. Programs using large numbers of animals such as toxicity testing or vaccine evaluations are not encouraged at the Center.

We can accommodate but 1,000 monkeys in our facilities and if a new species is needed for some special project, a priority must be established and a reduction made to accommodate the new research and the new species.

The 'administration' plays a negative role in the selection of species and the total number of primates used. For example, because there are limited facilities for apes and no facilities for observations of group behavior, research involving great apes or behavioral psychology will not be found at the Center.

Pertinent questions that must be satisfactorily answered by an investigator before beginning research at the Center are: Must a simian primate be used for this study and has the proper one been selected?

Each of the species on hand was chosen by an investigator for some special reason. For example *M. cyclopis* was chosen because it had been reported that this species was the best model on which to study trachoma [1, 2]. It was soon found that, insofar as this particular study was concerned, *M. cyclopis* was inferior to *Aotus trivirgatus* [3]. *M. radiata* were imported for use in reproductive studies because the cervical secretions during the menstrual cycles were superior to those of other animals for comparative studies related to human reproductive physiology [4].

A new project on the structure and function of neocortical area of Galago [RANDOLPH, unpubl.] will take advantage of the simpler cyto-architectural divisions of the somatic sensory cortex of the 'bush-baby' *(Galago crassicaudatus)* in comparison with studies of structure and function of more complex sensory areas in species reported by others (5).

Neurologists have found the simian primate species invaluable in experimental research on the brain and nervous system. DENNY-BROWN and YANAGISAWA [6] have reported an extremely interesting series, including the effects of complete decortication in the infant macaque. These studies of course necessitated the availability of infant macaques and special nursery facilities.

The experiments in clinical medicine involving transfusions in treatment of hepatic coma and transplantation of the liver by TREY *et al.* [7] and SLAPAK *et al.* [8] were most thoroughly studied in the larger species of simian primates before being successfully applied to man. Previous failures in this field were traceable to the use of unsuitable animal models.

Although a search for 'animal models for study of human diseases' is not an objective of the Center, it was recognized that the 'spin-off' of the research would be the identification of 'models'. A few of these identifications made by staff or associate staff members of the New England Regional Primate Research Center will be mentioned.

JONES *et al.* [9] found a difference in results of phenylalanine tolerance tests that indicated an inherited defect in metabolism.

CHRISTEN *et al.* [10] found that several species of New World monkeys have uric acid concentrations in serum and urine which are more comparable to those found in man than those found in prosimians or Old World monkeys.

In the field of oncology the possibilities opened by the work of MELENDEZ *et al.* [11,12] with the discovery in simian primates of oncogenic herpes viruses are numerous.

I believe these are sufficient illustrations to indicate how various species are selected for experimentation and why various species are to be found in one Center. At the same time it should be pointed out that although a great many 'basic principles' may be worked out on experimental species, we do not confuse the experimental procedure with a spontaneous disease of man. There is in every case a 'study model', some better than another, none the 'real thing'.

It may be of interest to some that all of the 26 species of simian primates at the Center thrive on the same basic food (a commercial monkey chow) supplemented with occasional morsels of fruit (apples and oranges).

All species having had an opportunity to breed in confinement (17 different species) have conceived and delivered live young at the Center, a few females of the macaque group conceived and raised at the Center, have matured and produced young. Other species have not had this opportunity. *Macaca fascicularis* have proved to be the most efficient species to reproduce in our Center. Adult females menstruate during the entire year; on the average, 80% of the females produce young each year and young are born during each month of the year. Distinct postfertilization menstrual bleeding is useful in this species to indicate early pregnancy.

The Center has acquired 4,411 simian primates since 1964. Of these, 650 were born at the Center, the rest purchased or transferred from other laboratories.

Approximately 2,450 monkeys were used in research and 850 died of natural causes (20% of total). The majority of these deaths occurred within our 45-day quarantine period and were attributed to a variety of causes. In Old World monkeys, colitis was the greatest single cause of death (23%) followed closely by pneumonia (17%). In New World monkeys, pneumonia accounted for 18% of deaths and colitis for 8.7%. Other major causes of death included septicemia (4.8% Old World, 11.3% New World); meningo-encephalitis (4.1% Old World, 5.5% New World); peritonitis (4.8% Old World, 6.8% New World); nephritis (2.0% Old World, 3.9% New World) and herpesvirus infections (5.2% New World).

Of the total number acquired 37 or 0.08% were destroyed because they were positive to tuberculin test. These were all Old World monkeys or chimpanzees.

Twenty-two of the 50 wild-caught pregnant females aborted. They were three to six months into gestation when purchased. On the other hand, abortion or neonatal deaths in the colony bred animals was less than 10% of pregnancies and in some species, the macaques for example, even less.

The total number of deaths include a colony of 119 tree shrews *(Tupaia glis)* all died or killed one another within four years. The original imported colony probably lived out their normal span of life. They reproduced regularly but most young were killed by their dams or other older animals. None of the surviving young produced. The colony was 'moved' three times, a procedure we consider undesirable as stability seems to be a requisite for success of some colonies, including the tree shrew [13].

It is unfortunate that meaningful direct comparisons of success or failure cannot be made between laboratories. For example, the figures of laboratories dealing with Old World monkeys are not meaningful when extropolated to figures from laboratories dealing principally with New World monkeys. Records of morbidity from laboratories purchasing directly from source are not meaningful when compared with those of laboratories which purchase 'conditioned' animals.

Finally, the total number of monkeys accumulated will have a proportional effect on the probability of diseases. Laboratories having the larger numbers usually have the more serious epizootics and the more exotic diseases, for example, Yaba disease or hemorrhagic fever.

References

1 WANG, S. P. and GRAYSTON, J. T.: Trachoma in the Taiwan monkey, *Macaca cyclopis*. Ann. N.Y. Acad. Sci. *98:* 177–187 (1962).

2 MORDHORST, C. H.: Experimental infections and immunogenicity of TRIC agents in monkeys. Amer. J. Ophthal. *63:* 1603–1615 (1967).

3 BELL, S. D. and FRASER, C.E.O.: Experimental trachomas in owl monkeys. Amer. J. trop. Med. Hyg. *18:* 568–572 (1969).

4 MCARTHUR, J. W.; OVADIA, J., and SMITH, A. W.: The menstrual cycle of the bonnet monkey *(M. radiata)*. Changes in cervical mucus secretion, vaginal cytology, sex skin and urinary estrogen excretion. Folia primat. *17:* 107–121 (1972)

5 HUBEL, D. H. and WIESEL, T. N.: Receptive fields and functional architecture of monkey striate cortex. J. Physiol. *195:* 215–251 (1968).

6 DENNY-BROWN, D. and YANAGISAWA, N.: The evolution of spastic paraplegia following cortical ablation in infant monkeys. Trans. amer. neurol. Ass. *96:* 169–172 (1971)

7 TREY, C.; KING, N. W., and GARCIA, F. G.: Reproduction of fulminant hepatic necrosis in rhesus monkeys and the reversal of hepatic coma by exchange transfusion. J. clin. Invest. *46:* 1125 (1967).

8 SLAPAK, M.; CHIR, M.; BADDELEY, M.; WEXLER, M.; SARAVIS, C.; SISE, H.; GARCIA, S.; GIOUARD, M., and MCDERMOTT, W. V., jr.: Extended preservation of the primate liver by simple cooling and orthotopic autotransplantation; in GOLDSMITH and MOOR-JANKOWSKI Medical primatology 1970, pp. 119–129 (Karger, Basel 1971).

9 JONES, T. C.; LEVY, H. L.; MAC CREADY, R. A.; SHIH, V. E., and GARCIA, F. G.: Phenylalanine tolerance tests in simian primates. Proc. Soc. exp. Biol. Med. *136:* 1087–1090 (1971).

10 CHRISTEN, P.; PEACOCK, W. C.; CHRISTEN, A. E., and WACKER, W. E. C.: Urate oxidose in primate phylogenesis. Europ. J. Biochem. *12:* 3–5 (1970).

11 MELENDEZ, L. V.; HUNT, R. D.; DANIEL, M. D.; GARCIA, F. G., and Fraser, C. E. O.: *Herpesvirus saimiri* II, experimentally-induced malignant lymphoma in primates. Lab. Anim. Care *19:* 378–396 (1969).

12 MELENDEZ. L. V. *et al.*: *Herpesvirus ateles,* a new lymphoma virus of monkeys. Nature New Biol. *235/58:* 182–184 (1972).

13 ELLIOT, O. S.: Bibliography of the tree shrews 1870–1969. Primates *12:* 323–414 (1971).

Author's address: Dr. BERNARD F. TRUM, New England Regional Primate Research Center of Harvard University, *Southborough, MA* (USA)

Medical Primatology 1972. Proc. 3rd Conf. exp. Med. Surg. Primates, Lyon 1972, part I, pp. 156–165 (Karger, Basel 1972)

Laboratory Breeding and Husbandry of Macaques, African Green Monkeys and Greater Bushbabies[1,2]

D. A. VALERIO, P. T. JOHNSON, W. E. LEVERAGE and G. E. THOMPSON

Litton Bionetics, Inc. Rockville, Md.

Introduction

Some of our experiences in the laboratory breeding of several non-human primate species (hereinafter referred to as simians) have been previously reported [5, 7, 8, 12]. Our multispecies simian breeding colony was established 11 years ago for the production of newborn and/or pregnant females for various biomedical research studies. For the past five years the colony population has averaged about 800 breeding primates. Macaque species, primarily *Macaca mulatta,* represent 85–90 % of the breeding colony. The intent of this paper is to summarize our experiences in the laboratory breeding of the rhesus monkey *(M. mulatta),* the cynomolgus monkey *(M. fascicularis)*, the bonnet monkey *(M. radiata),* the African green monkey *(Cercopithecus aethiops)*, and the greater bushbaby *(Galago crassicaudatus)*. The general husbandry practices employed will also be discussed as will the differences between the species.

Husbandry

All simians were purchased from a commercial vendor (Primate Imports Corp., Long Island, N.Y.). The galagos were purchased in the autumn

1 Conducted in part under contract No. NIH-NCI-71-2025, Special Virus Cancer Program NCI, NIH and USPHS.

2 The work was performed in an AAALAC accredited facility and according to the principles of animal care promulgated by the NAS-NRC.

of 1965, the African green monkeys were purchased in August, 1967, and the majority of the macaques were purchased between 1961 and 1967. All the simians described in this paper are housed in a two-story, windowless, environmentally controlled facility, which has been previously described [11].

The macaques and African green monkeys are fed a commercial pelleted monkey diet twice daily. The galagos are fed the pelleted diet only in the afternoon and receive bread slices in the morning. In addition, all primates are fed a special vitamin sandwich in the morning and sugar cubes containing Isoniazid each afternoon (8mg/kg body weight).

All simians underwent a three- to six-month quarantine and conditioning period prior to entry into the breeding colony. During quarantine, tuberculin testing was conducted every two weeks until the group had five consecutive negative tuberculin tests. An exception to the above was the galagos, which were tuberculin tested monthly during the quarantine period. Once in the breeding colony all primates are tuberculin tested every two months. The canine teeth have been surgically removed from all macaque and African green males. Greater detail on the husbandry and management of our breeding colony has been reported [11].

Macaque Breeding (M. mulatta, M. fascicularis, M. radiata)

For the successful breeding of macaques perhaps the most important factor is the proper monitoring of the menstrual cycle. In our colony this is done by subjecting all nonpregnant females to a daily vaginal swab. The detection of each day of menses is important to insure the optimal mating period. In 1969, VALERIO [7] and VALERIO *et al.* [12] reported on the length of menses and length of menstrual cycles recorded from 457 *M. mulatta,* 87 *M. fascicularis,* and 17 *M. radiata,* that conceived during a one-year period. The mean length of menses for *M. mulatta* and *M. fascicularis* was 3.9, while it was 3.8 for *M. radiata*. This is a remarkable similarity for these three macaque species. Furthermore, for these three species, very few conceptions resulted from cycles with menses longer than seven days suggesting that the incidence of ovulation was infrequent or that the endometrium was not prepared for implantation. In contrast to this, vaginal bleeding of only a one-day duration must be seriously considered as true menses because the mating of 39 *M. mulatta* based on this phenomenon resulted in conceptions.

The ability to predict the time of ovulation is extremely important for an efficient breeding operation. Practically speaking, for large breeding colonies this is best achieved by observation of menses and knowledge of previous menstrual cycle lengths.

The evaluation of menstrual cycles for these macaque species has also been previously reported [7,12]. The analyses included 3,561 menstrual cycles for *M. mulatta,* 372 for *M. fascicularis* and 76 for *M. radiata*. All short cycles were included since no cycles were shorter than 15 days. Menstrual cycle lengths greater than 50 days were not analyzed and arbitrarily were considered to be periods of amenorrhea, rather than long menstrual cycles. Menstrual cycle lengths that immediately preceded a conception were selected for analysis on the basis that they probably represented the most normal distribution and information for a species. *M. mulatta* cycle lengths had a bimodal distribution with peaks of 28 and 30 days with an additional smaller peak at 32. Menstrual cycle data for *M. fascicularis* was not as defined as that of *M. mulatta*. The most obvious peak occurred with menstrual cycles of 33 days. Although fewer cycles were analyzed for *M. radiata,* the distribution appeared to be bimodal with peaks of 28 and 30 days, which is comparable to that of *M. mulatta*. For all three species periods of amenorrhea, that is, 50 days or greater, could immediately precede a cycle in which conception occurred.

The breeding of macaques employed single pairs of animals in which the female was taken to the male's cage. The male to female ratio of 1:7–10 in our colony has been effective for providing sufficient rest for the males as well as permitting their maximal use. All matings are scheduled in advance using the data of menstrual cycle lengths and previous mating history. Males that had previously impregnated a certain female were generally reselected for at least three consecutive monthly matings. In general, the majority of the females had regular menstrual cycles (26–30 days) and were mated to the selected male beginning on day 11 of the cycle for a period of 72 h. Other 72-hour schedules were employed dependent on cycle length histories, as described in a previous report [12]. The second day of a three-day mating was taken as the estimated day of ovulation (EDO) for the calculations of lengths of gestation.

As previously shown in our work [7], once female *M. mulatta* have been in the laboratory at least three years, annual conception rates in excess of 100% can be expected (111%). For *M. fascicularis* females a comparable percentage can be expected, approximately 100% or more. Annual reproductive performance for *M. radiata* in our last analyses was approximately 70 % conceptions. This data for *M. radiata* is presently being reanalyzed. We have previously reported that fetal wastage averages approximately 15 %. The mean birth weights and gestational ages for live births of the three macaque species are shown in table I.

We have previously discussed breeding seasonality for our laboratory [7, 12]. It is sufficient to say that after females have been acclimated for a couple of years to the laboratory, breeding seasons do not exist as they do in their natural habitat, although conception and consequently delivery peaks do occur. However, we have conceptions and consequently births every month of the year for the three macaque species discussed.

African Green Breeding

Data concerning the laboratory breeding of the African green monkey *(Cercopithecus aethiops)* and published elsewhere *in extense* [5] will be summarized here.

Our colony of African greens consists of 30 females and six males. Publications on the reproductive characteristics of this species are limited as well as conflicting. Due to the difficulty reported by various laboratories that had attempted breeding of African green monkeys, we initiated this effort with some changes from that employed in our macaque breeding. It was believed there might be an advantage to housing the males in the same room with the females as opposed to separate rooms; females were still taken to the male's cage for breeding. The collection of semen from this species has been unsatisfactory. (See table II).

The detection of menses was performed by daily vaginal swabs, as with the macaques. The period of training the female to present her external

Table I. Mean birth weight and gestational age for live births of three macaque species

	M. mulatta	*M. fascicularis*	*M. radiata*
Number of live births	967	109	16
Mean birth weight, g			
combined sexes	488±76[1]	343±46	404±56
males	500±75	352±44	411±56
females	477±77	329±47	388±34
Gestational age, days			
combined sexes	164±6	160±7	162±4

1 ±SD.

Table II. Normal mean values of semen parameters

Species	Volume			Concentration/ml $\times 10^6$	Total concentration $\times 10^6$	Motility %	Abortion %	Dead %
	liquid	coagulum	total					
M. mulatta October to December	1.2	1.34	2.47	365	689	60	22	15.8
M. mulatta January to July	0.6	0.71	1.34	317	499	40	24	20
M. fascicularis	0.47	0.37	0.83	347	486	33	32	26
M. radiata	0.4	0.44	0.84	204	288	41	32	20
C. aethiops	1.0	0.24	1.24	145	298	58	24	13

genitalia for this process was longer and more tedious due to the nervous and excitable disposition of the African green. The majority of menses are detectable by vaginal swabs alone. However, the menstrual flow is not nearly as intense as it is in the *M. mulatta* and for those females that were amenorrheic or had too scant a menses an occult blood detection test (Hematest®) was utilized. The mean menses length of 583 menses analyzed was 3.7 days, which is slightly shorter than that for the macaque species discussed. It is significant that females with 1+ menses detected by the Hematest® and bred accordingly at approximately midcycle have often become pregnant.

The menstrual cycle length of African green monkeys is not well established in the literature. BUTLER [2] describes seven *Cercopithecus aethiops* as cycling at a mean of 72.3 ± 36.23 days with a range of 24–149 days. Our mean value for 392 cycles was 30.9 days. The difference between our data and BUTLER's [2] probably is that we have separated the *normal* cycle from the amenorrheic cycle which is arbitrarily defined as any cycle length greater than 50 days. After an initial 12-month attempt at breeding this species for a 72-hour period, and due to the fact that so little was known on individual matings of these animals, we have bred the regular cycling females for a six-day period as opposed to a 72-hour period for macaques. Normal cycling females are placed with the male on the tenth day following the onset of menses, for a six-day mating. Pregnancy is diagnosed by digital palpation of the uterus at approximately 25–30 days after breeding, as in macaques. Implantation bleeding, a phenomenon routinely observed in *M. mulatta* at 15–21 days postbreeding, has not been reported for *C. aethiops*. However, in 33% of the pregnancies in the colony, bleeding has been detected in this range of 15–21 days postbreeding, but the duration has been predominantly one day. Due to the excitability of this species, a stabilizing period of six to 12 months in the laboratory has been necessary before menstrual regularity is established. During the last two years, 75% of the females have conceived at least once per year and this required an average of five matings.

Breeding seasons concerning other colonies have been described by NAPIER and NAPIER [6] as nonfixed. CHASE and LONG [3], however, has reported a breeding season of August-March for an outdoor colony of *C. aethiops* in San Diego, Calif. Although peaks of productivity occur regularly in our colony, conceptions as well as the arrival of newborn have continued throughout each year. The length of gestation for *C. aethiops* as reported by NAPIER and NAPIER [6] is 180–213 days. However, in our colony, 38 live births have had a mean gestation length of 163 ± 6 days. Thirty-six months of breeding resulted in 75 pregnancies and 54 live offspring. A 28-percent fetal wastage observed in this species is higher than in the macaques. The African green monkey can be bred in a laboratory environment with reasonable success.

Greater Bushbaby Breeding (Galago crassicaudatus)

The greater bushbaby, *Galago crassicaudatus,* has been utilized in various biomedical studies. It is representative of the primate suborder Prosimii and due to their size they are ideal for housing and handling. They are nocturnal in habit, are easily bred in captivity and relatively simple to maintain. Our colony consists of about 36 galagos equally divided as to sex. Galagos, as opposed to other species reported in this paper, are bred as monogamous pairs. Body weights for adult males average 1,300 g and for females 1,100 g.

Estrus in this species is detected by the observation of changes that occur in the external membrane that nearly covers the vulvar orifice. As estrus approaches the vulvar area becomes hyperemic and the external membrane regresses. Mating is accomplished by placing the female with its selected male at the time of estrus and removing the female when the membrane returns to the anestrus position. The 17-day estrus length was the most frequently observed (range 3–104 days), especially in the last 12 months of breeding data reported. The mean estrous cycle length was 50.3 days.

Post partum estrus occurs in this species [1]. In our colony it was detected as early as two days after parturition. The mean value for initiation of *post partum* estrus was 10 ± 5 days for the 2½-years in which the offspring were immediately removed from their mothers. The mean length of the *post partum* estrus was 23 ± 20 days. Forty-three percent of *post partum* estrus matings resulted in conceptions.

Breeding seasons *per se* have been described for primates in their natural habitat as well as in captivity [1, 6]. For an outdoor colony of *G. crassicaudatus,* COOPER [4] has described a primary breeding season of January-February and a secondary season in June-July. Secondary seasons, however, appear to be heavily dependent on *post partum* estrus. In our laboratory, irrespective of the method of infant rearing, we have found no distinctive breeding seasons. During the years when the offspring were being hand reared, the number of conceptions per female was almost two per year. In our colony the mean gestation length was 128 ± 13 days for 69 pregnancies, which resulted in live births. Fifty-three months of breeding activity resulted in 69 pregnancies with either single or twin births for a total of 77 live offspring.

The overall fetal wastage rate was 29.8 % which included abortions and stillbirths Eight sets of twins were born for a twinning rate of 11.5 %. The mean birth weight of combined sexes is 47.4 $\pm$ 6 g. The greater bushbaby can be maintained and reproduced with ease in the laboratory environment (for more detail see a recently published article [8]).

Semen Characteristics and Artificial Insemination

The reproductive potential of males in our breeding colonies are monitored by semen evaluations. Semen evaluations are performed on these males routinely at 60-day intervals. Males with abnormal or unacceptable semen profiles are checked more frequently and removed from breeding until their evaluations return to acceptable ranges [10]. Briefly, the technique involves the collection of semen by electroejaculation into a one-ml solution of 2 % trypsin in 2.5 % lactated Ringer's. The ejaculate is allowed enzymatic liquifaction for 30 min at 37 °C at which time it is evaluated for semen volume (coagulum, liquid and total), sperm concentration, total sperm numbers, motility (at 10 % increments), abnormal and live: dead ratios.

Based on this method of evaluation we have found that *M. mulatta* males have seasonal semen profiles, the highest quality semen occurring during the months of October through December. Further, we have found that after acclimatization to our colony environment, males no longer are azoospermic during the months of February through July, although seasonality of impregnation still exists in a few males. Not enough information is available from our colony to make the same statistical evaluation of *M. fascicularis* or *M. radiata* semen. Normal mean values of semen parameters by species are shown in table II.

The use of trypsin in our laboratory is limited to semen evaluations since enzymatic damage to spermatozoa can occur after prolonged exposure. Semen used in artificial insemination is allowed to undergo autolysis for 30 min at 37 °C.

Initial studies in artificial insemination in our laboratory were necessarily concerned with developing the most successful technique. First, the coagulum was allowed to partially liquify and both the liquid and the coagulum were deposited into the vagina. Inseminations were performed either on days 11 through 14 of the menstrual cycle or on alternate days, generally, days 10, 12 and 14 of the cycle. The conception rate in *M. mulatta* by this technique was 4.03 %. The next technique attempted was intrauterine deposition of liquid semen [9]. An 18-gauge teflon catheter with a stainless steel stylus was used to traverse the cervix. Once the uterus was entered the stylus was removed and a one-ml syringe containing the semen was attached. Generally, 0.2–0.5 ml of semen was deposited. The conception rate from this technique was 20 %. Even though conception rates were comparable to natural mating, the intrauterine technique was difficult and time-consuming; therefore, an intracervical deposition technique was developed. The same insemination technique is used as with intrauterine insemination except that the semen is deposited in the cervix, generally, 0.1–0.2 ml of semen. Intracervical deposition has resulted in a 19-percent conception rate compared to an 18-percent conception rate for natural matings during the same period.

Most recently we have been studying the dilution of semen to derive an optimum concentration of motile spermatozoa necessary to impregnate females. In this study conceptions occurred when 7.5×10^6 motile cells were delivered in a 0.17-ml volume intracervically. Details of this study will be reported later this year.

References

1 BUETTNER-JANUSCH, J.: The breeding of galagos in captivity and some notes on their behavior. Folia primat. *2:* 93–110 (1964).

2 BUTLER, H.: Observations on the menstrual cycle of the grivet monkey *(Cercopithecus aethiops aethiops)* in the Sudan. Folia. primat. *4:* 194–205 (1966).

3 CHASE, J. E. and LONG, J. O.: A successful breeding colony of African green monkeys, *Cercopithecus aethiops*. Experimental breeding of subhuman primates. 5th Annu. Rep. Appendix IV, NIH Cancer Institute Contract PH43-63-56 (Institute for Comparative Biology, Zoological Society of San Diego, San Diego 1966).

4 COOPER, R.W.: Experimental breeding of nonhuman primates. 7th Annu. Rep., NIH Cancer Institute for Contract PH43-63-56 (Institute for Comparative Biology, Zoological Society of San Diego, San Diego 1969).

5 JOHNSON, P. T.; VALERIO, C. A., and THOMPSON, G. E.: Breeding the African green monkey, *Cercopithecus aethiops,* in the laboratory environment (in press).

6 NAPIER, J. R. and NAPIER, P. H.: A handbook of living primates (Academic Press, New York 1967).

7 VALERIO, D. A.: Breeding *Macaca mulatta* in a laboratory environment. Lab. Anim. Hb. *4:* 223–230 (1969).

8 VALERIO, D. A.; JOHNSON, P. T., and THOMPSON, G. E.: Breeding the greater bush-

baby, *Galago crassicaudatus*, in a laboratory environment. Lab. anim. Sci. *22:* 203–206 (1972).

9 Valerio, D. A.; Leverage, W. E.; Bensenhaver, J. C., and Thornett, H. D.: The analysis of male fertility, artificial insemination and natural matings in the laboratory breeding of macaques; in Goldsmith and Moor-Jankowski Medical primatology 1970, pp. 515–526 (Karger, Basel 1971).

10 Valerio, D. A.; Leverage, W. E., and Munster, J. H.: Semen evaluation in macaques. Lab. anim. Care *20:* 734–740 (1970).

11 Valerio, D. A.; Miller, R. L.; Innes, J. R. M.; Courtney, K. D.; Pallotta, A. J., and Guttmacher, R. M.: *Macaca mulatta.* Management of a laboratory breeding colony (Academic Press, New York 1969).

12 Valerio, D. A.; Pallotta, A. J., and Courtney, K. D.: Experiences in large-scale breeding of simians for medical experimentation. Ann. N.Y. Acad. Sci. *162:* 282–296 (1969).

Authors' addresses: David A. Valerio, DVM, Vice President, Biomedical Research Division, Litton Bionetics, Inc., 5541 Nicholson Lane, *Rockville, MD 20852;* Philip T. Johnson, DVM, MS, W. Edwin Leverage, BS and George E. Thomson, Department of Laboratory Animal Science, 5510 Nicholson Lane, *Kensington, MD 20795* (USA)

Medical Primatology 1972. Proc. 3rd Conf. exp. Med. Surg. Primates, Lyon 1972, part I, pp. 166–173 (Karger, Basel 1972)

Providing and Maintaining Primates for an Industrial Research Department

H. WEBER

Medical and Biological Research Division, Sandoz Ltd., Basel

In biological and medical research, the introduction of sophisticated experimental models and the application of statistical analysis have carried in their wake the development of a new branch of science pertaining to laboratory animals. Research in this field has led to the recognition of many variables that may influence the experiment, which may be grouped broadly under the headings experimental environment, the genetic background and the condition of the experimental animals. While it is possible for the investigator to standardize experimental conditions, the genetic background and condition of the test animal usually escape his direct control. The spectrum of animal models has been enlarged by the introduction of primate species into medical research. Compared with today's traditional laboratory animals such as mice, rats or dogs, the 'quality' of the new species has set back biological investigation by some 20 years to a time when experiments were done in laboratory animals of no specific genetic background or age, and in carriers of latent disease. This diminishes the advantage of greater relevance of experimental results regarding comparison with findings in man.

The common laboratory animals are bred for experimental use. They have undergone domestication to such an extent that their handling meets with little, if any, resistance. Monkeys in contrast are animals of the wild. They are usually anything but willing to co-operate with the investigator, and many are strong enough to inflict serious wounds. The situation is aggravated by the fact that many primate species may be carriers of disease potentially fatal to man.

In order to supply our laboratories according to their changing demands with healthy monkeys and to ensure minimal risk to the personnel handling the animals, we had to introduce strict quarantine procedures and develop safe caging facilities of optimal flexibility.

Regulations in Switzerland prohibit the importation of monkeys except for delivery to zoos and research centers. Quarantine is compulsory, but the measures to be taken during that time are left entirely in the hands of the veterinarian in charge. Animals arriving at our laboratories in Basel are first dipped in an insecticide bath and then placed in individual cages, separated from each other by Securit glass walls. Quarantine lasts at least 12 weeks during which time we undertake the following examinations and tests:

(1) clinical examination with special attention paid to the oral cavity and auscultation of the thorax;

(2) tuberculin test;

(3) thorax radiography;

(4) examination for Salmonella and Shigella in the feces;

(5) herpes simplex neutralization test;

(6) course of treatment against endoparasites;

(7) treatment of sick animals, where this is possible.

On arrival the monkeys are given Ringer's solution to drink. From the second day onward the animals receive a pelleted diet the formula of which has been previously published [WEBER and GRAUWILER, in press]. This diet has proved to be quite satisfactory as sole nutrition for macaques over periods of at least two years. It is also used for South American species which, however, usually receive orange juice, and fruits in addition. During the first week of quarantine the monkeys are gradually trained to our routine feeding regime by reducing the feeds from four to one per day. The monkeys soon learn not to drop the pellets into the fecal tray where they are irretrievable.

Clinical examination is done twice weekly for the first six weeks. Special attention is paid to ulcerations in the mouth, and to bronchial rales. With the smaller squirrel monkey this procedure is not usually carried out; clinical inspection is more superficial, which may partly account for the higher losses.

In the larger species (macaques) the tuberculin test is done intradermally, in the lower arm, using 500 IU of the purified protein derivative tuberculin (one part bovine type and two parts human type). With the squirrel monkey 100 IU are injected into an upper eyelid. Originally our quarantine period lasted eight weeks, and two tuberculin tests were made, one at the beginning, the other at the end of this period. However, in 1970 a tuberculous epizootic in two of our laboratory colonies of rhesus monkeys forced us to eliminate 57 animals and to reconsider our quarantine procedure. The epidemiologic trace-back pointed to four originally infected animals from one imported batch that had not shown positive reactions within the first eight weeks of quarantine. These animals had been distributed among three colonies in two of which the disease had had ample time to spread before the first infected animals were discovered. The course of the disease was typical and need not be described here. The agent, identified as a bovine type, was associated with only very mild tuberculin reactions at all stages of the disease. After consulting with Dr. SCHMIDT of the Southern Research Institute we decided that a 12-week quarantine period with three tuberculin tests at four-week intervals would give us better protection. According to Dr. SCHMIDT, no diminution of the tuberculin reaction due to short intervals of testing is to be feared. With some mycobacteria strains positive reactions may not develop before eight weeks after infection; however, excretion of bacteria can occur earlier than 12 weeks after infection. In order to exclude the possibility of missing animals in the late anergic stage all macaques are X-rayed before delivery to the laboratory. The animals are not treated with isoniazide, chiefly because we prefer not to interfere with drug metabolizing enzyme systems in our experiments. On the other hand, tuberculin testing is continued at eight-week intervals in macaques and 12-week intervals in South American species, even after quarantine.

For Salmonella and Shigella testing, rectal swabs are taken twice during quarantine, streaked on SS and McConkey agar plates and sent for diagnosis to an external bacteriological laboratory in Basel (Bakteriologisches Laboratorium Dr. Roth, Basel, Switzerland). Positive animals are treated with ampicillin.

Serum is taken at the beginning and at the end of quarantine and sent to the Institute of Virology at the University of Zurich for a neutralization test for herpes simplex virus. It is assumed that due to the cross-reaction between herpes B antibodies and herpes simplex antigen, a positive reaction will reflect prior exposure of the animal to B-virus [SABIN, 1934; KALTER *et al.*, 1967].

Table I gives the results of investigations and losses during quarantine in three imported monkey species in the past two years. The most startling finding is the high incidence of positive titers against herpes simplex virus in *Macaca arctoides*. If our aforementioned assumption is correct, this would mean that most of our *M. arctoides* could be possible carriers of herpes B virus. Until now we have only euthanized those monkeys showing clinical signs of herpes disease. This was the case with two of 250 imported rhesus monkeys. The infection, however, could not be confirmed definitely by histological evidence of inclusion bodies or by test injections in rabbits.

Repetition of the neutralization test on 20 rhesus monkeys that had previously been negative and had been having contact with positive animals since two years after quarantine showed no rise in antibody titers. This would indicate that individual isolation during quarantine can prevent the spread of infection. Nevertheless, according to the statement of HUNT and MELENDEZ [1969], animals once infected may remain life time carriers of viable virus, a fact that begs for caution.

The most common diseases encountered in rhesus monkeys during quarantine are endoparasite infestations, bronchopneumonia and diarrhea. Figure 1 shows the chronological appearance of the last two clinical signs during quarantine.

The losses due to bronchopneumonia occur usually in the first two or three days after arrival, when treatment for severely ill animals is too late. A comparison of the losses during quarantine in 1967, 1969 and 1971, with death rates of 26, 23 and 11%, respectively (table II), shows that the measures taken since 1969 (chiefly consisting in improved housing and caging) are effective in reducing mortality. Our experience, however, shows that the incidences of bronchopneumonia and diarrhea over the same time period were hardly altered.

With squirrel monkeys the chief *post mortem* findings in animals which die during quarantine are pneumonia, severe endoparasite infestations and degeneration of the liver. A strikingly high proportion of animals die without showing any clinical signs prior to death or apparent organic changes.

Table I. Results of necropsies and laboratory investigations during quarantine of three monkey species imported in the years 1970/1971

	M. mulatta	*M. arctoides*	*S. sciureus*
Total number of imported animals	251	97	222
Number of losses (%)	32 (13)	8 (8)	79 (35)
Number of deaths due to			
bronchopneumonia	7	0	14
diarrhea	19	5	0
liver degeneration	0	0	9
tuberculosis	4	0	0
herpetic lesions	2	0	0
parasites	0	2	16
tumors	0	1	0
'acclimatization stress'	0	0	39
Results of laboratory investigations			
salmonella positive, %	0	0	1
shigella positive, %	1	1.2	1
herpes positive, %	33	90	0

Clinical investigation in these animals is handicapped by their size. The term 'acclimatization stress' which we choose as diagnosis, is a poor explanation for deaths due possibly to simple mistakes in feeding methods as has been described by MORELAND [1970].

As soon as the quarantine period is over, the animals are put under the direct care of the investigator. Since it does not seem possible to exclude the presence of latent viral infections [MELNICK and BANKER, 1954] safe handling procedures are essential even after the monkeys are released from the quarantine area.

Table II. Losses in imported rhesus monkeys during quarantine in the years 1967, 1969 and 1971, and percentage of dead animals with diarrhea and bronchopneumonia. Not included are animals that died or were killed due to other reasons

	1967	1969	1971
Number of animals imported	42	90	171
% of animals which died during quarantine	26	23	11
% of dead animals showing diarrhea	63	83	53
% of dead animals showing bronchopneumonia	73	20	26

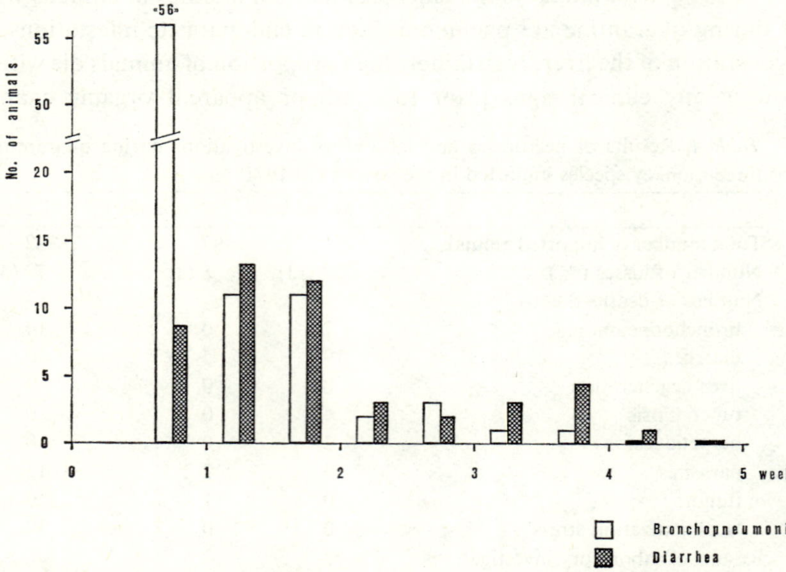

Fig. 1. Incidence of fresh cases of bronchopneumonia and diarrhea in rhesus monkeys as encountered in clinical investigations during the first weeks of quarantine. The figure represents the findings in a total of 251 animals imported in the years 1970 and 1971.

After some years of experience with various types of cages, both commercially available and constructed in our own workshops, we recommend the following conditions for caging subhuman primates:

(1) possibility for animal to adapt natural positions;

(2) prevention of fecal recontamination;

(3) visual but no physical contact between animals;

(4) safe catching device without necessitating the use of tranquillizers;

(5) possibility for minor manipulations, e.g. parenteral injections, without having to take the animal out of the cage;

(6) restriction of fixed room installations to a minimum;

(7) effective cleaning and disinfection;

(8) economy.

Until recently we have been using a large cage of 135 cm height 55 cm width and 65 cm depth for rhesus monkeys and stump-tailed macaques. The lower half of the front wall can be hinged backwards so as to confine the monkey to the upper half of the cage in which it can be squeezed to the front wall.

We have now, however, changed over to the cages used by the Laboratory for Experimental Medicine and Surgery in Primates [DAVIS *et al.*, 1969]. This cage has been modified by adding a hinged front door with widely-spaced bars through which an arm or leg of the monkey can be extruded. This facilitates getting hold of the animal and safely performing minor manipulations. The dimensions of this suspended cage are optimal for the individual caging of monkeys of the size of fully-grown macaques or the group caging of smaller species. For this size of animal – considering the minimal requirements for cage height – a double-storied caging system would be impracticable. The uneconomical use of room space due to the resulting single-story construction is somewhat compensated by keeping the cages narrow and deep. As an advantage for the monkey there is no necessity to limiting the height of the cage to the required minimum; the height is only restricted by the weight of the cage.

Feces and urine are collected from the floor below the cages which is covered with wood-granulate. The granulate is scraped daily into plastic bags and incinerated. The floor is then cleaned with a disinfectant. For periodic cleaning of the cages the monkey is removed to a clean cage. The dirty cage is then sterilized and washed.

For individual caging of the smaller monkey species the described system would be far too uneconomical. At the moment a mobile cage rack is in use which holds four cages sized 45 × 60 × 50 cm. The back and side walls, as well as the roof of these cages, are of solid stainless steel. Again the monkeys can be squeezed against the bars of the front door. Feces and urine are collected in cardboard trays below the floor grid lined with plastic. For cleaning, the complete rack is moved to the cleaning area.

For examination and experimentation, a restraining chair for macaques is used, as described by HARDY *et al.* [1965]. Various modifications for the specific needs of individual laboratories exist. A similar chair has also been constructed for metabolic experiments with squirrel monkeys. The chair is also used for repeated daily oral drug administration in toxicity studies, using either a stomach tube or gelatine capsules. We should emphasise that transfer of the animal from the cage to the chair is done by two assistants in the following way: The animal is pressed against and held tightly to the bars of the cage door by one person holding the monkey by an arm or the collar. The door is then swung open and the legs are grasped. The other person takes the animal holding the arms behind its back, and places it into the chair, the bars of which are screwed in place tightly. If necessary, the arms are strapped down. This restraining procedure works well with smaller animals. However, full-grown male macaques usually require to be given a tranquillizer beforehand. For this purpose Ketamin (10–20 mg/kg) has been found very useful.

Handling of the animals is rarely complicated by mishaps. Occasionally an animal may escape. For its recapture a net is used or the cage is baited. As a safety precaution, all monkey rooms are equipped with an extra inside wire-mesh door. Windows are also protected with wire-mesh. Within the last two years two persons have been severely bitten. In both cases, safety precautions (wearing of gloves, no single-handed handling of a nonanesthetized animal!) had been neglected.

Injury to the animals is rare, but has happened in one toxicity study. During the administration of capsules to a group of 18 rhesus monkeys, two to two and a half years old, injury of the pharynx occurred in six animals, with perforation in three and subsequent retropharyngeal and pleural abscess formation. This mishap was due to either the small size of the animals or the inexperience of the administrator.

Occurrence of this kind, and the danger of disease transmission to man, would in our opinion indicate the need for specially equipped centralized units and trained personnel for experimentation on primates. This would also facilitate species separation and the economic acquisition of usually expensive equipment for caging and restraining the animals. However, in a pharmaceutical research department with rather specialized research groups, the investigator prefers to have all the experimental models he is interested in grouped closely around him, no matter what species are involved. It is our hope that with the quarantine procedure and the caging and handling devices just outlined, it will continue to be possible to use the monkey with a reasonable degree of safety as an additional laboratory animal even in decentralized laboratories.

References

DAVIS, J. H.; MCPHERSON, B. R. and MOOR-JANKOWSKI, J.: Maintenance and handling of primate animals for medical research. Ann. N.Y. Acad. Sci. *162:* 329–335 (1969).

HARDY, J. R.; HIROTA, K.; HARKINS, H. N., and NYHUS, L. M.: A new chair for the study of restrained monkeys. Med. Pharmacol. exp. *12:* 336–344 (1965).

HUNT, R. D. and MELENDEZ, L. V.: Herpes virus infections of non-human primates. A review. Lab. anim. Care *19:* 221–234 (1969).

KALTER, S. S.; RATNER, J.; KALTER, G. V.; RODRIGUEZ, A. R., and KIM, C.S.: A survey of primate sera for antibodies to viruses of human and simian origin. Amer. J. Epidem. *86:* 552–568 (1967).

MELNICK, J. L. and BANKER, D. D.: Isolation of B virus (Herpes group) from the central nervous system of the rhesus monkey. J. exp. Med. *100:* 181–194 (1954).

MORELAND, A. F.: Nutritional status and problems of feeding newly imported primates; in HARRIS Feeding and nutrition of non-human primates, pp. 29–43 (Academic Press, New York 1970).

SABIN, A. G.: Studies on the B-virus. I. The immunological identity of a virus isolated from a human case of ascending myelitis associated with visceral necrosis. Brit. J. exp. Path. *15:* 248–268 (1934).

WEBER, H. and GRAUWILER, J.: Experience with a breeding colony of *Macaca arctoides* used for the teratological testing of drugs; in BEVERIDGE Breeding primates. Int. Symp. on Breeding Non-Human Primates for Laboratory Use, Berne 1971 (Karger, Basel, in press).

Author's address: Dr. HEINZ WEBER, Sandoz AG, *CH-4000 Basel* (Switzerland)

Medical Primatology 1972. Proc. 3rd Conf. exp. Med. Surg. Primates, Lyon 1972. part I, pp. 174–175 (Karger, Basel 1972)

Journal of Medical Primatology

E.I. GOLDSMITH and J. MOOR-JANKOWSKI

Laboratory for Experimental Medicine and Surgery in Primates (LEMSIP), New York University School of Medicine, New York, N.Y.

The Journal was founded in 1971 as a forum for medical scientists and biologists interested in approximating the health and disease conditions of man in his phylogenetically closest relatives, the apes and monkeys.

This field of research has grown during the past decade much faster than most biomedical research areas[1] thus creating the need for a specialized journal to publish papers refereed and edited from the viewpoints of the human disease under study as well as of the appropriate animal husbandry, correct taxonomy and due concern for humane experimentation and conservation of endangered species.

The fifteen editors and fifty consultants (referees) of the Journal represent an international cross-section of biomedical scientists interested and knowledgeable in experimental use of simians. All papers submitted are evaluated by at least two referees.

The Journal is listed in *Biological Abstracts, Current Contents, Excerpta Medica* and *Index Medicus*.

While the main interest of the Journal lies in human biology and medicine, equal priority is awarded to simian biology because of its importance for experimental design and the dearth of knowledge in this area, and to papers on new methods of simian husbandry and methodology of experimentation, particularly important for the new discipline of Medical Primatology.

Papers submitted should be original and written in English. As a rule,

1 E.P. FRIDMAN: 'Recent Developments of Biomedical Research in Primates with a Short Historical Review', Sukhumi Primate Information Center, Institute of Experimental Pathology and Therapy of the USSR Academy of Medical Science, Sukhumi, USSR. J. med. Prim. *1:* 220–229 (1972).

the papers should not exceed 8 printed pages, including illustrations and references; however, longer papers will be occasionally accepted. Short papers in form of case histories, clinical, pathological, epidemiological, etc. observations of 1–3 printed pages are also invited, as are book reviews and announcements as well as reports of meetings.

Papers should be submitted in triplicate, with one set of original illustrations and two sets of copies, to expedite handling. They should be sent to one of the Editors-in-Chief of the Journal.

Authors' addresses: E.I. GOLDSMITH, MD, Editor-in-Chief, Journal of Medical Primatology, New York Hospital – Cornell Medical College, 525 East 68th Street, *New York, NY 10021;* and J. MOOR-JANKOWSKI, MD, Editor-in-Chief, Journal of Medical Primatology, New York University School of Medicine, 550 First Avenue, *New York, NY 10016* (USA)

Medical Primatology 1972. Proc. 3rd Conf. exp. Med. Surg. Primates, Lyon 1972, part I, pp. 176–178 (Karger, Basel 1972)

Series 'Primates in Medicine'

E.I. GOLDSMITH and J. MOOR-JANKOWSKI

Laboratory for Experimental Medicine and Surgery in Primates (LEMSIP), New York University School of Medicine, New York, N.Y.

Vol. 1
First Holloman Symposium on Primate Immunology and Molecular Genetics
Volume Editor: C.H. KRATOCHVIL
XVI + 99 p., 33 fig., 28 tab., 1968

Vol. 2
Using Primates in Medical Research
Part I: Husbandry and Technology
Volume Editor: W.I.B. BEVERIDGE
XIV + 132 p., 5 fig., 5 tab., 1969

Vol. 3
Using Primates in Medical Research
Part II: Recent Comparative Research
Volume Editor: W.I.B. BEVERIDGE
VI + 162 p., 3 fig., 15 tab., 1969

Vol. 4
Chimpanzee: Central Nervous System and Behavior
Volume Editor: H.H. REYNOLDS
VIII + 160 p., 11 fig., 19 plates, 1969

Vol. 5
Conservation of Non-human Primates in 1970
Author: BARBARA HARRISSON
VI + 97 p., 3 tab., 1971

Vol. 6
Chimpanzee: Immunological Specificities of Blood
Volume Editor: C.H. KRATOCHVIL
VI + 150 p., 6 fig., 27 tab., 1972

Vol. 7
Transplantation in Primates
Volume Editor: G.P. MURPHY
VIII + 148p., 36 fig., 27 tab., 1972

Vol. 8
1. Microbiology, Clinical Chemistry and Some Aspects of Hematology
Volume Editor: S.S. KALTER
(in press)

This series was started in 1968 to provide short to medium-size volumes concerned with the use of primate animals in human health research. It is directed mainly to the medical profession and, initially, it carried the subtitle 'Experimental Medicine and Surgery in Primates' to emphasize the value of simians as models for research in human health. This subtitle was recently abandoned because of the rapid progress in the recognition by health professions of the importance of primate experimentation.

The first volume of the series reviewed the considerable amount of basic biological information on simians obtained mainly by the group brought together by CLYDE H. KRATOCHVIL at the Holloman U.S. Air Force Base.

Volumes 2 and 3 were the outcome of a symposium on methodology of primate experimentation organized by the Merieux Institute in Lyon, France, under the auspices of World Health Organization. The two volumes represent a handbook on the methodology of human health related primate experimentation.

Volume 4, devoted to the central nervous system and behavior, initiated the series of reports on experiences gained in chimpanzee experimentation at the Holloman A.F.B. laboratory, where the U.S. Air Force pioneered in the use of chimpanzees not only for space research but also for basic biological studies and human health research. The volume was edited by H.H. REYNOLDS who initiated or directed much of the studies described.

Volume 5 is devoted to the conservation of primates and lists the most complete information available at this time. It is intended as a handbook and reference source for the international effort of conservation which, until the publication of this volume, was handicapped by the lack of a reliable up-to-date status report.

Volume 6 continues the series begun in volume 4, on chimpanzee experimentation carried out at the Holloman chimpanzee laboratory under the leadership of CLYDE H. KRATOCHVIL during the period 1963–1970. It may be safely stated that it was during that time at Holloman that the foundation was laid for the modern biomedical research use of chimpanzees. The work at Holloman concerned numerous aspects of basic and clinical research for which a completely new methodology of husbandry and experimentation on these powerful and potentially dangerous animals was developed. The data obtained as well as the experimental designs were either classified at that time or available only in internal technical reports of the U.S. Air Force. The aim of the chimpanzee volumes is to make this information available to the general readership. Moreover, while the Holloman Air Force laboratory was phased out in 1970 because of budget consideration, a considerable number of Holloman animals and research programs have been transferred to other institutions. Thus, an important part of knowledge on chimpanzee immunology reported in volume 6 was acquired on the basis of continued studies on the Holloman Chimpanzee Panel for Immunological Studies presently located in New York.

Volume 7 reviews succinctly the considerable amount of information on the use of simians in transplantation experiments as well as in organ transplantation therapy for human patients. The editor of the volume, G.P. MURPHY has had the numerically largest experience in simian transplantation in addition to his considerable background in human transplantation surgery.

Volume 8 is the first of a series of monographs on the baboon as the experimental animal in health research. It originates from the Southwest Foundation for Research and Education, San Antonio, Texas, which has been one of the pioneers of the use of baboons in medical research and which has particularly advanced our knowledge of simian microbiology by the studies of S.S. KALTER and co-workers in their Simian Virus Reference laboratory sponsored by the U.S. National Institutes of Health and the World Health Organization. The volume reviews in detail the large bulk of information on baboon microbiology and clinical chemistry obtained in the laboratory of the author and contains complete bibliographical information on work by others.

Several new monographs are in preparation and the series is intended to be pursued by publication of two volumes per year.

Authors' addresses: E.I. GOLDSMITH, MD, Editor, Primates in Medicine, New York Hospital – Cornell Medical College, 525 East 68th Street, *New York, NY 10021;* J. MOOR-JANKOWSKI, MD, Editor, Primates in Medicine, New York University School of Medicine, 550 First Avenue, *New York, NY 10016* (USA)

Medical Primatology 1972. Proc. 3rd Conf. exp. Med. Surg. Primates, Lyon 1972, part I, pp. 179–181 (Karger, Basel 1972)

Keeping Pace with Active Research on Experimental Medicine and Surgery in Primates

D. F. Hersey

Smithsonian Science Information Exchange, Washington, D.C.

I shall present here rather briefly the program of the Smithsonian Science Information Exchange in the United States. The Exchange is somewhat unique among information systems in that it does not deal with published research in either journals or the report literature. It does, however, cover that important segment of ongoing research which has been funded but may not have reached a stage where publication or presentation at meetings has been accomplished.

For more than 20 years the Exchange has been providing research scientists and managers in both the United States and abroad with information on who is currently working in a given area of research, and where such research is being carried out. In addition, information is provided on who supports the research, the period of support and in many cases the level of support. A short (approximately 200-word) summary is also provided for each of the more than 90,000 active research projects registered with the Exchange each year.

To better understand the Exchange one must have some understanding of the types of research projects registered with the Exchange and where they come from. The Exchange's coverage includes basic and applied research in all areas of the life (including agricultural and medical sciences), physical, social and engineering sciences. Information on grants, contracts and other awards as well as in-house research carried out by US Federal Agencies and other organizations is registered with SSIE. In addition, a large portion of non-federally supported research supported by private foundations, fund raising agencies such as the American Cancer Society and American Heart Association, universities, and to a lesser extent industrial laboratories and overseas organizations is also registered at SSIE. Although the Exchange does not contain a complete record of all ongoing research in the United States, it is the most complete data bank of active research currently available anywhere, offering the advantage of a single source of query, uniform indexing and retrieval, regardless of the source of support of research.

The value and usefulness of the Exchange's data bank is amply demonstrated by a constant increase in the demand for user services. The Exchange fulfills an important function in helping research scientists and management personnel avoid unwarranted duplication, as well as maintain a current awareness in fields of rapidly changing research efforts. The Exchange attempts, with a considerable degree of success, to have the individual research scientists prepare the abstracts of their own research and to have them updated on an annual basis to reflect changes in the direction of their research.

The Exchange's computerized files are set up in such a way that its scientific staff can respond fairly quickly and accurately on research in all areas of research including primatology. There are at present over 1,000 projects in the area of primatology registered with the Exchange. Of that number, there are some 246 dealing with neurophysiology, 236 that deal with psychology, 186 and 263 projects dealing with cardiovascular studies and pharmacology and toxicology, respectively. I have selected these broad areas at random only for the purposes of giving you a feel for the range of coverage. The Exchange provides information on both broad areas, such as those I have just mentioned, or in very specific areas. As an example of our information service on specific research areas, I have chosen a typical package of ongoing research records one would receive from the Exchange in response to each of the following inquiries for all active research on

(1) psychotropic drug research in primates;

(2) the limbic system of primates;

(3) primate hemodynamics, including blood flow and volume studies in localized tissues and organs.

The first of these contained some 37 projects, the second 28 and the third 74 projects. Each of the projects represents an individual research project. The packages, which are typical of those actually received by users, would show a wide range of supporting organizations, and of performing organizations, many of which you may not be familiar with. The intent of our Notices of research is to provide a simple, rapid mechanism for research workers to contact one another with the intent of exchanging more detailed information on problems of common interest.

There are several reasons why the Exchange is able to provide information about ongoing research which is designed to fit the requirements of its scientific users. The Exchange's system of indexing is interdisciplinary in nature and both indexing and retrieval are carried out by a staff of scientists and engineers including doctorate level personnel. The indexing system itself is hierarchical and open-ended in nature which allows the Exchange the flexibility of answering questions in the user's own words, regardless of the specificity of the request without the need for key words. The open-endness of the system, as you might guess, allows us to add or delete terminology necessary for indexing or retrieval to reflect ongoing research interests without force fitting documents into inappropriate categories of terminology. It also allows us to maintain a more modest size vocabulary than one can achieve with a thesaurus and provides a more realistic mechanism for updating terminology which can be done almost daily if required under the SSIE system. Conceptual indexing is also an important part of the SSIE system in that indexing is not dependent upon key words in the research summaries submitted to the Exchange. This type of indexing provides for a much greater depth in both indexing and retrieval. There is no substitute, of course, for the value any system gains by using professional scientists as a means of communication between a computerized data bank and an inquiring scientist.

In summary, the Smithsonian Science Information Exchange serves as a useful mechanism in (a) avoiding unwarranted duplication of effort on the part of either the scientist or the sponsoring organization of research projects, and by (b) serving as a referral center of active research and providing a mechanism for scientists to keep up with what their colleagues are working on even before it reaches the publication stage. Previous studies of SSIE users indicated approximately 90% learned of research projects they did not know about prior to their inquiry to SSIE.

Author's address: Prof. DAVID F. HERSEY, Smithsonian Science Information Exchange, *Washington, DC* (USA)

Medical Primatology 1972. Proc. 3rd Conf. exp. Med. Surg. Primates, Lyon 1972, part I, pp. 182–184 (Karger, Basel 1972)

Primate Zoonoses Surveillance Report

A. F. Kaufmann

Epidemiology Program, Center for Disease Control, Health Services and Mental Health Administration, US Department of Health, Education, and Welfare, Atlanta, Ga.

The Primate Zoonoses Surveillance Program was initiated in 1970 by the Center for Disease Control (CDC) to obtain data on spontaneous disease incidence in captive nonhuman primates and the extent of risk to human health associated with handling these animals. The program is based on monthly reports submitted by centers in the United States which utilize large numbers of nonhuman primates. These reports contain information on spontaneous disease due to all causes in individual animals, a colony population summary, and disease in personnel having contact with nonhuman primates.

Nine centers participated in 1970 and 1971: Delta Regional Primate Center, Tulane University, Covington, La; National Center for Primate Biology, University of California, Davis, Calif.; New England Regional Primate Research Center, Harvard University, Southborough, Mass.; Oregon Regional Primate Research Center, University of Oregon, Beaverton, Oreg.; Washington Regional Primate Research Center, University of Washington, Seattle, Wash.; Wisconsin Regional Primate Center, University of Wisconsin, Madison, Wisc.; Yerkes Regional Primate Research Center, Emory University, Atlanta, Ga.; Laboratory Division, Center for Disease Control, Atlanta, Ga.; and Research Animal Branch, USAF School of Aerospace Medicine, Brooks Air Force Base, Tex.

The Primate Zoonoses Surveillance Report, which serves as a feedback mechanism of the surveillance program, summarizes the information received from the participating centers and other contributors. Currently, the report is issued quarterly. Approximately two-thirds of each report is composed of discussion and summarization of the surveillance data contributed by the participating centers, and the remainder consists of informal notes on interesting cases, serologic surveys, necropsy series, and other information derived from a variety of sources.

The Primate Zoonoses Surveillance Report should not be considered a formal scientific journal. A primary purpose of the report is rapid dissemination of information on diseases of current interest. For this reason, much of the information is preliminary. All informal notes in the surveillance report are edited to be as concise as possible, with little detailed discussion of materials and methods. Anyone wanting to quote the report should contact the original investigator for confirmation and interpretation.

Because a brief summary of the primate zoonoses surveillance data will be presented elsewhere in this volume [2], my discussion will be limited at this time to other information.

Each issue of the Primate Zoonoses Surveillance Report contains several short notes. As mentioned earlier, these range from case reports and descriptions of disease outbreaks to reports of serosurvey results. The following summarized notes are examples of information contained in the report.

Rabies in Nonhuman Primates

Rabies affects nonhuman primates much more frequently than is commonly thought [3, 4]. In the period 1929–1970, rabies was diagnosed in seven monkeys recently imported into the United States. In 1967 and 1968, 21 countries in Africa, Asia, Europe, and South America reported to the World Health Organization human exposure to rabid monkeys necessitating Pasteur treatment. In 1968, an additional 14 monkeys not associated with human exposure had laboratory-confirmed rabies. In addition, the Rabies Surveillance Report of the Pan American Zoonoses Center lists 11 laboratory-confirmed cases of monkey rabies occurring in seven Middle and South American countries in the one-year period July 1, 1969 through June 30, 1970.

Reports from India and Indonesia, where rabies is epizootic, support the concept that rabies is not a rare disease in monkeys. The Pasteur Institute of Southern India, Coonoor, reported that 1,354 people received anti-rabies treatment at the Institute after being bitten by proved or presumably rabid monkeys in the period 1908–1970. At least one fatal human case of rabies from a monkey bite has occurred in India.

The Bio-Farma laboratory in Indonesia documented rabies in 83 cynomolgus monkeys *(Macaca fascicularis)* in the period 1953–1968. All 83 cases involved human exposure, but fortunately no human disease resulted.

The incubation period of rabies in monkeys may be prolonged. One rhesus monkey *(Macaca mulatta)* challenged at CDC survived 105 days before signs of disease developed. Until found comatose in its cage, the monkey appeared normal in all respects. Others in its group developed the furious form of the disease in 15–35 days.

Epidemic Neurologic Disease in Rhesus and Vervet Monkeys

In 1970, an outbreak of an obscure neurologic disease occurred in the CDC monkey colony [1]. The disease first occurred in a group of rhesus monkeys being conditioned in a quarantine area. The outbreak eventually spread to three other groups and resulted in the death of 74 monkeys.

Most of the involved monkeys were simply found dead with no antecedent signs of clinical illness. Approximately 20% of those affected had signs of neurologic abnormalities prior to death. During routine handling, some animals began having convulsions, which would start as fine muscular tremors and progress to spasmodic contractions of entire muscle groups and limbs. In a few animals, opisthotonus was observed. The seizures would last for approximately one minute, after which the animals were comatose for periods ranging from a few minutes to an hour. Partial paralysis was observed in two monkeys.

Eleven affected animals were completely necropsied. Morphologic changes were surprisingly few. Mild edema, congestion, and perivascular hemorrhages were observed in the brain of ten monkeys, and a nonsuppurative meningoencephalitis mainly confined to the brain stem was present in one animal. Interstitial myocarditis was observed in three of the 11 monkeys. A picornavirus, which has not been completely identified to date, was isolated from seven of nine monkey brains cultured.

In closing, the Primate Zoonoses Surveillance Report is a quarterly report of the Center for Disease Control dealing with spontaneous disease of nonhuman primates and their handlers. It is distributed without charge to anyone having an interest in nonhuman primate diseases and their public health significance.

References

1 BRODERSON, J. R.; GARY, G. W.; KAUFMANN, A. F.; KISSLING, R. E.; PERL, D. P., and QUIST, K. D.: An epidemic of encephalomyelitis in recently imported rhesus monkeys. Primate Zoonoses Surveillance Report No. 3 (Center for Disease Control, Health Services and Mental Health Administration, Atlanta, Ga. 1971).

2 KAUFMANN, A. F.: Nonhuman primate zoonoses surveillance in the United States; in GOLDSMITH and MOOR-JANKOWSKI Medical primatology 1972, part III, pp. 58–67 (Karger, Basel 1972).

3 RICHARDSON, J. H.: Rabies in nonhuman primates. Primate Zoonoses Surveillance Report No. 4 (Center for Disease Control, Health Services and Mental Health Administration, Atlanta, Ga. 1971).

4 RICHARDSON, J. H.: Rabies in nonhuman primates. Primate Zoonoses Surveillance Report No. 5 (Center for Disease Control, Health Services and Mental Health Administration, Atlanta, Ga. 1971).

Author's address: Dr. ARNOLD F. KAUFMANN, Center for Disease Control, Health Services and Mental Health Administration, US Department of Health, Education and Welfare, *Atlanta, GA 30333* (USA)

Medical Primatology 1972. Proc. 3rd Conf. exp. Med. Surg. Primates, Lyon 1972, part I, pp. 185–190 (Karger, Basel 1972)

An Information Service in Support of Biomedical Research

A Subset on Primatology

H. E. KENNEDY

Biosciences Information Service of Biological Abstracts, Philadelphia, Pa.

Biosciences Information Service of Biological Abstracts (BIOSIS) is the world's largest English language abstracting and indexing service in the biological and biomedical sciences. The BIOSIS information data base (literature reference data) consists of bibliographic citations and/or abstracts of more than three million research reports accumulated from 1926 through 1971. The data base is increased by almost one-quarter million additional items annually. BIOSIS' coverage of the scientific literature includes an estimated 85% of the world's total published primary basic research literature on biology- and biologically-related topics. This coverage represents approximately 8,000 primary source journals from 100 countries which are published in 55 of the world's languages [PARKINS, 1969].

Biomedical Literature Coverage

References and abstracts to basic biomedical research reports constitute approximately 65% (156,000 papers) of the coverage by BIOSIS each year. Basic biomedical research is emphasized. Clinical reports, case reports and reports concerned with the surgical arts are deliberately not covered to an appreciable extent in order that basic research may be covered as exhaustively as possible. Because all of the life sciences are related by the common quality that constitutes life itself, every report included by BIOSIS can be of potential interest to the biomedical research investigator. Thus, many users of *Biological Abstracts (BA)* and other BIOSIS publications and services discover material of interest to them from peripheral sources outside the scope of their usual definition of primary interest and emphasis.

Primatology Research Literature

Of BIOSIS' total coverage in 1971 of 240,000 research reports, 75,158 dealt with studies involving primates. Most of these reports (71,953) involved human beings and were not concerned with nonhuman primates. The remaining 3,205 employed nonhuman primates as the predominant subject for study of the body systems, behavior, as host organisms for infectious disease studies, or as tools or test organisms for other types of research. The nonhuman primate papers reported by BIOSIS in 1971 originated in 742 journals. These journals represent basic primary research publications, exclusive of newsletters and other types of publications which do not report research data. A frequency analysis of research reports according to the nonhuman primate families studied is presented in table I. More than 45% of the papers dealt with members of the family Cercopithecidae and approximately one-third were concerned with nonhuman primates in general. In the latter case, the particular primate under study was not specified in the paper, so it could not be indexed taxonomically to a level more specific than order. Approximately 10% of the papers were concerned with the Cebidae and slightly fewer involved the Pongidae. Each of the families Callithricidae, Lorisidae, Lemuridae and Tupaiidae was studied in at least 40 but not more than 70 reports. Less than 10 papers each were concerned with the Daubentoniidae, Tarsiidae and Indridae.

The subject headings (concepts) to which papers may be assigned for publication in *BA* and cross-indexed in the BIOSIS machine-readable indexes consists of 623 topics. Research studies on every nonhuman primate

Table I. Number of papers covered in 1971 according to family

Family	Number of papers
Nonhuman primates, general	941
Callithricidae	70
Cebidae	309
Cercopithecidae	1,468
Daubentoniidae	8
Indridae	2
Lemuridae	43
Lorisidae	49
Pongidae	270
Tarsiidae	4
Tupaiidae	41
Total	3,205

Table II. Frequency of subject emphasis of papers according to primate studied

	Callithricidae	Cebidae	Nonhuman primates, general	Cercopithecidae	Pongidae
Aerospace physiology				+	
Allergy			+		
Anthropology			+		+
Behavior, animal		+	+	+ + +	+
Behavior, comparative			+	+	
Behavior, conditioning		+		+ +	
Blood, lymph				+	
Blood cells				+	
Blood vessel pathology			+	+	
Cardiovascular physiology				+ +	
Developmental biology				+ +	
Endocrine system			+	+ + +	
Evolution			+		
Animal genetics				+	
Immunology			+	+ +	
Immunopathology			+	+ +	+
Systematic mammalogy			+ +	+	+
Medical bacteriology			+	+ +	+
Medical virology	+		+ +	+ + +	+
Carcinogenesis			+	+ +	
Nervous system anatomy		+	+ +	+ +	
Nervous system pathology			+ +	+ +	
Nervous system physiology		+ +	+ + +	+ + +	
Dental physiology				+	
Parasitology		+	+	+ +	
Pharmacology, cardiovascular				+	
Drug metabolism			+	+	
Neuropharmacology			+	+	
Psychopharmacology			+	+	
Radiation effects			+	+ +	
Reproductive physiology			+	+	
Sense organs, physiology			+	+ +	
General biology			+		
Temperature regulation				+	
Toxicology, general			+	+	
Toxicology, pharmacological			+	+	
Animal virology			+	+ + +	

+ = >12 papers; + + = >24 papers; + + + = >48 papers.

family, with major subject emphasis on 216 of the subject headings, were indexed in 1971. Again, a larger number of subject concepts were represented by the Cercopithecidae than any other family. General primate studies were reported on a wider range of topics than any family other than Cercopithecidae.

A scattergram of the distribution of the nonhuman primate literature according to subject concepts was studied to determine the nature of reports most frequently encountered in the literature and the organisms involved in them. The results of this study are represented in table II. The most frequent occurrence of organism and subject coincidence is Cercopithecidae in studies of animal behavior, animal virology, endocrine system, medical virology and nervous system physiology. The nervous system, especially physiology, was the subject of more studies than any other topic and involved particularly Cercopithecidae and Cebidae, and nonspecific (general) nonhuman primates. Pongidae were more frequently used in studies of anthropology, behavior, immunopathology, systematic mammalogy, medical bacteriology and virology than in studies of other topics. The Cebidae were used frequently in studies of behavior, nervous system and parasitology, while the chief subject of studies with the Callithricidae was medical virology. The predominant use of Cercopithecidae and reference to primates in general results in a long list of high-frequency subject areas, as noted in table II.

File Structure

The current BIOSIS file consists of magnetic tape and hardcopy publications of bibliographic information and indexes for all articles selected for coverage. About 60% of the total file is published with abstracts, i.e. that portion which constitutes the semimonthly publication, *Biological Abstracts*. The remainder is published as bibliographic citations and indexes in the monthly publication *Bioresearch Index*.

The machine-readable file is derived from indexing techniques which result in a set of computer-manipulatable indexes either in natural language or in numeric codes. The words used by the author in the title of his paper, augmented to ensure that the composite 'title' adequately represents the item, constitute the source of the natural language subject index, B.A.S.I.C.

Title augmentation is an intellectual effort by biologists on BIOSIS' staff. In general, augmentation is not controlled by a fixed thesaurus, but rather by natural language relative to the need of each individual abstract, e.g. organisms, instrumentation, methodology, etc., when such fundamental aspects of the paper are not encompassed in the original title. Retrieval of a given paper without specifying exactly the words used by the author is facilitated further by the Cross Index. Any paper (on the average) will be indexed under six subject headings in addition to the one under which it is published in *BA*. The name of any of the 623 subject headings and subheadings of *BA* under which the subject article might appropriately be published is used as another index tag for that item.

The third major indexing technique is based upon the hierarchical taxonomic positions of the organisms under study. Called the Biosystematic Index, it is created as the editor analyzes the paper and classifies it according to the organism(s) involved. Usually, the levels of taxonomic hierarchy used are phylum, class, order and family. When seeking papers describing work on a particular organism, or the effects of a substance on a given organism, the user is best advised to search the Biosystematic Index first. The general nature of the study in which he is interested is revealed by the brief truncation of subject heading which follows the taxonomic category.

Beginning January, 1974, an organism index will be published in addition to the Biosystematic Index. This index will consist of an alphabetical listing of all organisms according to genus and species, followed by a truncated notation of the general subject of the paper. Each will be keyed to the appropriate abstract and reference number.

In addition to the indexes discussed above, author indexes are prepared. Annual cumulations of each index are prepared and published as separate issues.

Advantages of a Broad Scope Information Bank for the Subject Specialist

A large broad-based information bank offers many advantages for the subject specialist whose interest may be narrow but wishes thorough coverage. Some of the advantages of a broad-based data bank, i.e. one that includes wider subject scope than a less broad subdiscipline of biology or biomedicine (primatology) are:

(1) Many interdisciplinary journals or journals of only peripheral interest to the specialist (e.g. primatologist) are covered for other areas of biological or biomedical interests and yield occasional papers of importance to a seemingly distant special interest.

(2) Journals of remote geographic origin or those which are published in less universal languages are covered, and while the mainstream journals serve most of the user's needs, articles from other journals may be the source of important contributions to the literature or may suggest ideas which trigger productive research.

(3) Single input to the information base, i.e. the abstracting and indexing of a given paper and its inclusion in a machine-readable information bank, provides for multiple output, effecting reuse of the input for users with a wide range of interests and needs. Specialist user groups can be served, as well as special subject-oriented publications and services derived according to the needs of individual users (SDI) or groups of users (such as the membership of a given society).

(4) Input requires intellectual interfacing only once, regardless of output or user specialty. This principle offers sound economic advantages for the user when compared with the cost of collecting only specialized information that is of importance to a specific group while attempting to cover all of the information with which it is concerned.

(5) Carefully structured indexes (machine-readable) enable computers to search, sort and retrieve specific information from complex and massive files.

A large information bank which attempts to cover the entire scope of biology and biomedicine also has its disadvantages, but the advantages more than compensate for them. A major disadvantage is the ever-increasing quantity of literature that must be processed. This contributes to bulk of secondary publications (an impediment to their use) and rising costs.

References

PARKINS, P.V.: Biosciences Information Service of Biological Abstracts. Encyclopedia of Library and Information Science, vol. 2, pp. 603–621 (Dekker, New York 1969).

Author's address: H. E. KENNEDY, Biosciences Information Service of Biological Abstracts, 2100 Arch Street, *Philadelphia PA 19103* (USA)

Medical Primatology 1972. Proc. 3rd Conf. exp. Med. Surg. Primates, Lyon 1972, part I, pp. 191–192 (Karger, Basel 1972)

Experience of Organization and Activity of the Sukhumi Primate Information Center

B.A. LAPIN and E.P. FRIDMAN

Institute of Experimental Pathology and Therapy of the USSR Academy of Medical Sciences, Sukhumi

In connection with the rapid development of medicobiological research on primates during 1966–1967, the Sukhumi Primate Information Center (SPIC) was established at the Institute of Experimental Pathology and Therapy of the USSR Academy of Medical Sciences (IEPT). The following tasks were set for the Center:

(1) To accumulate and provide facts and bibliographical information on all aspects of biology and experimental use of primates in medical and biological investigations.

(2) To inform systematically the staff of the IEPT and other concerned institutions of the literature on experimental primatology.

(3) To collect the entire appropriate literature of primate publications widely scattered in hundreds of publications.

Initially the staff of the SPIC consisted of eight persons, but it has grown to 15 people at the present time.

During 1967, in collaboration with the All-Union Institute of Scientific and Technical Information (ISTI), a search system on cards K-5 with marginal perforation (perfocards) published in 'Scientific and Technical Information' was devised which now functions in six scientific fields concerning primates.

To facilitate a maximal coverage of primate literature, 90% of which, as it is known, is published abroad, the Center established contacts with the Primate Information Center of the Washington Regional Primate Research Center (Seattle, USA). The Sukhumi Center systematically sends bibliography of Russian works to the Washington Center, and receives from the USA a weekly bulletin of the world literature with the authors' addresses. The SPIC is now engaged in the search of literature from seven other established sources.

The SPIC requests literature directly from the authors or institutions, ISTI and IMI (Institute of Medical Information) among them, and also through interlibrary exchange.

As it has been shown by the analysis of publications for 1968, the Sukhumi PIC received more than 85% of the scientific literature (including more than 60 separate issues) of the total number of publications it identified (almost 3,200). In 1970 SPIC received 2,914 titles of publications.

All the literature arriving at SPIC is processed as follows: the title and, as a rule, the summary are translated into Russian (in rare cases a full translation is made); the summary is typed on the perfocards and codified; the perfocards are duplicated on the alphabetical card index.

At least once a month (as a rule, more often) SPIC issues its own bulletin of the literature arriving at the Institute with titles in Russian and in the original languages. Fifteen bulletins were issued in 1970, listing the total of 3,273 publications. The bulletin is electrographically reproduced.

Having met with difficulties of translation (over 3,000 works per year), SPIC worked out a system of so-called waiting information carriers which permits us to use successfully summaries from ISTI and IMI short assay journals and to reduce the work of translators by more than one-third.

Scientific research work of SPIC consists of annual analysis of primate publications, of the trends developing in world research, of the comparison of world literature as reflected by the USSR short assay journals with the picture provided by the SPIC data, as well as of analysis of other problems found in literature received at Sukhumi (research in primate systematics, similarity and difference of man and other primates; generalization of the data on medical and biological problems which can be investigated only in simians). Since 1968, six scientific papers have been published based on SPIC materials.

By the first of June, 1971, SPIC had about 11,000 perfocards. Over 1,500 publications are now in the processing stage. This represents a rather impresssive collection of world literature for 1967–1970 which enables us to provide information on any aspect of experimental primatology. In SPIC there are about 6,700 addresses of the first authors and more than 1,650 addresses of the institutions engaged in primate research.

The information presented here indicates that SPIC has fulfilled the task set out by its planners. At present the Center is trying to find ways to switch its work over to the computer memory.

Authors' address: Dr. B.A. LAPIN and Dr. E.P. FRIDMAN, Institute of Experimental Pathology and Therapy of the USSR Academy of Medical Sciences, *Sukhumi* (USSR)

Medical Primatology 1972. Proc. 3rd Conf. exp. Med. Surg. Primates, Lyon 1972, part I, pp. 193–195 (Karger, Basel 1972)

Laboratory Animal Science

A. M. LEASH and E. C. MELBY, jr.

School of Medicine, Case Western Reserve University, Cleveland, O. and School of Medicine, The Johns Hopkins University, Baltimore, Md.

Laboratory Animal Science was originated in 1950 as *Proceedings of the Animal Care Panel,* an organization devoted to the improvement of standards for management and use of animals utilized in biomedical research. The *Proceedings* rapidly evolved to quarterly and then bimonthly publication.

In 1963, the title was changed to *Laboratory Animal Care,* and publication was managed by the Williams & Wilkins Company. In 1968 publication was assumed by the sponsoring organization, now known as the American Association for Laboratory Animal Science (AALAS). The title of the journal was changed once again in 1971 to *Laboratory Animal Science.*

Laboratory Animal Science is the bimonthly publication of the American Association for Laboratory Animal Science. Publication, advertising, and subscription offices are: P.O. Box 10, Joliet, IL 60434 (USA). The editor is AARON M. LEASH, DVM, School of Medicine, Case Western Reserve University, Cleveland, OH, 44106 (USA). Circulation is approximately 3,500 copies. The journal is mailed to every state in the United States and to at least 50 foreign countries.

Laboratory Animal Science is a refereed journal of approximately 150 pages per issue (100 pages of text and 50 pages of advertising). A wide range of papers are published which fall into the general categories of morphological and biological values, breeding, management, use, and diseases of experimental animals. They are classified as 'original articles': comprehensive reports of definitive or experimental studies, and review articles; or as 'technical notes': reports of observations, case reports, clinical laboratory findings, newly developed technics, and innovations in housing and handling of experimental animals.

The volume contents are indexed by author and subject in each December issue.

Advertisements by major domestic suppliers of laboratory animals, equipment, and supplies make this journal an excellent reference for purchasing agents. The advertisers are indexed in each issue.

All papers presented at the annual meeting of the American Association for Laboratory Animal Science become the property of the journal. Many are eventually published in Laboratory Animal Science, although permission is freely given to publish elsewhere.

Laboratory Animal Science is indexed in *Index Medicus, Biological Abstracts,* and *Current Contents.*

Since 1950, approximately 300 papers have been published in Laboratory Animal Science in the field of primatology. Subjects have included: breeding, nutrition, normal value for blood cerebrospinal fluid and other fluids and tissues, bacterial, mycotic, viral, and parasitic diseases, metabolic disorders, neoplasia, and other pathological conditions, anesthesia, experimental surgical procedures, procurement, housing, husbandry, handling and restraint, devices, and technics.

Two supplements dealing with primates have been published within recent years; both were proceedings of workshops at Southwest Foundation for Research and Education, San Antonio, Tex.: Primate Parasitology Workshop, 1970, and Primate Virology Workshop, 1971.

The American Association for Laboratory Animal Science invites membership applications. Members receive Laboratory Animal Science, as well as other publications of the Association. Subscriptions to the journal, without membership, are also available.

Because of its spectrum of subject matter published, Laboratory Animal Science enjoys a wide range of readership and an increasing circulation. For these reasons, in part, it has been and continues to be a popular journal for authors of papers on primates.

Primate species have included:
Alouatta caraya, A. villosa;
Aotus trivirgatus;
Ateles fusciceps, A. geoffroyi;
Callicebus moloch, C. sp.;
Callithrix jacchus (Hapale jacchus);
Cebus albifrons, C. capucinus;
Cercocebus atys, C. torquatus;
Cercopithecus aethiops, C. cephus, C. pygerythrus;
Erythrocebus patas;

Galago senegalensis;
Gorilla gorilla;
Hylobates lar entelloides;
Lagothrix sp.;
Macaca arctoides (speciosa), M. cyclopis, M. fuscata, M. fascicularis (irus, philippensensis), M. mulatta, M. nemestrina, M. radiata;
Pan troglodytes, P. sp.;
Papio anubis, P. cynocephalus, P. papio;
Saguinus fuscicollis, S. geoffroyi, S. nigricollis (Tamarinus nigricollis), S. oedipus (Oedipomedas oedipus);
Saimiri sciureus;
Tarsius syrichta;
Tupaia glis, T. g. chinensis.

Authors' addresses: Dr. AARON M. LEASH, School of Medicine, Case Western Reserve University, *Cleveland, OH 44106;* Dr. EDWARD C. MELBY, jr., School of Medicine, The Johns Hopkins University, *Baltimore, MD 21205* (USA)

Medical Primatology 1972. Proc. 3rd Conf. exp. Med. Surg. Primates, Lyon 1972, part I, pp. 196–197 (Karger, Basel 1972)

Comments on the Laboratory Primate Newsletter

A. M. SCHRIER

Brown University, Providence, R.I.

My nine-year-old daughter, Marya, overheard me expressing some uncertainty about what I should say in this paper. She immediately sat down to write a short speech, which I will now use to introduce my comments. I quote:

'Ladies and gentlemen, I would like to talk to you about the Laboratory Primate Newsletter. We need more material. The Newsletter has been going on for ten years and has been interesting to all readers. But we need more facts and more ideas. We want the Newsletter to keep on doing what it has been doing.' Although I would not quite word it exactly the way my daughter did, she did touch on some of the points that I had planned to make.

As many of you know, four major functions of the Newsletter have evolved since I founded it in 1962. The first of these is to provide information on care, breeding, and procurement of nonhuman primates for laboratory research; the second function is to disseminate general information about the world of primate research (such as announcements of meetings and seminars, and descriptions of research projects); the third function is to help meet the special research needs of individual investigators by publishing requests for research material or for information related to specific research problems; and the fourth function is to serve the cause of conservation of nonhuman primates by publishing information on that topic.

I should emphasize that we do not adhere rigidly to guidelines concerning material submitted for publication in the Newsletter. Most articles, notes, and notices fall in one or another of the categories just outlined. However, we are willing to publish any article so long as it contains useful ideas or data and relates in some way to research on nonhuman primates. We try to give special consideration to articles containing data that is not conveniently publishable elsewhere and to articles of an editorial nature.

I founded the Newsletter in 1962 because I felt that there was a serious lack of up-to-date information concerning care of nonhuman primates. The many problems I encountered in setting up my own primate laboratory a few years earlier contributed strongly to that opinion. At that time, there were few sources of information of any kind that one could use in setting up a primate laboratory and in maintaining the animals. Since then, there has been an explosive growth in the number of primate laboratories in many countries, as well as in the number of fields of primate research. As might be expected, concomitantly, there has been a great growth in the number of sources of information about primates and primate research. This has taken the form of books and conference proceedings, specialized journals, and pamphlets and booklets. Some of the current sources of information have already been or will be described by the others on this program.

Despite this growth in other sources of information, the feedback that we have gotten from many of our readers, both by mail and by word of mouth, strongly suggests that the Newsletter continues to fill a real need. However, there is now some question about the present nature of this need. Has the Newsletter become too much like a journal, or is it not enough like a journal? Are the abstracts of published articles useful? What other regular features might be useful? And so on. I hope, therefore, to call on our readers to help define the direction the Newsletter should take in the future. I would greatly appreciate any suggestions or comments from readers about these questions. We plan to circulate a questionnaire about this in the near future.

I do not know whether we can continue to issue the Newsletter free of charge indefinitely[1], for the growth of our mailing list and size of the issues continue to strain our budget. In any event, with the Newsletter now in its second decade, we plan to issue it as long as we continue to get the support of our readers.

1 The major financial support for the Newsletter comes from the Division of Research Resources of the United States, National Institutes of Health, grant No. RR-00419.

Author's address: Dr. ALLAN M. SCHRIER, Psychology Department, Brown University, *Providence, RI 02912* (USA)

Medical Primatology 1972. Proc. 3rd Conf. exp. Med. Surg. Primates, Lyon 1972, part I, pp. 198–200 (Karger, Basel 1972)

Primate Information Center as a Tertiary Communication Service

MARYEVA W. TERRY

Primate Information Center, Regional Primate Research Center, University of Washington, Seattle, Wash.

In the parlance of scientific communication, the Primate Information Center is a 'tertiary service'. In other words, at least two forms of formal communication of the information have occurred before the Information Center enters the communications network: publication of the article by a 'primary' journal and publication of the citation or an abstract by the major indexing services such as *Biological Abstracts, Index Medicus* and *Excerpta Medica* (the so-called secondary services). A fair question would thus seem to be why is yet a third service needed to insure communication to interested scientists. There are several possible answers to this question. In the case of the Primate Information Center, four of the more important ones are as follows.

The first is to provide a single point of access to the work by preceding services. For example, in 1971 we published in *Current Primate References* citations of articles appearing in 823 journals and other serial publications. To find these citations we searched, or had searched for us, 10 standard secondary services. In addition, the library of the Oregon Regional Primate Research Center, The Primate Information Centre at Sukhumi and our own staff contributed citations not available from other sources. In other words, each of the secondary services has its own view of the interests of its users and none of these views wholly encompasses primate research.

A second answer is to provide different kinds of information for different uses. The secondary services are very large, dealing with articles by the hundreds of thousands each year. Thus, they have time and space to gather and store only a finite amount of information about each article. In contrast, the Information Center deals with only a few thousand articles, so that it has time to derive a variety of information packages from each article to service a variety of needs. For example, in addition to bibliographic services, we can provide extracts of normal values and extracts of data on drug metabolism. Our activities in these areas are slowly expanding, and we hope to see this kind of information repackaging become the greater proportion of our service.

A third answer, which is not unrelated to the previous, is individualized attention to information needs. Because we deal with a limited literature and a relatively small clientele, we are not bound by our systems. If we receive an interesting and challenging question which requires available information beyond what we have already extracted from the literature, we can – and do – go back and search out the answer.

The fourth answer is to maintain long-term access, so renewal of interest in a topic that has been dormant for some years does not involve a difficult literature search. Although we are presently limiting the scope of our searches on current topics to the recent literature, we still have access to the primate literature published since 1940 and, if the current files do not provide the answer, we extend the search back in time immediately. Detailed information on the content of our bibliographic files and suggestions on how to approach them will be found in the proceedings of the previous Conference [TERRY, 1971].

Addendum

To those persons interested in the primate literature *per se,* the following data in table I may be of passing interest. They appear to be fairly typical of the publication

Table I. Citations of journals in 1971 *Current Primate References*

Number of articles per journal	Number of journals	% of total journal articles cited
1	365	11
2	151	9
3	72	7
4	46	6
5	34	5
6	34	6
7	18	4
8	14	3
9	17	5
10	5	1
11–20	47	20
20	20	21

Table II. Journals cited at least 20 times in 1971 *Current Primate References*

American Journal of Obstetrics and Gynecology
American Journal of Physical Anthropology*
Anatomical Record*
Brain Research
Dissertation Abstracts, International*
Electroencephalography and Clinical Neurophysiology*
Endocrinology
Experimental Neurology
Federation Proceedings*
Folia primatologica
Journal of American Veterinary Medical Association*
Journal of Comparative and Physiological Psychology
Journal of Dental Research*
Laboratory Animal Science
Nature, London
Pharmacologist*
Physiologist*
Primates
Proceedings of Society for Experimental Biology and Medicine
Psychonomic Science
Science
Transactions of the Royal Society for Tropical Medicine and Hygiene*

* Asterisks indicate those for which publication of abstracts, particularly society proceedings, contributed substantially to the total citations.

patterns, but since only one year's input is represented here, no firm conclusions should be drawn regarding communication in primatology.

Table II lists the journals that were cited 20 or more times. Since the publication year of *Current Primate References* does not correspond in any consistent way to the publication years of journals, they are listed in alphabetical order rather than ranked by number of citations.

Reference

TERRY, M.W.: The primate information center; in GOLDSMITH and MOOR-JANKOWSKI Medical primatology 1970, pp. 980–985 (Karger, Basel 1971).

Author's address: Dr. MARYEVA W. TERRY, Primate Information Center, Regional Primate Research Center, University of Washington, *Seattle, WA 98195* (USA)

Reproduction and Perinatal Studies

Medical Primatology 1972. Proc. 3rd Conf. exp. Med. Surg. Primates, Lyon 1972, part I, pp. 202–205 (Karger, Basel 1972)

Cyclic Desquamation of the Endocervical Epithelium of the Bonnet Monkey *(Macaca radiata)*

J. Bashir-Farahmand, Priscilla D. Taft and Janet W. McArthur

Departments of Gynecology and Pathology, Massachusetts General Hospital and Harvard Medical School, Boston, Mass.

We have utilized the method of exfoliative cytology to obtain a quantitative description of the desquamative cycle in the cervical epithelium of the bonnet monkey *(Macaca radiata)*. This species was chosen in preference to the human for such studies because of the profuseness of its cervical mucus secretion, which exceeds by at least tenfold the quantity produced by the human female during adult reproductive life.

Materials and Methods

Two adult female bonnet monkeys were isolated from contact with males and were followed throughout two complete menstrual cycles. The following indices were employed: urinary excretion of total estrogens [Brown *et al.*, 1968], cervical mucus weight [McArthur *et al.*, 1972] and cytological measurements as will be detailed subsequently. The ovulatory character of these four cycles was confirmed by demonstrating the presence of a fresh corpus luteum at laparotomy.

The cytological studies were carried out upon the sediment obtained by daily vaginal irrigations performed according to the technique of Davis [1962a and b]. The washings were prepared for microscopic examination by Millipore filtration and Shorr staining. Two observers working independently determined the ratio of endocervical to vaginal cells, and made a differential count of the endocervical cells. The ratio was based upon a count of all of the cells in a sufficient number of fields to yield a total count of 400 cells of both vaginal and endocervical type. The differential count was obtained by extending the counting process until a total of 400 endocervical cells had been identified. The vaginal cell types were defined according to the system of classification recommended by the International Academy of Gynecological Cytology [1958]. Three types of endocervical cells were distinguished: secretory (or mucoid), ciliated, and reserve cells.

Results

The proportion of cervical cells among the desquamated cells in the vaginal washings during a representative cycle is shown in table I, together with the urinary excretion of total estrogens and the cervical mucus weight. It is evident that there are two maxima in urinary estrogen excretion: a pronounced rise during midcycle and a lesser increase during the premenstrual phase. The quantity of cervical mucus was seen to increase gradually during the follicular phase and to reach a maximum which is concomitant with the midcycle surge in urinary estrogen. After a sharp postovulatory

Table I. The urinary estrogen excretion, cervical mucus weight and percentage of endocervical cells in the vaginal washings during menstrual cycle I in bonnet monkey Debbie

Day of cycle	Urinary estrogens mg/24 h	Cervical mucus weight, mg	Endocervical cells %
1	1.80	592	27.5
2	1.55	810	35.0
3	1.80	887	31.5
4	1.49	1030	18.5
5	1.75	499	8.0
6	2.25	1566	2.5
7	1.70	2125	33.0
8	1.82	2456	21.5
9	2.52	3764	28.0
10	2.78	3877	5.0
11	6.29	5003	23.5
12	5.06	5103	1.5
13	2.60	3020	23.5
14	2.19	1608	60.0
15	2.91	856	25.0
16	2.02	452	68.0
17	1.57	360	33.5
18	1.52	250	36.0
19	1.42	130	11.0
20	1.38	159	47.0
21	1.49	160	22.0
22	1.70	168	2.0
23	2.00	124	20.0
24	2.35	135	35.0
25	1.52	130	77.5
26	1.41	162	8.5
27	3.51	249	41.5

decline, mucus secretion remains minimal until the late luteal phase, when the quantity again begins to increase. This terminal rise is more evident in the three cycles not shown. The proportion of endocervical cells tends to be minimal during the follicular phase of the cycle and to increase abruptly as the urinary estrogen and the cervical mucus weight decline from their midcycle peak. Variable proportions characterize the midluteal phase, with a second phase of desquamation occurring premenstrually.

The differential count of the endocervical cells during the same cycle is depicted in table II. It can be seen that secretory cells constitute the bulk of the desquamate throughout the cycle, and that insignificant variations occur in the proportion of ciliated cells. The reserve cells, on the other hand,

Table II. The relative proportion of different endocervical cell types in the vaginal washings during menstrual cycle I in bonnet monkey Debbie

Day of cycle	Secretory, %	Ciliated, %	Reserve, %
1	70	15	15
2	75	15	10
3	77	13	10
4	78	12	10
5	80	8	12
6	78	8	14
7	84	11	5
8	87	10	3
9	95	5	0
10	96	4	0
11	95	5	0
12	98	2	0
13	97	3	0
14	98	2	0
15	95	5	0
16	97	3	0
17	98	2	0
18	95	5	0
19	70	4	26
20	90	3	7
21	90	2	8
22	85	4	11
23	80	5	15
24	87	4	9
25	63	7	30
26	66	14	20
27	63	13	24

disappear completely from the desquamate during the midcycle estrogen surge, only to reappear in maximal numbers during the latter part of the luteal phase and during menstruation.

Histological evidence to corroborate the cytological data was obtained by removing the uterus of an animal in which active desquamation of the cervical epithelium was taking place. This disclosed the fact that denudation of the epithelium was effecting the exposure of the reserve cell layer. Desquamation of the reserve cells could then occur.

Conclusions

We have shown by means of cytological techniques that there is periodic desquamation of the cervical epithelium and, to a lesser extent, of the subjacent layer of reserve cells. Desquamation is minimal during the follicular phase of the cycle, when ovarian estrogen secretion is increasing. It is maximal during the immediate postovulatory phase, when urinary estrogen levels are falling and when the presumed antiestrogenic effect of progesterone upon the cervical epithelium is being exerted. The physiologic significance of these observations remains to be determined.

References

Brown, J. B.; MacLeod, S. C.; MacNaughtan, C.; Smith, M. A., and Smith, B.: A rapid method for estimating oestrogens in urine using a semi-automatic extractor. J. Endocrin. *42:* 5–15 (1968).

Davis, H. J.: The irrigation smear. Amer. J. Obst. Gynec. *84:* 1017–1023 (1962a).

Davis, H. J.: The irrigation smear. Accuracy in detection of cervical cancer. Acta cytol. *6:* 459–467 (1962b).

International Academy of Gynecological Cytology, Terminology Subcommittee. Opinion polls on cytological definitions and terminology. Acta cytol. *2:* 25–139 (1958).

McArthur, J. W.; Ovadia, J.; Smith, O. W., and Bashir-Farahmand, J.: The menstrual cycle of the bonnet monkey *(Macaca radiata)*. Folia primat. *17:* 107–121 (1972).

Author's address: Dr. Jaleh Bashir-Farahmand, Department of Gynecology, Massachusetts General Hospital, *Boston, MA* (USA)

Medical Primatology 1972. Proc. 3rd Conf. exp. Med. Surg. Primates, Lyon 1972, part I, pp. 206–215 (Karger, Basel 1972)

Physiological Studies on the Baboon Uterus by Means of Uterocutaneous Fistulae

W. G. Breed, Anne Fraser, P. Eckstein and P.V. Peplow

Department of Anatomy, Medical School, University of Birmingham, Birmingham

Introduction

Repeated sampling of endometrial tissue and uterine secretions in non-human primates is difficult owing to the inaccessibility of the uterus. If specimens can be obtained at all through the cervix, as for instance in baboons, they are likely to be contaminated with blood, cervical mucus and cells from the vagina.

In an attempt to overcome this difficulty van Wagenen and Morse [1940] prepared uterocutaneous fistulae in rhesus monkeys by removing a wedge-shaped piece from the uterine fundus and suturing the opening to the abdominal wall. Such preparations were later also carried out by Sturgis [1942] and Good and Moyer [1966], adopting a very similar surgical technique [see also Kraemer *et al.*, 1971]. Although these fistulae allowed regular biopsies of the endometrium to be obtained [see Good and Moyer, 1968] they have several disadvantages: (a) washings of the uterine cavity are liable to be lost, at least partially, through the cervix and vagina, (b) menstrual bleeding will probably occur through both the fistula and vagina, and (c) if intrauterine devices are inserted through the fistula they may be expelled into the vagina unobserved [unpubl. observations]. We set out, therefore, to investigate the possibility of preparing a second type of uterine fistula which would avoid these complications.

Material and Methods

The baboons used were seven adult, regularly cyclic, probably parous females *(Papio cynocephalus)*. Their maintenance, together with the method used for routinely inspecting for menstruation and sexual skin activity has been described by BREED *et al.* [1972] except that once the fistula had been prepared, the abdomen, as well as the perineum, were inspected for signs of menstrual bleeding. The method used for obtaining peripheral blood and for the progesterone assay (by competitive protein binding) is given in the same paper.

Preparation of Type-1 (Fundal) Fistulae (fig.1)

After anaesthesia with Sernylan (phencyclidine hydrochloride, 0.5 ml/10 kg, 20 mg/ml; i.m.) and Nembutal (sodium pentobarbital, 2.5 ml/10 kg, 60 mg/ml) three animals were prepared in a similar manner to that already described for rhesus monkeys [VAN WAGENEN and MORSE, 1940; STURGIS, 1942; GOOD and MOYER, 1966]. Laparotomy was performed and a conical piece of myo- and endometrium was removed from the top of the uterus. The serosa surrounding the opened fundus was then elevated to the level of the skin incision and the adjacent peritoneum stitched to it both ventrally and dorsally with 3–0 chromic catgut. The remainder of the peritoneum was then closed and the rectus sheath and abdominal skin near the fistula sutured to the serosa and myometrium with silk thread. Great care was taken not to occlude the opening. The rest of the abdominal wall was then closed.

Preparation of Type-2 (Uterocervical) Fistulae (fig.2)

Since this type of preparation requires a much longer operation than the type-1 fistula, halothane was preferred to nembutal, following the initial tranquillization with Sernylan. Four baboons were used. When full surgical anaesthesia had been induced, a midline abdominal incision was made. After displacing the bladder to one side, the body of the uterus and upper cervix were palpated to determine the position of the internal os.

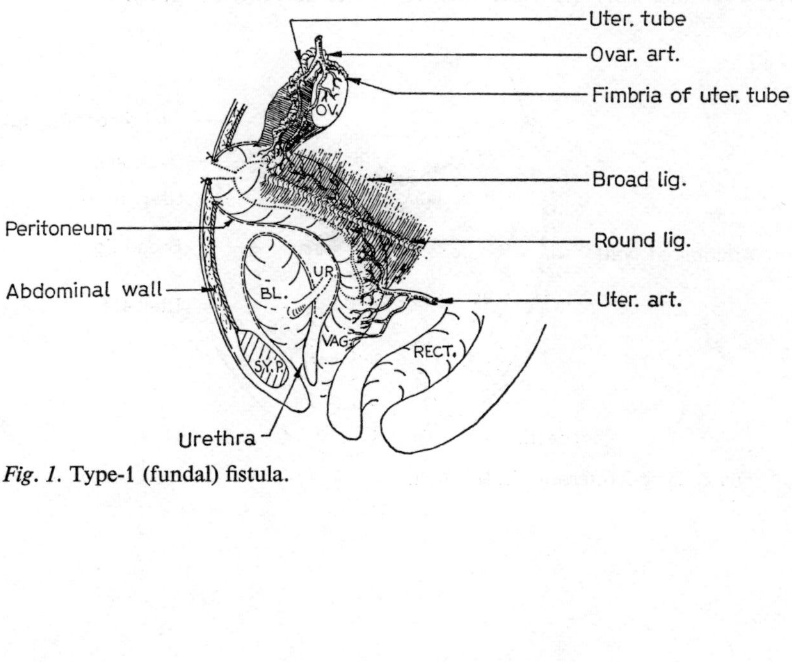

Fig. 1. Type-1 (fundal) fistula.

Part of the broad ligament was then carefully separated from the lateral border of the lower part of the body of the uterus and a complete cut made by electrocautery just cranial to the internal os. Some bleeding always occurred from the division of the uterine arteries and their branches along the lateral borders of the cervix and uterus, but this was surprisingly slight. After completion of the incision and cessation of the bleeding, the body of the uterus was swung round and the peritoneum circumferentially attached to the serosal coat about 1 cm below the cut surface. After closing the rest of the peritoneal cavity the adjacent parts of the rectus sheath and the abdominal skin were stitched to the body of the uterus near its cut surface.

All fistulae were probed once a week, and could usually be kept patent indefinitely in this way. Sometimes a membrane formed over the stoma, but this could be broken down with artery forceps. Endometrial biopsies were taken weekly. These were subsequently fixed, sectioned at 7μm and stained with HE. Five-millilitre samples of blood were collected regularly throughout the cycle and the resulting plasma stored until required.

Results

The following observations indicate that in baboons the presence of either type of uterine fistula does not effectively interfere with the ovarian, endometrial or sex skin cycle.

Menstrual Cyclicity and Menstruation

In baboons with fundal (type-1) fistulae, menstruation occurred both through the abdominal opening and through the vagina, but in the case of type-2 fistulae only onto the surface of the abdomen (fig. 3).

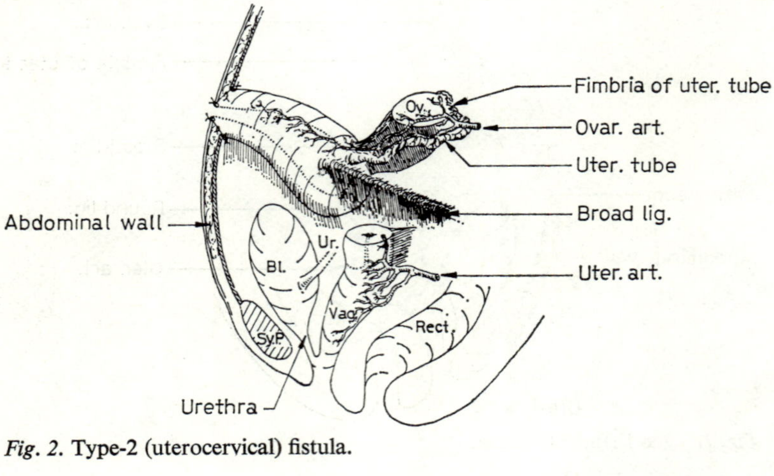

Fig. 2. Type-2 (uterocervical) fistula.

Table I lists cycle lengths and duration of bleeding for both groups of animals from three months before to 10–12 months after establishing the fistula, and shows that neither was affected.

In the three animals with fundal (type-1) fistulae there was regular slight protrusion of endometrium from the wound during the secretory phase of the cycle, as previously observed by VAN WAGENEN and MORSE [1940] intwo rhesus monkeys. This did not occur in any of the females with type-2 fistulae.

Table I. Length of menstrual cycle and menstruation

Type of fistula	Number of baboons	Cycle length, mean ± SE in days				
		before fistula made	after fistula made (months)			
			1–3	4–6	7–9	10–12
Fundal	3	38.3±0.97 (9)	37.2±1.2 (9)	35.5±0.8 (9)	37.0±1.3 (9)	35.9±1.1 (9)
Uterocervical	4	31.5±0.71 (12)	32.6±1.9 (12)	32.2±1.4 (12)	31.0±1.0 (12)	31.6±1.1 (10)
		Duration of menstruation, mean ±SE in days				
Fundal	3	3.33±0.37 (9)	3.33±0.68 (9)	3.55±0.55 (9)	3.44±0.52 (9)	3.44±0.64 (9)
Uterocervical	4	2.83±0.27 (12)	2.25±0.30 (12)	2.33±0.39 (12)	2.50±0.19 (12)	3.27±0.40 (11)

( ) = Number of cycles.

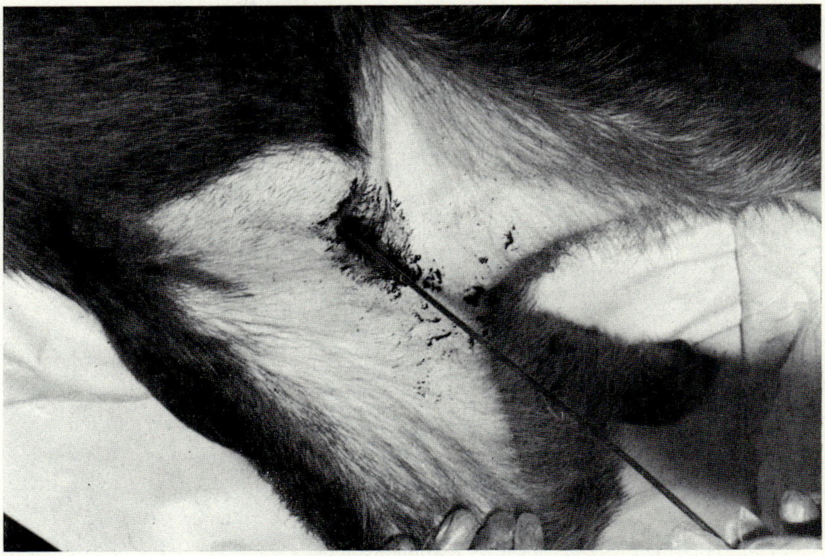

Fig. 3.

Sexual Skin Activity

Normal changes in the sexual skin, comparable to those of intact females [see BREED *et al.,* 1972] continued in all animals with fistulae throughout the post-operative period of observation (fig. 4). The regular sequence of swelling (turgescence) followed by shrinkage of the sexual skin (detergescence) suggested that ovarian hormonal activity was not impaired by the presence of a fistula.

Plasma Progesterone Levels

The concentration of plasma progesterone in the fistulous animals, when plotted in relation to the stage of the menstrual cycle, was similar to

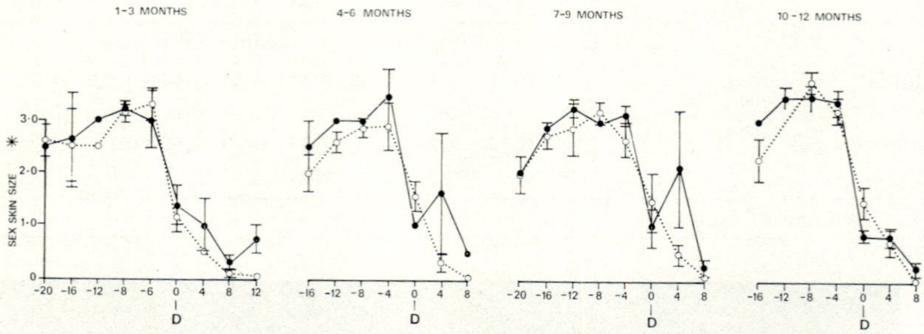

Fig. 4. Sex skin changes after making fistulae. D = onset of deturgescence; ○ = utero-cervical fistula; ● = fundal fistula. (*in arbituary units)

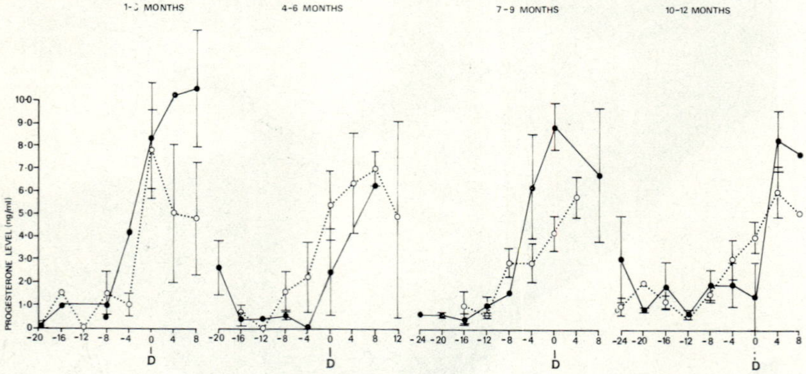

Fig. 5. Plasma progesterone after making fistulae (±S. E.). D = onset of deturgesence; ○ = uterocervical fistula; ● = fundal fistula.

that in intact controls for the three-month periods of observation (fig. 5). This confirms, more directly, that ovarian luteal function continued normally, regardless of the length of time after establishing the fistulae.

Endometrial Changes (fig. 6–9)

Histological examination of uterine biopsies, obtained at weekly intervals from both types of fistula, revealed no marked alteration of the cyclical changes in the endometrium, either immediately or up to 16 months after operation. They resembled the endometrial changes reported by NOYES *et al.* [1950] in the human uterus.

The endometrial cycle in the baboon has recently been briefly described by MACLENNAN and WYNN [1971]. Since some of our observations were not mentioned by these workers, an outline of the histological changes as observed by us in endometrial biopsies is given below.

In the early (proliferative) stage of the cycle just after the cessation of bleeding some remnants of the endometrium from the previous cycle can still be discerned, i.e. glands with large lumina and the presence of coiled

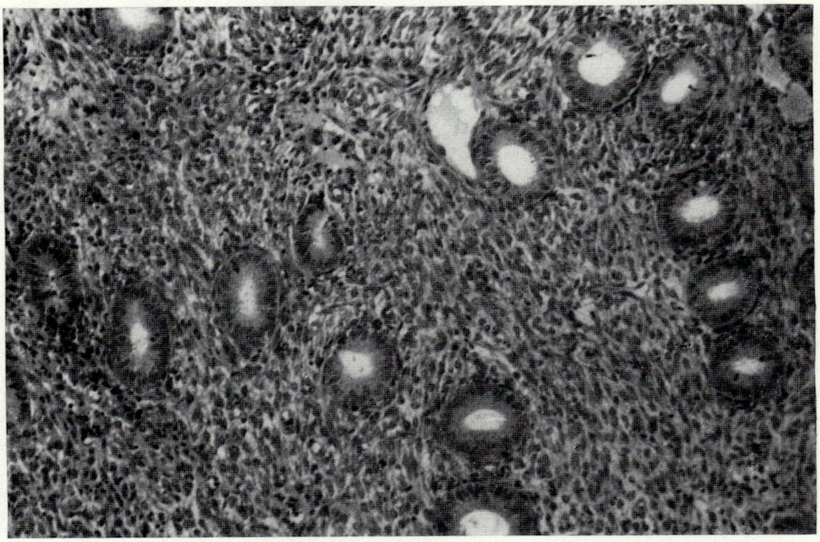

Fig. 6. Biopsy taken from a uterocervical fistula nine months after its preparation and 16 days before onset of deturgescence. Typical proliferative endometrium with compact stroma, small glands with basal nuclei. × 105.

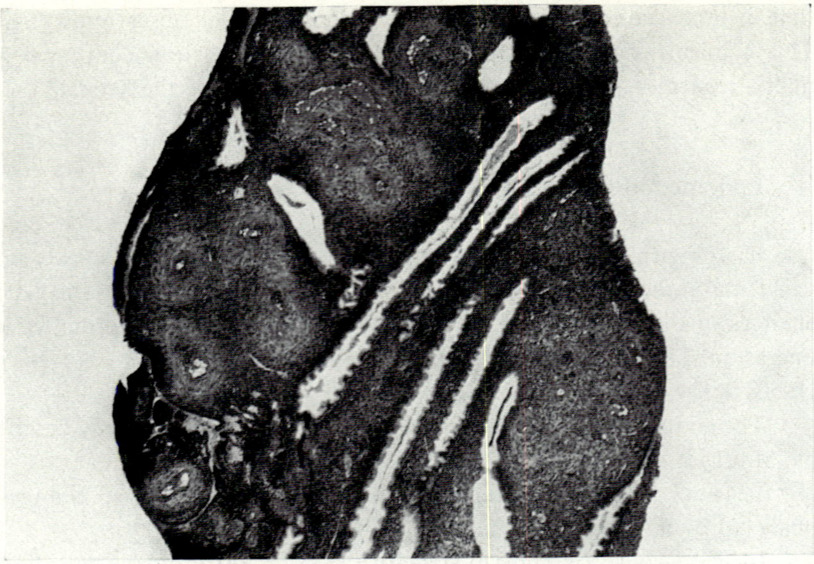

Fig. 7. Biopsy taken from a uterocervical fistula 6½ months after preparation and ten days after onset of deturgescence. Typical secretory endometrium with large glands containing secretion and prominent coiled arterioles. × 22.

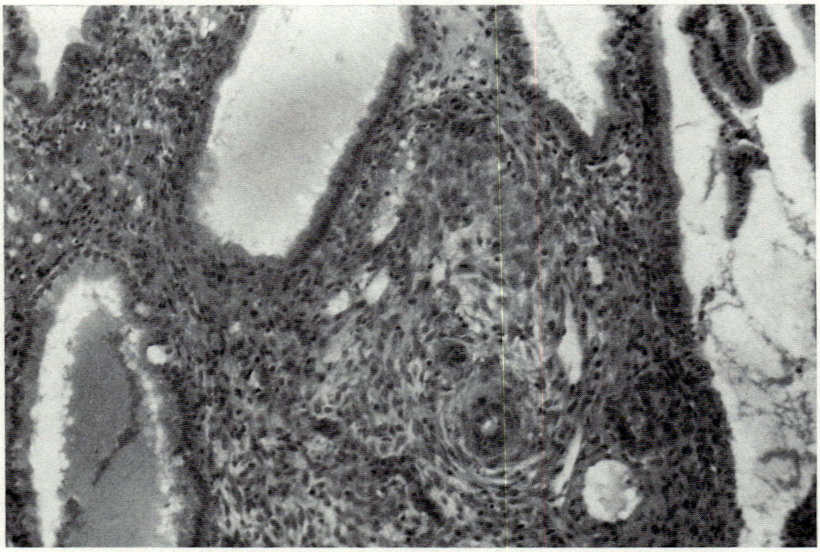

Fig. 8. Biopsy taken from a fundal fistula 16 months after preparation and 11 days after onset of deturgescence. Large glands of typical secretory endometrium and presence of a coiled arteriole. × 105.

arterioles. Typical early proliferative endometrium, however, has small glands with large columnar cells and basal nuclei amongst which occasional mitoses occur. The stroma consists of cells with vesicular nuclei and very little cytoplasm.

From the mid-proliferative phase to the onset of deturgescence the lumen of the glands somewhat increases, and there is marked pseudostratification of the epithelium lining the glands. Mitoses can often be seen in the stromal cells at this time.

After the onset of deturgescence the coiled arterioles become larger and more conspicuous. At the same time there is a reduction in size of the epithelial gland cells and the pseudostratification of their nuclei disappears. In spite of some variation within and between different biopsies, an increase in the glandular lumen is usually apparent at this stage, the nuclei either becoming centrally situated or remaining essentially basal in position, while the luminal aspects of the glands frequently develop a 'whispy' fringe. Stromal oedema can frequently be seen in secretory endometrium and predecidual changes also occur occasionally.

Thus it may be concluded that since typical endometrial changes, similar to those described for the human uterus, continue, endometrial cyclicity and function are not impaired as a result of making the fistulae. Presumably any reduction in blood supply resulting from the division of the uterine arteries is compensated for through the anastomotic branches of the ovarian arteries.

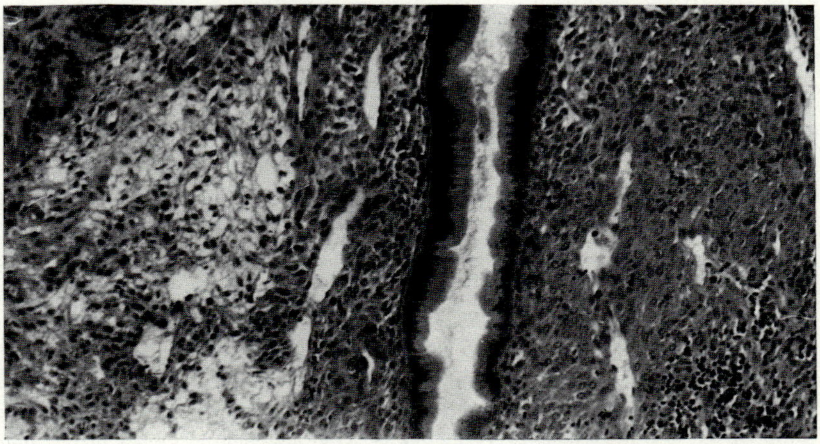

Fig. 9. Biopsy taken from a fundal fistula 8½ months after preparation and three days after onset of deturgescence. Large gland with secretion and the presence of precidual tissue. × 105.

Discussion

One of the potential problems of this type of preparation is the risk of bacterial contamination. Swabs taken from the fistulae soon after they had been prepared showed that, not surprisingly, a typical skin flora occurred in the wounds. Treatment with a suitable antibiotic, however, resulted in virtual elimination of all organisms other than *Staphylococcus aureus*. Presumably as a result of the weekly probing and tissue sampling a bacterial flora was sometimes reintroduced, but it could always be controlled with antibiotics. Such contamination, however, did not cause any noticeable deterioration in the health of the animals or affect the endometrial cycle (see above).

It should be stressed that in the second type of fistula in which the uterus is freed from the cervix, all cervical elements are completely eliminated and all menstrual bleeding occurs onto the abdominal wall. Such a preparation lends itself to a variety of investigations of uterine physiology and biochemistry. Our own objectives are to study the effects of plastic intrauterine devices on the endometrium and its function. Preliminary histological findings suggest that no alteration in the cyclical endometrial changes occurs after insertion of an IUD. Work on the biochemical composition of both menstrual blood and uterine flushings, and on possible effects of a device on them, is in progress.

Acknowledgements

The financial support of the Ford Foundation of New York is gratefully acknowledged. We also wish to thank Mrs. JUDITH STEPHENSON for performing the progesterone assays, Mr. J. BAZELEY for drawing figures 1 and 2, and Messrs. H. SMITH, F. JAMES and G. TONKS for their care of the animals.

References

BREED, W. G.; STEPHENSON, J. M.; ECKSTEIN, P.; PEPLOW, P. V., and BUTT, W. R.: Effect of an intrauterine device on menstrual cyclicity and luteal function in the baboon. J. Reprod. Fertil. *28:* 249–257 (1972).

GOOD, R. G. and MOYER, D.L. Technique of serial endometrial biopsy in the monkey through a uterocutaneous fistula. J. Reprod. Fertil. *22:* 573–574 (1966).

GOOD, R. G. and MOYER, D. L.: Estrogen-progesterone relationships in the development of secretory endometrium. Fertil. Steril. *19:* 37–49 (1968).

KRAEMER, D.C.; MOORE, G.T., and VERA CRUZ, N.C.: in HAFEZ Comparative reproduction of non-human primates (Thomas, Springfield 1971).

MAC LENNAN, A. H. and WYNN, R. M.: Menstrual cycle of the baboon. I. Clinical features, vaginal cytology and endometrial histology. Obstet. Gynec. *38:* 350–358 (1971).

NOYES, R. W.; HERTIG, A. T., and ROCK, J.: Dating the endometrial biopsy. Fertil. Steril. *1:* 3–21 (1950).

STURGIS, S. H.: Method for obtaining uterine fluid from the monkey. Effect of pilocarpine, atropine, physiological salt solution and adrenaline. Endocrinology *31:* 664–672 (1942).

WAGENEN, G. VAN and MORSE, A. H.: Cyclic changes in the exteriorized uterus. Endocrinology *27:* 268–273 (1940).

Authors' address: Dr. W.G. BREED, Mrs. ANNE FRASER, Prof. P. ECKSTEIN and Dr. P. V. PEPLOW, Department of Anatomy, The Medical School, University of Birmingham, *Birmingham, B15 2TJ* (UK)

Medical Primatology 1972. Proc. 3rd Conf. exp. Med. Surg. Primates, Lyon 1972, part I, pp. 216–226 (Karger, Basel 1972)

Breeding Results from an Experimental Colony of *Macaca arctoides*

SIGRID BRÜGGEMANN and J. GRAUWILER

Medical and Biological Research Division, Sandoz Ltd., Basel

Introduction

In June, 1970, an experimental breeding colony of *Macaca arctoides* for drug testing was started [GRAUWILER and BRÜGGEMANN, 1973]. This paper is to report our experience and breeding results obtained during the past two years. The colony was started with 36 females and 11 males. It consists now of 64 females and seven males.

A. Cycle Determination and Mating Procedures

1. Cycle Determination

Accurate follow-up of the menstrual cycle in the female monkey is essential for mating at the optimal time of the cycle. In teratological experiments timed breeding is indispensable because the exact day of conception must be known if the dam is to be treated during a specific phase of gestation.

(a) Because menstrual bleeding in *M. arctoides* is slight and of short duration, daily vaginal swabbing is necessary. This procedure is easy because the stump-tailed macaque can be trained to show its vulva. The first day on which blood is detected is considered as day 1 of the menstrual cycle (and the last day of the previous cycle).

(b) Evaluation of the ovarian cycle has been attempted in our laboratory in various ways. Measurements of vaginal temperature and determinations of the ratio of sialic acids to chlorides in vaginal (cervical) mucus, as mentioned elsewhere [WEBER and GRAUWILER, 1972] and cytological examination of vaginal smears were done according to HENDRICKX [1971 a].

2. Mating

The females in our colony are mated only once during one menstrual cycle. The time is chosen as described for other macaque species [STENGER, 1968; HAFEZ, 1971; MACDONALD, 1971]. If possible, the female is mated two days before the expected middle of the following cycle, a formula proposed by HENDRICKX [1972].

The stump-tailed macaques cannot be left alone without supervision when mated in relatively narrow cages because the male tends to fight as soon as mating is terminated [MACDONALD, 1971]. Therefore, the females in our colony are usually removed from the male's cage after 30 min or 1 h at most.

Mating is observed and successful ejaculation is verified by vaginal sampling with a saline-filled pipette. Number and viability of sperms are estimated and given a point score. The day of mating is considered as day 0 *post copulationem* (p.c.).

3. Diagnosis and Follow-Up of Pregnancy

(a) Pregnancy is verified by means of a bioassay in virginal mice based on qualitative proof of chorionic gonadotrophin in the maternal serum [WILSON *et al.*, 1970]. The test is performed between days 18 and 25 p.c. It is repeated five to seven days later, if the results are questionable. Attempts are being made to apply this technique for urine as well.

(b) Follow-up of pregnancy is done by rectal palpation of the uterus, using a scale proposed by WILSON *et al.* [1970]. Pregnancy can be ascertained as early as day 25, but more accurately after day 30. Normal development of uterine size is checked in a biweekly examination, and in pregnant animals daily vaginal swabbing is performed. If bleeding is detected, rectal examination is done immediately to check uterine size and consistency in order to recognize signs of impending abortion as early as possible.

(c) Evaluation of offspring: Fetuses have been obtained for the most part by cesarean section on days 70 or 100 p.c. In addition, for evaluation of other developmental stages, cesarean sections have been performed in untreated dams on days 30, 50, 60, 80 and 140 p.c. The fetuses are examined externally, fixed in formalin and later dissected.

(d) Gestation period: Four dams have been allowed to go to term to provide data on the normal duration of gestation and postnatal development. One of the dams was allowed to raise her young, and two young were raised by hand from birth following VALERIO's method [VALERIO *et al.*, 1969]. The fourth young was stillborn.

B. Breeding Results

1. Cycle

a) Menstrual Cycle

From 393 menstrual cycles in 42 animals, 319 lasted between 26 and 40 days, with a mean cycle length of 30.5 days (MACDONALD [1971] found a mean cycle duration of 30.7 days for the same species). Twenty cycles were shorter than 26 days and 54 cycles longer than 40 days. It was arbitrarily decided that a cycle length of 26–40 days might represent the norm; according to several investigators [HAFEZ, 1971], this time span corresponds with the normal cycle lengths of other nonhuman primates, especially other macaques and baboons [HENDRICKX, 1971a; VAN WAGENEN, 1966]. The distribution of the cycle length shows one peak around 30 and another around 60 days. A third peak might be suspected around 90 days. This might suggest that one or two menstrual bleeds could have been missed occasionally (fig. 1).

b) Ovarian Cycle

None of the different approaches for evaluating and timing the ovarian

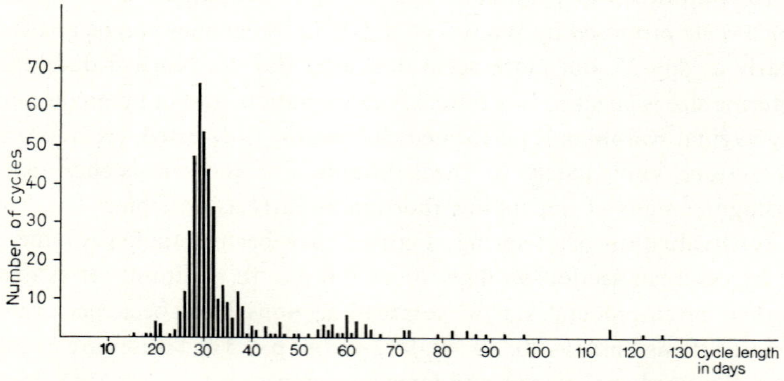

Fig. 1. Distribution of 427 menstrual cycles of *Macaca arctoides* showing a peak around 30 days (mean length of cycles between 26 and 40 days: 30.5). A second peak shows up around 60 days, and a third could be expected near 90 days.

cycle and diagnosing the time of ovulation gave satisfactory results as yet. With the exception of vaginal cytology, they have been abandoned because of the considerable variation and uncertainty of results.

2. Pregnancies

a) Normal Pregnancies

Uterine size, as ascertained by rectal palpation, begins to increase from day 25 p.c. onwards. There are no significant differences in comparison with the development of uterus size found by WILSON *et al.* [1970] in rhesus monkeys.

Implantation bleeding was observed in 48 of 70 pregnancies (68.6%). Bleeding began between days 16 and 24 p.c. in 44 of these (>90%) approximately 28–37 days after the last menstrual bleeding. There was one case of implantation bleeding as early as day 13 p.c. and one each beginning on days 28, 29 and 39 p.c. Implantation bleeding lasted from one to 14 days, but in most cases up to six days. As menstrual bleeding lasted approximately one to three days, half of the implantation bleeds could not be distinguished from menstruation by the duration of bleeding. As with menstrual bleeding, in most cases implantation bleeding was not detectable except by vaginal swabbing (table I).

Table I. Onset and duration of implantation bleeding. Bleeding set in between days 27 and 42 of menstrual cycle or between days 16 and 25 p.c. in most cases. Duration is usually relatively short

Day of menstrual cycle	27–35	36–42	52
Number	42	5	1
%	88	10	2

a) Onset of bleeding

Day p.c.	13	16–25	28–29	39
Number	1	44	2	1
%	2	92	4	2

b) Duration of bleeding

Days	1	2–3	4–6[1]	7–10[1]	11–14[1]
Number	12	11	13	6	6
%	25	23	27	12.5	12.5

1 Discontinuous.

b) Pregnancy Rates

From June, 1970, to April, 1972, 81 pregnancies occurred in our colony. Nineteen females had one pregnancy, 11 had two, eight had three and four had four pregnancies. The total number of matings was 418, i.e. 19.4% or every fifth of the matings were successful. Successful matings occurred throughout the year with the exception of November, 1971 (fig. 2).

c) Abortions

During the reported time span one of 26 untreated controls aborted while 11 abortions of 25 pregnancies of thalidomide-treated animals occurred.

3. Fetuses

a) Fetal Weight and Size

Fetal weight increased between days 50 and 60 p.c. from 2 to 10 g. It reached roughly 20 g by day 70 p.c. and approximately 40 g by day 80 p.c. The weights of the 100-day p.c. fetus varied between 110 and 160 g. The weight of one 140-day p.c. fetus was 258 g and the birth weights of three of our four young delivered naturally were 405 g (162 days p.c.) and 520 and 543 g (both 180 days p.c.). The fourth young was born on day 178 p.c. and weighed 505 g on the third day p.p. (fig. 3).

Crown-rump lengths were approximately 40 mm in the 50-day-old fetus, 50 mm at 60 days, 70–80 mm at 70 days, 100 mm at 80 days and

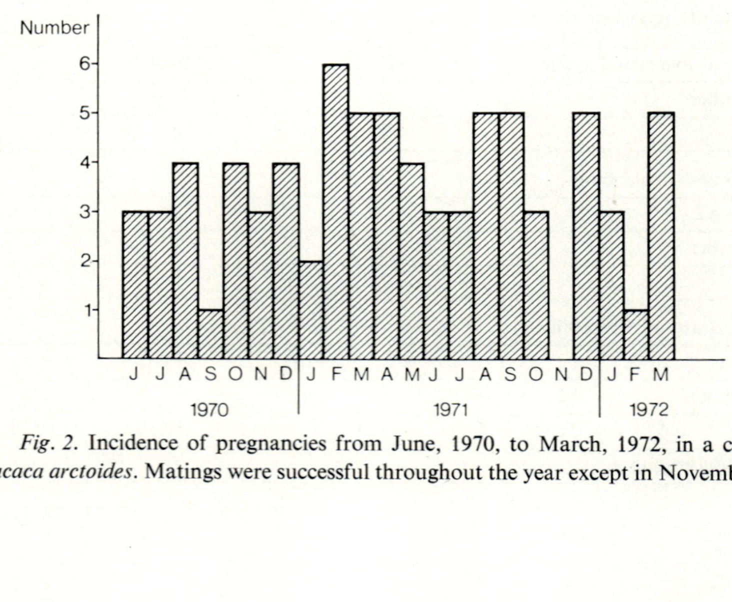

Fig. 2. Incidence of pregnancies from June, 1970, to March, 1972, in a colony of *Macaca arctoides*. Matings were successful throughout the year except in November, 1971.

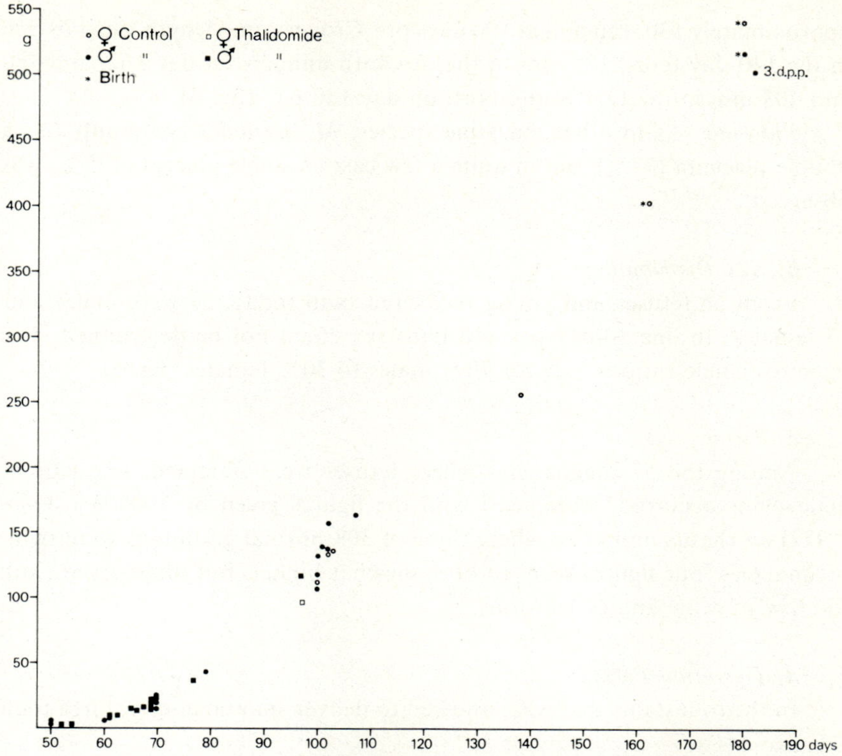

Fig. 3. Increase in body weight in *Macaca arctoides* fetuses from day 50 p.c. to birth. No difference between weights of fetuses from untreated dams and dams treated with thalidomide can be observed.

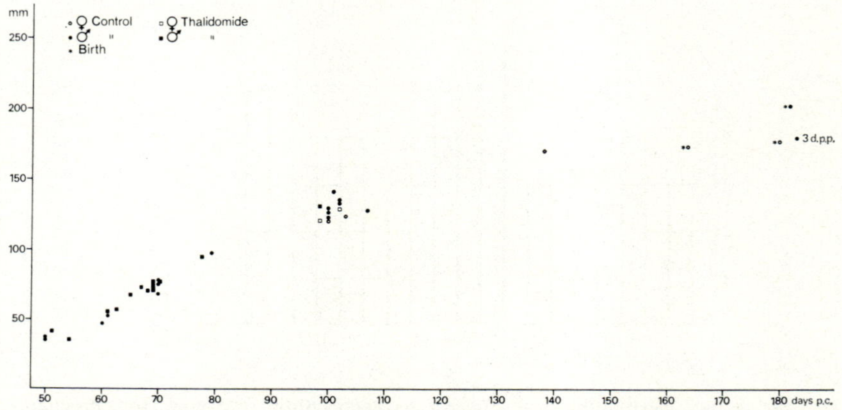

Fig. 4. Increase in crown-rump length in *Macaca arctoides* fetuses from day 50 p.c. to birth. After day 100 p.c. an increase in length seems to slow down.

approximately 130–150 mm at 100 days p.c. Crown-rump length was 170 mm in the 140-day fetus, 175 mm in the newborn animals on day 162, and 180 and 203 mm in the two young born on day 180 p.c. (fig. 4).

Placenta: As in other macaque species, *M. arctoides* commonly has a double placenta (75 %), but in quite a few cases a single placenta (25 %) was observed.

b) Sex Distribution

From 56 fetuses and young recovered until today, 39 were males and 17 females; in one 30-day p.c. old fetus sex could not be determined. The male to female ratio is 2.29, or 70 % males to 30 % females (fig. 5).

c) Twins

Among the 57 pregnancies where fetuses were obtained, one pair of male twins occurred. Compared with the figures given by VAN WAGENEN [1972] on rhesus monkeys, where three of 308 normal gestations were twin-pregnancies, our figures seem to be somewhat higher, but numbers are still too low to draw any conclusions.

d) Gestation Period

In the four dams that were allowed to deliver spontaneously, birth took

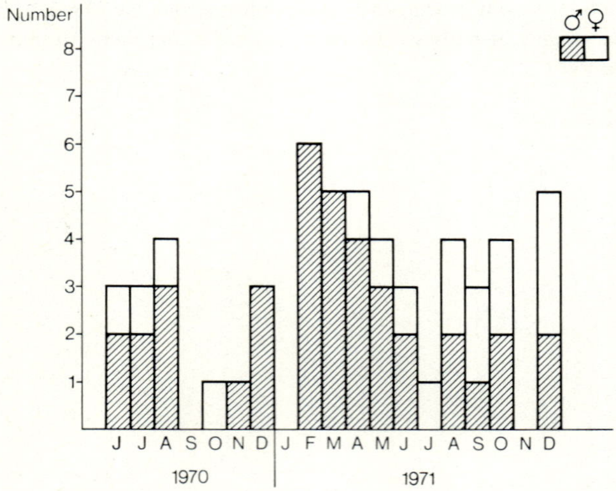

Fig. 5. Sex distribution of 56 *Macaca arctoides* fetuses recovered from pregnancies established between June, 1970, and December, 1971. After a predominance of the male fetuses up to March, 1971, a shift in favour of females seems to develop.

place after 162–180 days p.c. Three of the four animals were born around 180 days p.c., so possibly gestation length is somewhat longer than in the rhesus monkey, as postulated by STENGER [1972] and TRUM [1972].

e) Thirty-Day Embryo

As yet in the two 30-day control embryos available, only arm buds are present. In the rhesus monkey, HEUSER and STREETER [1941] observed both arm and leg buds by the 27th day p.c., so that leg bud formation may be somewhat later in *M. arctoides* (fig. 6).

C. Hand- and Mother-Reared Infants

One male, born on the 178th day p.c., was mother-reared and two females, born on the 162nd and 180th day p.c., were hand-reared from day

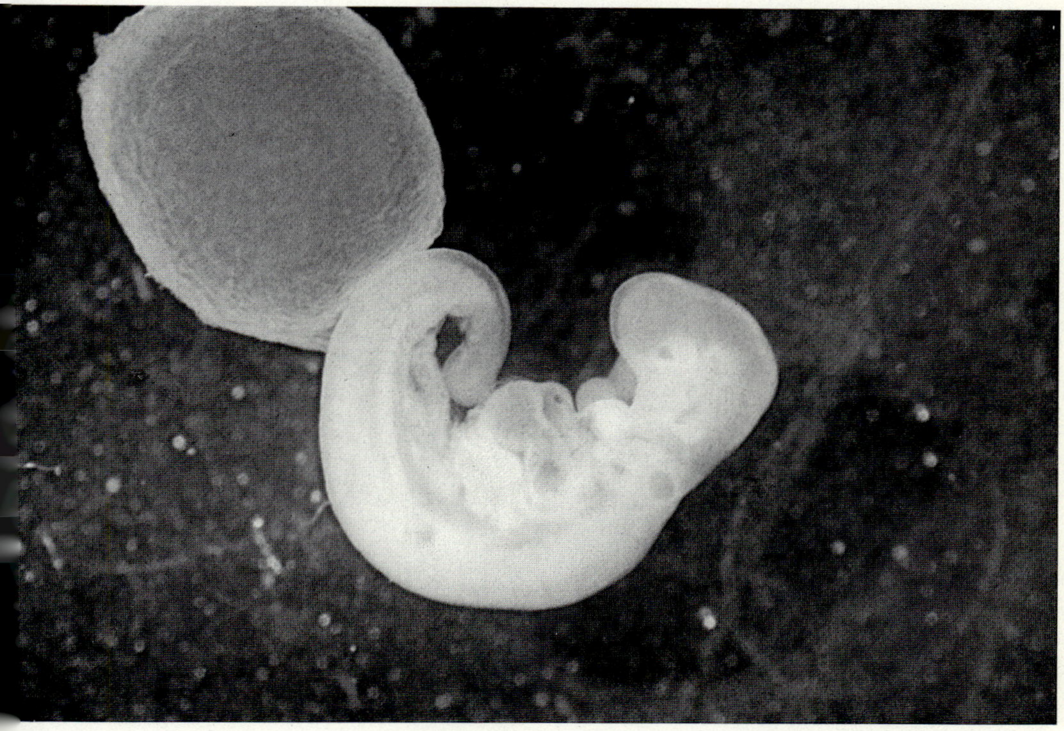

Fig. 6. 30-day embryo of *Macaca arctoides* showing full development of arm buds, but no hindlimbs are visible at this stage. Note the large vitelline sac.

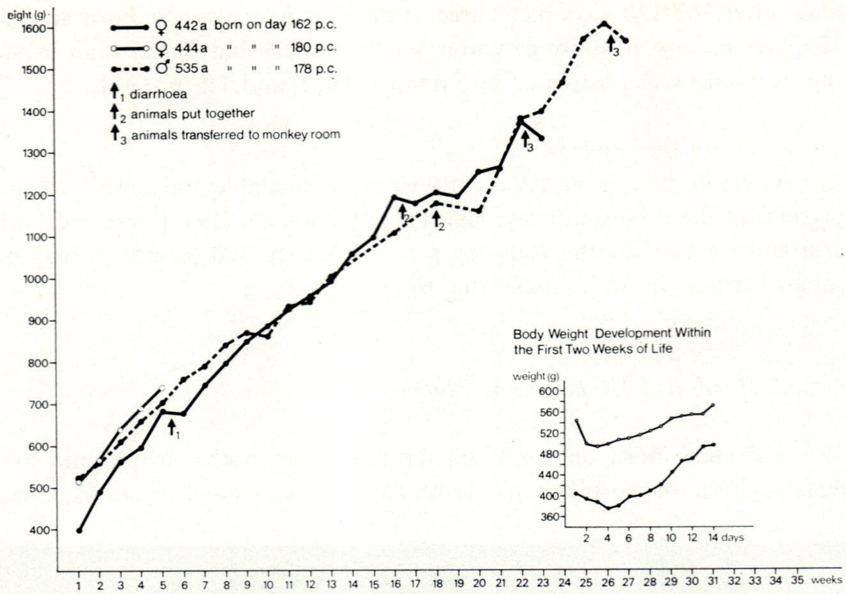

Fig. 7. Body weights of three juvenile *Macaca arctoides* from birth to 27 weeks. Weight gains of males and females do not show significant differences. Weight loss during the first two to four days was compensated by the tenth day.

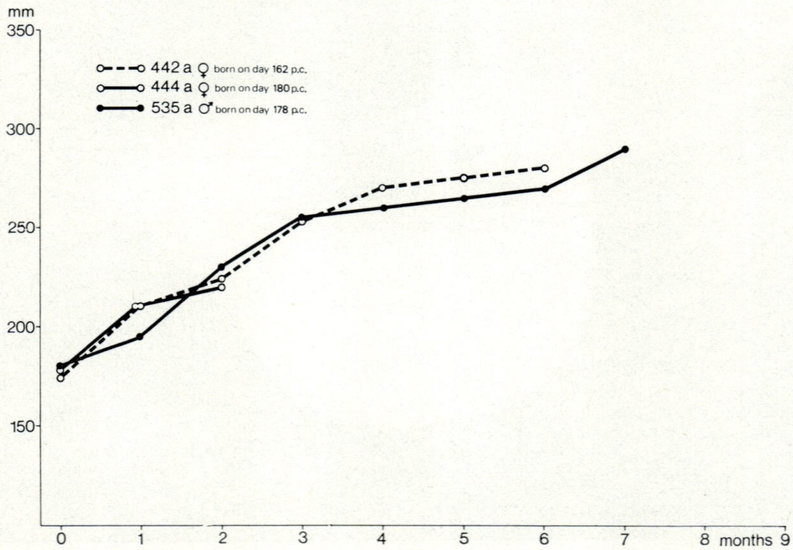

Fig. 8. Increase in crown-rump length of two juvenile *Macaca arctoides* from birth to six to seven months.

1 p.p. with baby food (following the technique of VALERIO *et al.* [1969]). The early born female weighed 405 g and the other two weighed 543 (female) and 505 g (male). Weight loss during the first week (amounting to 30–50 g) was seen in the two hand-reared females. The daily average weight gain during the first six weeks was 8.2 g in the early born female, 7.6 g in the other female, and 6.7 g in the male (fig. 7).

The crown-rump length increased at the same rate with both females, namely 40 mm in the first and 15 mm in the second month. The increase in the male amounted to 10 mm in the first and 20 mm in the second months (fig. 8). The time and sequence of dentition was not the same in all infants; i_1, i_2 and the upper molars appeared first, in that order usually. These were followed by the lower molars and the lower canines or vice versa. The last to appear were the upper canines. Teething started between the first and fifth week and was completed by the 13th to 14th week.

The different duration of intrauterine development seemed not only to have an effect on birth weight but also behaviour. For the first few days the early born female slept most of the time, crawled only a little and showed no facial expressions. On day 4 she walked on four legs. The 'teeth-chattering face' and the ability to recognise objects was observed on day 8 p.p. The second female was much more alert, walked on four legs and showed 'teeth-chattering face' by day 1 p.p. She could recognise objects on the sixth day.

The first female learned to climb only on day 31 p.p. when she had the opportunity to observe the four-weeks older male. The second female, on the other hand, started to climb on day 17 p.c. It is possible that she also learned to do so by observation, as she was caged next to the two older young who were then four and five months old.

Acknowledgements

The authors are gratefully indebted to Drs. H. WEBER and J. MESZAROS for performing the cesarean sections.

References

GRAUWILER, J. and BRÜGGEMANN, S.: Two year's experience with *Macaca arctoides* as an experimental animal for teratologic drug testing; in GOLDSMITH and MOOR-JANKOWSKI Medical primatology 1972, part III, pp.250–258 (Karger, Basel 1972).

HAFEZ, E.S.E. (ed.): Comparative reproduction of non-human primates (Thomas, Springfield 1971).

HENDRICKX, A.G.: Embryology of the baboon (University of Chicago Press, Chicago 1971a).

HENDRICKX, A.G.: Teratogenicity of thalidomide in the baboon *(Papio cynocephalus)*, bonnet monkey *(Macaca radiata)* and cynomolgus Monkey *(Macaca irus)*; in BIEGERT and LEUTENEGGER Proc. 3rd Int. Congr. Primatology, Zurich 1970, vol. 2, pp. 230–237 (Karger, Basel 1971).

HEUSER, C.H. and STREETER, G.L.: Development of the macaque embryo. Contrib. Embryol. Carneg. Inst. *29:* 17–55 (1941).

MACDONALD, G.J.: Reproductive patterns of three species of macaques. Fertil. Steril. *22.* 373–377 (1971).

STENGER, V.G.: A preliminary report on the use of the pregnant *Macaca speciosa* as an experimental subject; in VAGTBORG Use of non-human primates in drug evaluation, pp. 102–117 (University of Texas Press, Austin 1968).

STENGER, V.G.: Reproductive capabilities of the *Macaca speciosa* in captivity. 6 years' experience; in BEVERIDGE Breeding primates, pp. 100–104 (Karger, Basel 1972).

TRUM, B.F.: Research on reproduction at the New England Primate Research Center; in BEVERIDGE Breeding primates, pp. 198–202 (Karger, Basel 1972).

VALERIO, D.S.; MILLER, R.L.; INNES, J.R.M.; COURTNEY, K.D.; PALLOTTA, A.J., and GUTTMACHER, R.M.: *Macaca mulatta*, management of a laboratory breeding colony (Academic Press, New York 1969).

VAN WAGENEN, G.: Studies in reproduction *(Macaca mulatta)*. Proc. FDA Conf. on Non-human Primate Toxicology, pp. 103–113 (FDA-Administration, Warrenton 1966).

VAN WAGENEN, G.: Vital statistics from a breeding colony. Reproduction and pregnancy outcome in *Macaca mulatta*. J. med. Prim. *1:* 3–28 (1972).

WEBER, H. and GRAUWILER, J.: Experience with a breeding colony of stumptail macaques for teratological testing of drugs; in BEVERIDGE Breeding primates, pp. 92–99 (Karger, Basel 1972).

WILSON, J.G.; FRADKIN, R., and HARDMAN, A.: Breeding and pregnancy in rhesus monkeys used for teratological testing. Teratology *3:* 59–72 (1970).

Authors' address: Dr. SIGRID BRÜGGEMANN and Dr. JULES GRAUWILER, Medical and Biological Research Division, Sandoz Ltd., *CH-4000 Basel* (Switzerland)

Medical Primatology 1972. Proc. 3rd Conf. exp. Med. Surg. Primates, Lyon 1972, part I, pp. 227–231 (Karger, Basel 1972)

The Use of Radiologic Techniques for the Study of Placental Circulation in the Rhesus Monkey

M. W. Donner, S. I. Margulies and Elizabeth M. Ramsey

Department of Radiology, The Johns Hopkins Medical Institutions and Department of Embryology, Carnegie Institution of Washington, Baltimore, Md.

In our radiographic studies two modalities were utilized [1]. In some studies a system was employed which provides standard X-rays made in two planes simultaneously at a rapid rate. Our studies with this equipment were conducted at a speed of two films/sec. In other studies we used cineradiography, in which a motion picture camera photographs the intensified fluoroscopic image at a speed of 30 frames/sec. The rapid serial X-rays provide excellent definition and the cineradiography permits observation of circulatory dynamics. Thus the two techniques are complementary.

To demonstrate maternal arterial inflow to the placenta the femoral artery is exposed under anesthesia and a catheter is introduced into it to convey the radiopaque material into the maternal circulation. The contrast material is injected as a bolus with a pressure syringe [2]. It progresses from the femoral artery into the aorta, rising to the level of the renal arteries. The endometrial spiral arteries are filled and the medium enters the intervillous space in fountain-like spurts. These spurts grow as the contrast material spreads in the intervillous space of the placenta (fig. 1A).

To demonstrate venous drainage from the placenta, contrast material is injected into the intervillous space [3]. In approximately 80% of rhesus monkeys the placenta is bidiscoid. The location of the placental discs is ascertained by transillumination of the uterus and a catheter is introduced into the intervillous space of one of the discs. Through this the radiopaque medium is injected with a constant speed infusion pump. The contrast material slowly disperses throughout the intervillous space and drainage channels in the uterine wall appear in increasing numbers as the space fills (fig. 2).

The preceding illustrations have shown uteroplacental circulation during uterine relaxation. The effects of contractions upon circulation have

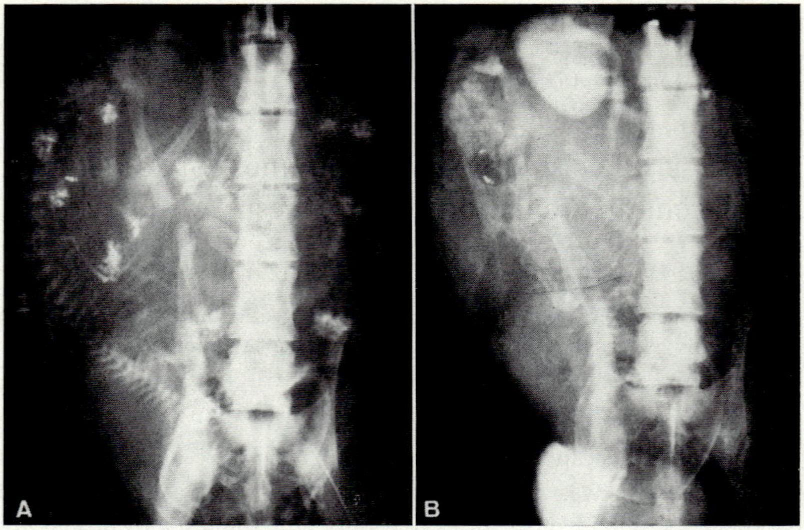

Fig. 1. Angiographs made 8 sec after injection of contrast medium into the femoral artery. *A* Injection during uterine relaxation. *B* Injection during a contraction of 42 mm/Hg. Monkey 62/30. SP No. 38, 141st day of pregnancy.

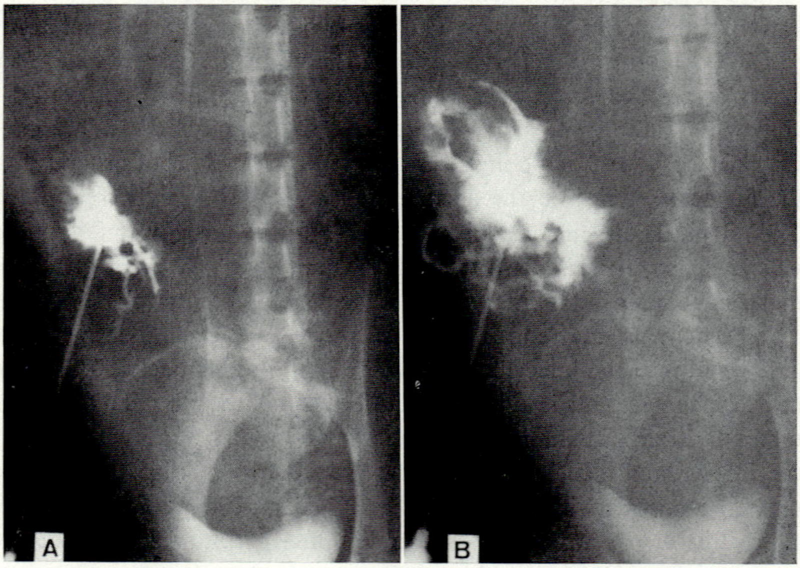

Fig. 2. Angiographs made at 2 and 6 sec after rapid injection of contrast material into the intervillous space during uterine relaxation. *A* The pool of opaque medium indicates the intervillous space. *B* Mural veins are also seen.

also been studied. In order to inject the contrast medium at the desired time, it is necessary to monitor uterine activity continuously. This is accomplished by the introduction of an open-end catheter into the amniotic cavity through the abdominal wall and the wall of the uterus. The catheter is attached to a pressure transducer and the latter to a multichannel physiological datarecording system [4]. The X-ray procedures are similar to those already described.

During uterine contraction the arterial inflow is reduced or even halted, depending upon the intensity of the contraction. In figure 1B, showing an experiment in which the medium was injected during an intense contraction, there are no spurts and there is only partial filling of the uterine vessels.

Uterine activity also affects venous drainage of the intervillous space. As the uterus begins to contract, the venous drainage channels disappear or become narrower. However, the pool of contrast material within the intervillous space remains essentially the same throughout the contraction. This suggests that contraction does not empty the intervillous space. As the uterus begins to relax, venous channels reappear.

In the foregoing, only half of the placental circulation has been considered. No mention has been made of the contribution of the fetus. Demonstration of the fetal circulation in the placenta of the monkey can be accomplished in one of two ways, either by injection of contrast material into the fetal femoral artery [5] or by direct injection into an interplacental vessel [6]. Both of these routes have been employed.

To inject radiopaque material into the femoral artery of the fetus laparotomy is performed on the pregnant monkey. The uterus is transilluminated to outline the margins of the placental discs so that these may be avoided. The uterine wall is then incised. The membranes are separated from the wall in the region of the incision. A fetal leg is manipulated through this opening, with the membranes intact when possible. The fetal groin is exposed and an incision is made over the femoral vessels. These are identified and the femoral artery is isolated and incised to permit the introduction of a small polyethylene catheter. When this has been accomplished, there is prompt return of blood under the head of fetal arterial pressure.

Radiopaque material injected into the fetal femoral artery returns to the fetal hypogastric arteries and the umbilical arteries. It rises to the level of the chorionic plate and enters the subchorionic vessels. Then, it slowly fills the capillaries of the fetal cotyledons, which appear as discrete, cottony puffs. The cotyledons are independent structures. There is no capillary communication from one to the other. From the cotyledons the blood drains into the umbilical vein and is returned to the fetal circulation.

The interplacental fetal vessels located by transillumination of the uterus are dissected free through a hysterotomy incision and are catheterized to permit injection of contrast material directly into the subchorionic circulation. This technique provides superior visualization of the fetal cotyledons and of the placental disc which has been injected. Contrast material enters the interplacental vessel and appears in the capillary network of the fetal cotyledons (fig. 3A).

In a final demonstration these two independent circulations are shown simultaneously to illustrate their relationship. The radiopaque material is first injected into the fetal circulation and, immediately thereafter, a maternal injection is begun. In figure 3A the X-ray on the left was made a few seconds after a fetal injection. The radiopaque material appears in the fetal capillary system of the cotyledon. In figure 3B is shown an early stage of an immediately subsequent maternal injection. Well-formed spurts from the maternal circulation directly overlie fetal cotyledons in the ratio of 1:1.

From the accompanying illustrations it is apparent that the technique of radioangiography is suitable for direct visualization of placental circulation. Employing it we have been able to demonstrate the pathways of inflow of maternal blood to the intervillous space and of drainage of blood back into the maternal systemic circulation. We have shown that myometrial

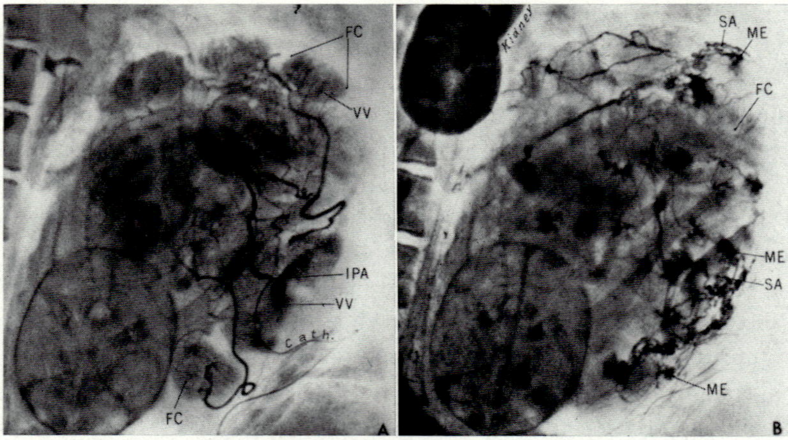

Fig. 3. Spot films made during a combined fetal and maternal injection study. Contrast medium injected, proximally, into an interplacental artery. MP 59. Monkey No. 65/80. 152 days pregnant. *A* Taken 3 sec after fetal injection. *B* Two seconds after maternal injection. SA = endometrial spiral artery; ME = maternal entry in association with a cotyledon; FC = fetal cotyledon; kidney = maternal kidney containing contrast medium; PA = interplacental artery; VV = villous vessels.

contractions curtail or halt both arterial inflow and venous drainage. Finally, we have demonstrated the basic relationship between maternal and fetal circulation in the placenta of the rhesus monkey.

References

1 DONNER, M. W.; RAMSEY, E. M., and CORNER, G. W. jr.: Maternal circulation in the placenta of the rhesus monkey. A radiographic study. Amer. J. Roentgenol. *90:* 638–649 (1963).

2 RAMSEY, E. M.; CORNER, G. W. jr., and DONNER, M. W.: Serial and cineradioangiographic visualization of maternal circulation in the primate (hemochorial) placenta. Amer. J. Obstet. Gynec. *86:* 213–225 (1963).

3 RAMSEY, E. M.; MARTIN, C. B., jr.; MCGAUGHEY, H. S. jr.; KAISER, I. H., and DONNER, M. W.: Venous drainage of the placenta in rhesus monkeys. Radiographic studies. Amer. J. Obstet. Gynec. *95:* 948–955 (1966).

4 CORNER, G. W. jr.; RAMSEY, E. M., and STRAN, H. M.: Patterns of myometrial activity in the rhesus monkey in pregnancy. Amer. J. Obstet. Gynec. *85:* 179–185 (1963).

5 MARTIN, C. B. jr.; RAMSEY, E. M., and DONNER, M. W.: The fetal placental circulation in rhesus monkeys demonstrated by radioangiography. Amer. J. Obstet. Gynec. *95:* 943–947 (1966).

6 RAMSEY, E. M.; MARTIN, C. B. jr., and DONNER, M. W.: Fetal and maternal placental circulations. Amer. J. Obstet. Gynec. *98:* 419–423 (1967).

Authors' addresses: Dr. M. W. DONNER and Dr. S. I. MARGULIES, Department of Radiology, The Johns Hopkins Medical Institutions, *Baltimore, MD;* Dr. ELIZABETH M. RAMSEY, Department of Embryology, Carnegie Institution of Washington, *Baltimore, MD* (USA)

Medical Primatology 1972. Proc. 3rd Conf. exp. Med. Surg. Primates, Lyon 1972, part I, pp. 232–236 (Karger, Basel 1972)

Natural and Artificial Control of Ovulation in Nonhuman Primates [1,2]

W. R. DUKELOW, R. M. HARRISON, J. M. R. RAWSON and M. P. JOHNSON

Endocrine Research Unit, Michigan State University, East Lansing, Mich.

Introduction

The placement of feral primates in captive conditions has usually resulted in extended periods of infertility. Interest in the reproduction of these species (both for species preservation and for the production of research animals), has resulted in extensive studies to improve or modify the ovulation detection processes for these animals. This work has recently been reviewed by DUKELOW [1970b]. Our efforts have been primarily directed towards characterization of the developing follicle as observed by laparoscopy and efforts to induce ovulation in a species which exhibits no menstrual cycle and which shows a strong seasonal effect even after several years of captivity *(Saimiri sciureus)*.

Material and Methods

Thirty-one mature female *Macaca fascicularis* were maintained under controlled temperature and lighting conditions in the primate facility of the Endocrine Research Unit. All animals had been in captivity for at least five years and many were born in captivity. The mean cycle length in the colony was 30.8 days, the median 33 days, and the mode 28 days. Cycles ranging from 27 to 32 days in length comprised 66.4% of all cycles.

The techniques employed in laparoscopic observation of the ovaries have been described previously [DUKELOW *et al.*, 1971].

Laparoscopic examinations were begun on the 10th day of the menstrual cycle and were continued for three to four days, at intervals of 4–12 h, depending on the follicular

1 Michigan Agricultural Experiment Station Journal Article No. 5939.

2 Supported by NIH contract No. 70-2061, USPHS grant No. 5-PO6-RR 00366-04 and NIH Research Career Development Award No. 1-K4-HD35, 306-01.

development. As the ovulation approached laparoscopies were made every 4 h. In three cases, observations were made every hour for 9 h with the animal under continuous anesthesia in an attempt to observe actual ovulation.

To avoid laparotomy no attempts were made to recover ova from the reproductive tract. Previous studies in our laboratory have demonstrated that laparoscopy can be used to predict ovulation time and, coupled with short-term mating sessions of 20–30 min duration, pregnancies can be obtained [JEWETT and DUKELOW, 1971].

In *Saimiri sciureus* ovulation was induced by the technique of DUKELOW [1970a], employing a sequential regime of progesterone (5 mg daily for five days) follicle stimulating hormone (FSH), (1 mg daily for four days), and 500 IU of human chorionic gonadotropin (HCG) to induce ovulation. To test the effect of 17α-acetoxy-6-methylpregna-4,6 diene, 3,20-dione (Megestrol Acetate, Mead Johnson) this compound was administered at levels of 50, 100, 250 and 500 μg daily by the technique of HARRISON and DUKELOW [1971], (i.m. injection) for a five-day period beginning with the first day of FSH. The effectiveness of the megestrol acetate in inhibition ovulation was measured by laparoscopic observation of the ovaries 12 and 36 h after HCG. In some trials, silastic implants of megestrol acetate were placed in *Saimiri* for up to 29 days prior to ovulation testing. These capsules (two per animal) released approximately 50 μg megestrol per day.

Results and Discussion

The characteristic changes that occur in the follicular morphology in *Macaca fascicularis* are illustrated in figure 1.

The site of the developing follicle could be identified two days prior to ovulation by, (a) a generalized swelling and darkening of a portion of the ovarian surface, and (b) an increase of approximately 35% in the size of the ovary (fig. 1a).

Within 30 h of ovulation, a stellate pattern of blood vessels is often seen on a portion of the ovary which will eventually form the follicular wall (fig. 1b). Within 8–10 h of ovulation the pattern of vessels became more pronounced and the follicular cone is established. This pattern often showed a network of smaller vessels around the base of the follicular cone with a larger vessel across the center of the follicular wall which bifurcated at the follicular border (fig. 1c). The vessel across the center of the follicle divided it into two hemispheres, each displaying a localized area of avascular tissue (stigmata). At the same time the fimbria approximates the follicle to prepare for recovery of the ovum (fig. 1d).

The most significant postovulatory changes (fig. 1e) are: (a) a generalized hemorrhagic appearance, (b) the occlusion of the stigmata, (c) a flattening or irregularity of the follicular wall, and (d) occlusion of the follicle (luteinization). These characteristics could be observed as early as 10 h following ovulation, but the corpus hemorrhagicum (CH) generally could not be reliably distinguished from the preovulatory follicle until 20–40 h following ovulation. In the majority of cases the vasculature present in preovulatory follicles remained intact through the remainder of the cycle (fig. 1f).

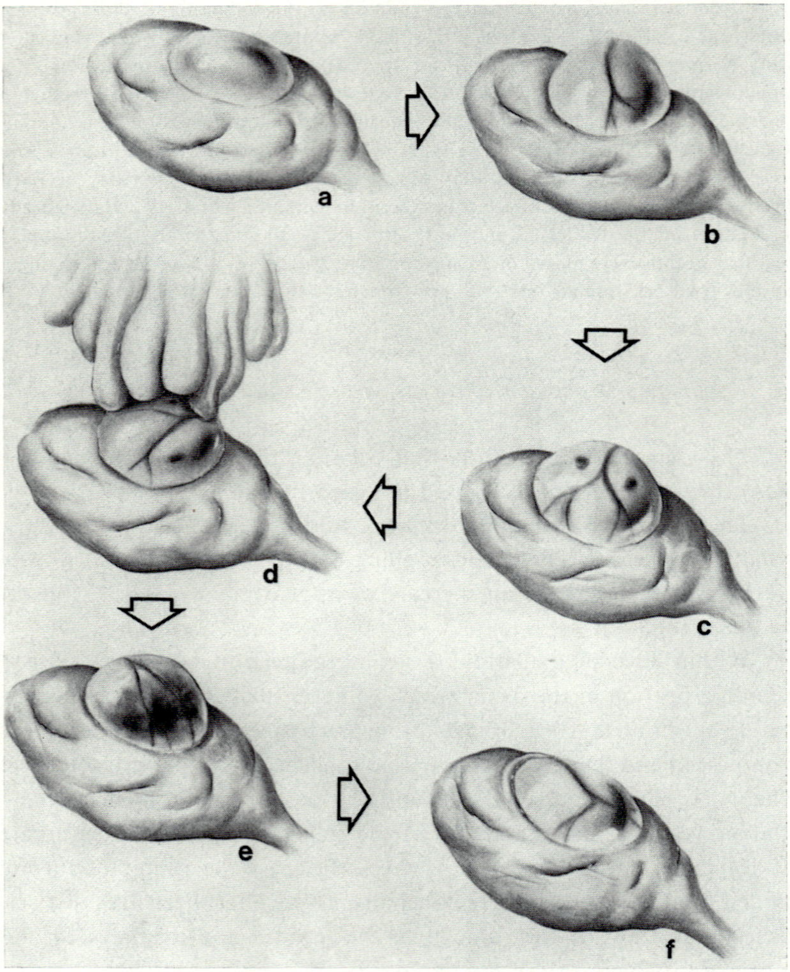

Fig. 1. Progressive development of the follicle in *Macaca fascicularis*. a Day 10 of the cycle, blister-like appearance. b Day 11, appearance of stellate blood vessel pattern. c Day 12, formation of stigmata or translucent areas. d Day 12 (8–10 h before ovulation), approximation of fimbria to follicle. e Day 14, post-ovulatory follicle with hemorrhagic appearance. f Day 24, persistance of corpus luteum with vasculature.

The identification of the morphological development of the follicle in *M. fascicularis* has provided a valuable tool which can be used to identify early developing follicles, predict ovulation times, assist in the timing and recovery of ova by follicular aspiration and evaluate the effects of pharmacological agents on ovarian function.

The effectiveness of the ovulation regime and its inhibition by megestrol acetate are indicated in table I.

The ovulation induction regime consistantly allows induction of single or double ovulations (other regimes will cause superovulation in this species [BENNETT, 1967; DUKELOW, 1970a]). A daily injected dose of 50 μg of megestrol acetate did not reduce the rate of induced ovulation when compared to controls subjected to the same ovulation regime. A level of 500 μg completely blocked ovulation while levels of 100 and 250 μg daily had dose-related but nonlinear effects.

Megestrol acetate, administered by silastic implant, inhibited but did not completely suppress response to the ovulation regime. Of the 16 animals in this group, eight had the implant for 11 days and eight for 29 days prior to testing. In the first group, three animals of eight (37.5 %) ovulated, whereas in the second group, one of eight (12.5 %) ovulated suggesting a relationship between the time the implant is in place and its effectiveness. Megestrol acetate is being extensively studied as a low-dose oral contraceptive in man. Its exact mode of action is unknown, but various workers have suggested that the mechanism of this effect is: (1) inhibition of sperm capacitation, (2) an increased rate of gamete transport through the reproductive tract, (3) an increased spermicidal effect within the cervical mucus, or (4) a combination of the above. Evidence has been presented in different species to substantiate each of these mechanisms. While the contraceptive action of megestrol acetate in squirrel monkeys has not yet been adequately demon-

Table I. Induced ovulation, and megestrol acetate inhibition of ovulation in *Saimiri sciureus*

Treatments	Number of observations	Number of animals ovulating	% ovulating	Mean number of ovulations per animal
Control	108	53	49	1.4
MA, 50 μg	10	6	60	1.3
MA, 100 μg	20	7	35	1.0
MA, 250 μg	20	4	20	1.5
MA, 500 μg	10	0	0	–
MA, implants	16	4	25	1.0

strated, the present work does indicate that a level of 100 to 250 μg of megestrol acetate administered daily for five days allows ovulation to be induced in these subjects. Its effects on the other reproductive phenomena mentioned above have not been determined.

What the exact mechanism of action of the low dose of megestrol acetate is remains to be investigated. At present, the combined techniques of controlled, induced ovulation and laparoscopic examination establish a reliable test system whereby the effects of various antifertility, pharmacological agents can be studied in primates.

Acknowledgement

The authors wish to express their appreciation to Mr. J. ITO, medical illustrator at the Oregon Regional Primate Research Center, for producing figure 1.

References

DUKELOW, W. R.: Induction of single and multiple timed ovulation in the squirrel monkey *(Saimiri sciureus)*. J. Reprod. Fertil. *22:* 303 (1970a).

DUKELOW, W.R.: Reproduction physiology of primates. Proc. Amer. Ass. Zoo Veterinarians, p. 51. (1970b).

DUKELOW, W. R.; JAROSZ, S. J.; JEWETT, D. A., and HARRISON, R. M.: Laparoscopic examination of the ovaries in goats and primates. Lab. anim. Sci. *21:* 594 (1971).

HARRISON, R. M. and DUKELOW, W. R.: Megestrol acetate. Its effect on the inhibition of ovulation in squirrel monkeys *(Saimiri sciureus)*. J. Reprod. Fertil. *25:* 99 (1971).

JEWETT, D. A. and DUKELOW, W. R. Laparoscopy and precise mating techniques to determine gestation length in *Macaca fascicularis*. Laboratory Primate Newsletter *10:* 16 (1971).

Author's address: Dr. W. RICHARD DUKELOW, Endocrine Research Unit, Michigan State University, *East Lansing, MI 48823* (USA)

Medical Primatology 1972. Proc. 3rd Conf. exp. Med. Surg. Primates, Lyon 1972, part I, pp. 237–247 (Karger, Basel 1972)

The Physiology of Physiologic Hyperbilirubinemia of the Newborn[1]

L. M. GARTNER[2] and DONNA L. LANE

Department of Pediatrics and the Rose F. Kennedy Center for Research in Mental Retardation and Human Development, Albert Einstein College of Medicine, Bronx, N.Y.

Introduction

Virtually every newborn human infant develops unconjugated hyperbilirubinemia during the first week of life. Although in most cases this physiologic jaundice is benign, in some infants it results in a type of brain damage, known as kernicterus [3]. Infants with high serum concentrations of unconjugated bilirubin, premature infants and infants having suffered from anoxia and acidosis are particularly susceptible to the neurotoxic effects of bilirubin, resulting either in death or permanent neurological disability [3]. In order to understand the mechanisms responsible for the regular accumulation of bilirubin in the newborn infant, we have undertaken a detailed study of the formation and excretion of bilirubin in the newborn rhesus monkey. Newborn simians are the only animals which develop physiologic jaundice and are, therefore, the only suitable models for study of this developmental disorder. Physiologic jaundice has been well characterized in the newborn rhesus monkey with mean peak concentrations of unconjugated bilirubin of approximately 5 mg/100 ml of sera occurring at about 24 h of age [5]. These peak levels are similar to those observed in full-term human neonates, although in the human the maximal level occurs during the third day of life.

It has usually been stated that physiologic jaundice results from deficient hepatic glucuronyl transferase activity, the enzyme catalyzing the con-

1 Supported by grants from USPHS, NIH, NICHD and from the Gail I. Zuckerman Foundation for Research in Chronic Liver Diseases in Children.

2 Career Development Awardee of the National Institute for Child Health and Human Development.

version of bilirubin from a lipid soluble compound to a water soluble diglucuronide of bilirubin [6]. The formation of this water soluble compound in the liver cell permits excretion into bile. Other possibilities exist, however, to explain the accumulation of bilirubin in the newborn. Thus, there may be increased bilirubin production from either erythrocyte or nonerythrocyte sources; diminished hepatic uptake of bilirubin resulting either from deficient hepatic blood flow or from developmentally abnormal cellular uptake and transport mechanisms; or from enhanced reabsorbtion of unconjugated bilirubin from the intestine with return of the pigment to the circulation. Each of these mechanisms may be superimposed on the demonstrated deficiency in the conjugation of bilirubin. The object of the present study is to simultaneously determine the relative contribution of each of these mechanisms to the development of physiologic jaundice and to determine the maturation of each of these processes in the newborn monkey.

Methods

Using the techniques of bile duct catheterization for bile collection, i.v. infusion of saline and unconjugated bilirubin, and repeated liver biopsy and arterial blood sampling, it was possible to quantitatively determine endogenous hepatic excretion, maximal cumulative hepatic uptake, maximal

Table I. Newborn *Macaca mulatta*

Animal	Age, h	Weight, g	% Liver weight
Q	5 FT (C-section)	485	3.68
L	7 FT	546	2.83
E	19 FT	405	2.73
B	27 FT	447	2.64
D	54 FT	351	3.08
P	80 FT	499	3.05
C	120 FT	494	2.63
A	300 FT	511	3.01
O	456 FT	556	2.81
J	57 prem. (3 weeks)	300	3.04
K	4 post. (2 weeks)	647	3.99

Ages in hours at time of study, duration of gestation, type of delivery, weight at time of study and liver weight as percent of body weight of each newborn *Macaca mulatta* studied. FT = delivery of animal at term; prem. = delivery of animal prematurely; post. = delivery of animal postmaturely; C-section = delivery of animal by cesarean section.

conjugation and maximal hepatic excretion of bilirubin in the intact anesthetized newborn rhesus monkey. The surgical, chemical and mathematical techniques used to arrive at these figures have been verified previously in adult male rhesus monkeys in studies performed in collaboration with Dr. CORNELIUS at the Institute for Comparative Biology of the San Diego Zoological Society in California [2].

On each figure the newborn data is presented in relation to age and in comparison with values obtained previously in adult animals using identical techniques [2].

Eleven newborn monkeys ranging in age from four to 456 h have been studied, including one animal born three weeks prematurely, one born two weeks postmaturely by cesarean section, and one delivered by cesarean section at term (table I). All animals were from the Laboratory for Experimental Medicine and Surgery in Primates of New York University Medical Center, (LEMSIP), New York. Except for the two cesarean section delivered animals, liver weights were approximately 3 % of body weight. The cesarean section delivered animals had liver weight of nearly 4 %, perhaps reflecting increased glycogen stores.

Results

Native serum bilirubin concentrations prior to infusion of bilirubin reached a maximum of between 4 and 5 mg/100 ml of serum at about 24 h of age and then rapidly declined to nearly normal levels by about 48 h (fig. 1). Not until 300 h, however, did the levels become entirely normal or 0.1 mg/100 ml. The initial period of high serum bilirubin concentrations is designated as phase I, while the second period of low grade hyperbilirubinemia is called phase II. The two cesarean section delivered animals, one postmature and one term, had very low serum bilirubin concentrations, probably reflecting their very young age (fig. 1).

The excretion of bilirubin in bile prior to infusion of bilirubin has been designated as endogenous bilirubin excretion (fig. 2). During the first few hours of life, when hepatic transport and metabolism of bilirubin is markedly restricted, endogenous excretion does not represent the load of bilirubin presented to the liver. After 48 h of life, when the physiologic limitations have largely been overcome, endogenous bilirubin excretion in bile does reflect the quantity of bilirubin presented to the liver, whether from new bilirubin synthesis or from intestinal reabsorbtion of bilirubin. Of particular

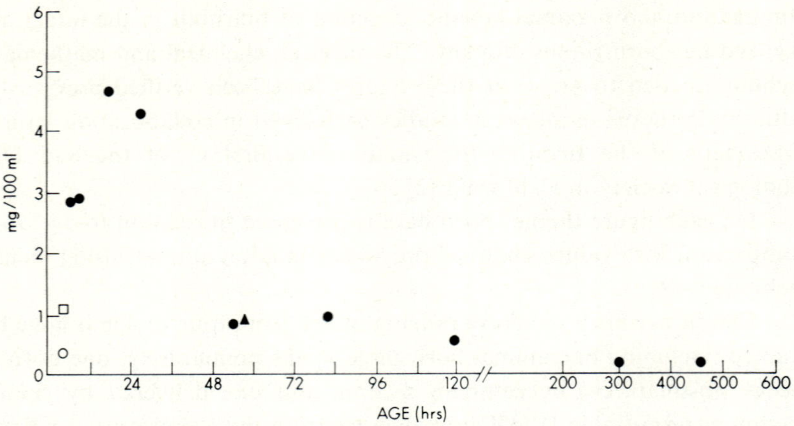

Fig. 1. Total serum bilirubin concentrations in mg/100 ml of each newborn *Macaca mulatta* at time of study as indicated on abscissa. ● = Full-term animals; ▲ = premature animals; ○ = postmature animals; □ = full-term animal delivered by cesarean section.

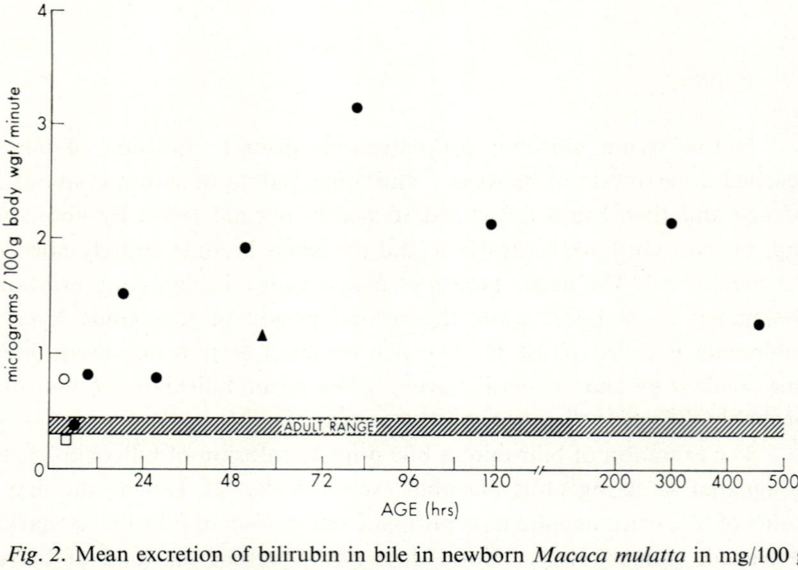

Fig. 2. Mean excretion of bilirubin in bile in newborn *Macaca mulatta* in mg/100 g body weight/min during one-hour period of i.v. saline infusion prior to infusion of unconjugated bilirubin (endogenous). Ages of animals in hours at time of study are indicated on the abscissa. ● = Full-term; ▲ = premature; ○ = postmature; □ = full-term delivered by cesarean section. Hatched bar indicates range of endogenous excretion of bilirubin in bile in adult *Macaca mulatta* [2].

note is the fact that endogenous excretion of bilirubin in the newborn period, up to at least 19 days of age is from four to ten times greater than that observed in adult animals (fig. 2). Thus, newborn liver is presented with a bilirubin load of approximately 2 μg/100 g body weight/min, as compared with 0.3–0.4 μg/100 g body weight/min in the adult. This marked increase in bilirubin load in the newborn period, as compared with the adult, is in marked contrast to the situation in the newborn guinea pig, an animal in which endogenous bilirubin excretion is the same in the newborn as in the adult [1]. The newborn guinea pig does not develop physiologic jaundice. The increased load of bilirubin presented to the liver in the newborn monkey is of such magnitude that it is unlikely that it is derived from accelerated destruction of circulating erythrocytes, but is more likely to be derived from either exaggerated nonhemoglobin heme turnover, pre-erythrocyte heme turnover or from a brisk enterohepatic re-absorbtion of bilirubin. The relative contribution of each of these to the increased bilirubin load of the newborn remains to be investigated.

Maximal cumulative hepatic uptake of bilirubin following infusion of bilirubin was found to be deficient during the first 72 h of life, but its capacity was approximately four- to sixfold greater than the rate of endogenous bilirubin excretion (fig. 3). This suggests that hepatic uptake of bilirubin was

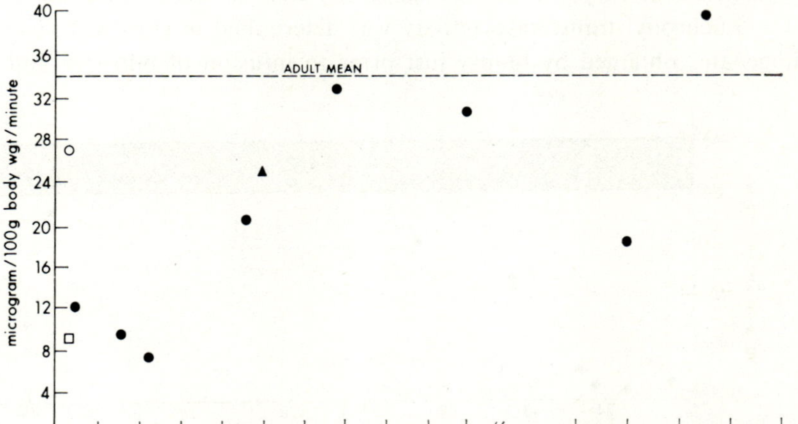

Fig. 3. Maximal mean cumulative hepatic uptake of bilirubin in mg/100 g body weight/min during i.v. infusion of unconjugated bilirubin in newborn *Macaca mulatta.* Ages of animals in hours at time of study are indicated on the abscissa. ● = Full-term; ▲ = premature; ○ = postmature; □ = full-term delivered by cesarean section. Dashed line indicates mean cumulative hepatic uptake of bilirubin in adult *Macaca mulatta* [2].

not the rate-limiting step in hepatic transfer of bilirubin in the newborn. This is confirmed by the observation that the concentration of unconjugated bilirubin in the liver at each point in the experiment was similar to that in the adult liver indicating that there is no significant impairment in the entrance of bilirubin into the hepatocyte at these relatively *high* serum concentrations of unconjugated bilirubin. Since uptake of bilirubin by the liver cell may be a process in which the rate of transfer is relative to the serum concentration of bilirubin, at much lower serum bilirubin concentrations uptake may partially restrict the entrance of bilirubin into the liver cell. This mechanism of deficiency of relative hepatic uptake of bilirubin may explain the low grade hyperbilirubinemia of phase II observed in the animals from 48 to 120 h of age. Preliminary analysis of the data indicates that the rate of transfer of bilirubin into the liver cell at lower serum bilirubin concentrations is less in newborns during phase II than in the adult animals. With higher serum bilirubin concentrations this difference is eliminated. ARIAS has demonstrated relatively poor hepatic uptake of sulfobromophthalein during this same period and correlated this poor dye uptake with deficiency of the hepatic cytoplasmic binding protein 'Y' [4]. Further study of the development of relative hepatic uptake is currently underway.

Of particular interest is the observation that maximal hepatic uptake in the postmature monkey at four h of age is very close to that of the adult (fig. 3).

Glucuronyl transferase activity was determined *in vitro* with liver homogenates obtained by biopsy just prior to infusion of bilirubin and was

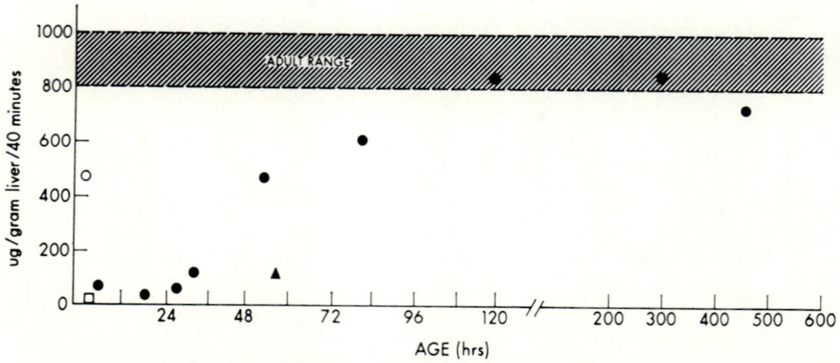

Fig. 4. Mean glucuronyl transferase activity in liver from newborn *Macaca mulatta* expressed as mg/g liver wet weight/40 min of incubation with optimal concentrations (13.2×10^{-2} M) of uridine diphosphoglucuronic acid (UDPGA) *in vitro*. Ages in hours at time of study are indicated on the abscissa. ● = Full-term; ▲ = premature; ○ = postmature; □ = full-term delivered by cesarean section. Hatched bar indicates range of hepatic glucuronyl transferase activity in adult *Macaca mulatta* [2].

markedly deficient during the first 24 h of life (fig. 4). After this period of time, enzyme activity gradually increased to reach adult levels by 120 h (fig. 4). If the units of activity are converted from μg of bilirubin glucuronide formed/g of liver/40 min to μg/100 g body weight/min, the units used for expression of the *in vivo* data, then the conjugating capacity of the adult is from 64 to 80 μg/100 g body weight/min, while that of the newborn is between 1 and 2 μg/100 g/min during the first 24 h of life. Such conversions must be viewed with caution but estimates of bilirubin glucuronide formed *in vivo* compare very well with the estimate from the *in vitro* data. Thus, whereas the rate of glucuronide conjugation is approximately 200 times greater than endogenous bilirubin excretion in the adult monkey, the rate of conjugation of bilirubin in the newborn during the first 24 h of age is *less* than endogenous bilirubin production, suggesting that deficient glucuronide conjugation *is* the rate-limiting step responsible for phase I of physiologic jaundice. Glucuronyl transferase activity is the rate-limiting step, however, only because the newborn animal has a markedly increased bilirubin load presented to the liver (fig. 2). It is hypothesized that if the neonate did not have an increased bilirubin load it would not have phase I of physiologic jaundice. Similarly, if glucuronyl transferase activity was normal, even with the increased bilirubin load, there would be no early hyperbilirubinemia. Thus, it is a combination of both factors, increased load and deficient conjugation, which appears to produce phase I of physiologic jaundice.

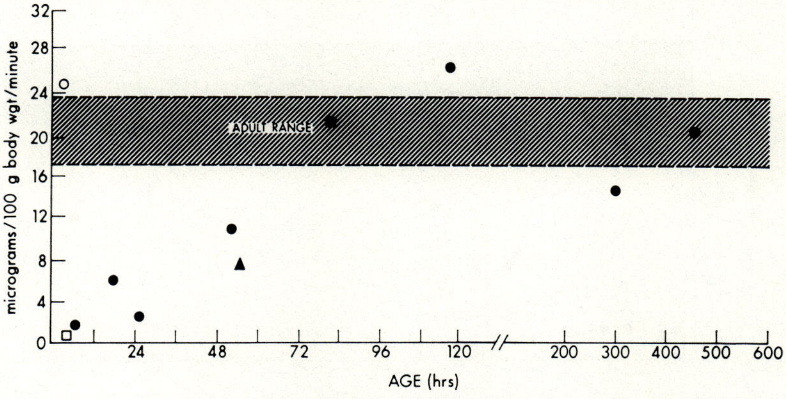

Fig. 5. Maximal mean hepatic excretion of bilirubin in bile by newborn *Macaca mulatta* during i. v. infusion of unconjugated bilirubin expressed as mg/100 g body weight/min. Ages of animals in hours at time of study are indicated on the abscissa. ● = Full-term; ▲ = premature; ○ = postmature; □ = full-term delivered by cesarean section. Hatched bar indicates range of maximal hepatic exretion of bilirubin in adult *Macaca mulatta* [2].

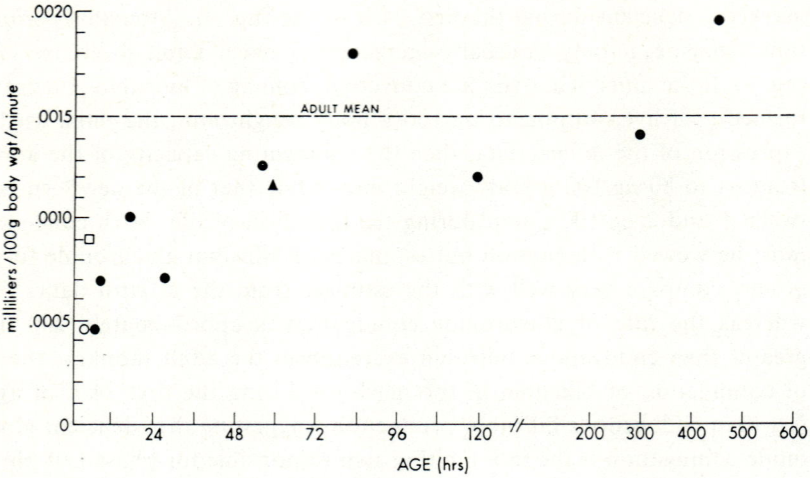

Fig. 6. Mean bile volume secreted by newborn *Macaca mulatta* during one-hour period of i.v. saline infusion expressed as ml/100 g body weight/min. Ages of animals in hours at time of study are indicated on the abscissa. ● = Full-term; ▲ = premature; ○ = postmature; □ = full-term delivered by cesarean section. Dashed line indicates the mean volume of bile secreted by adult *Macaca mulatta* [2].

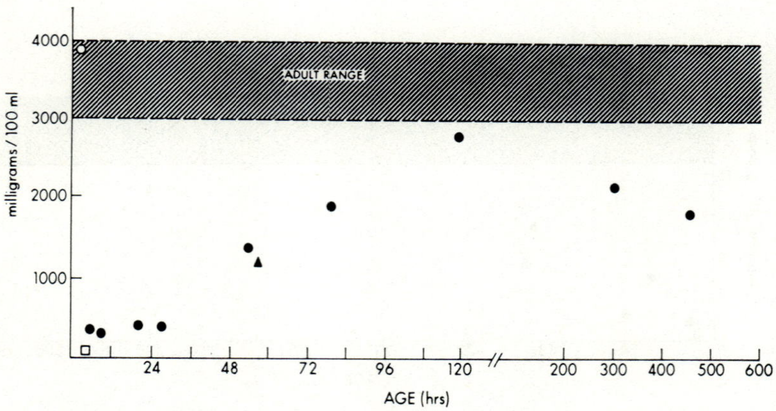

Fig. 7. Mean maximal concentrations of bilirubin in bile of newborn *Macaca mulatta* during i. v. infusion of unconjugated bilirubin expressed as mg/100 ml of bile. Ages of animals in hours at time of study are indicated on the abscissa. ● = Full-term; ▲ = premature; ○ = postmature; □ = full-term delievered by cesarean section. Hatched bar indicates range of maximal bile bilirubin concentrations in adult *Macaca mulatta* [2].

Of note again is the observation that the postmature monkey had glucuronyl transferase activity at four h of life which was equivalent to that of animals three days old (fig. 4). In contrast, the premature animal was deficient as compared to term animals of comparable age (fig. 4).

Maximal hepatic excretion was also markedly deficient during the first 24 h of life and rose progressively during the next three days to reach adult levels by 96 h (fig. 5). Excretion of conjugated bilirubin by the liver cell is not the rate-limiting step during the first 36 h of the newborn period. This conclusion is supported by the observation that the pigment which accumulated in the liver and in the serum during infusion of animals less than 36 h of age was all of the unconjugated or indirect-reacting type. Presumably all bilirubin which was conjugated was subsequently excreted without retention in the liver or circulation.

Of note once again is the markedly accelerated maturation of excretion in the four-hour-old postmature monkey (fig. 5).

Bile flow in the newborn monkey was only modestly restricted in comparison with the marked limitation noted in the other functions described during the newborn period (fig. 6). Bile flow in the postmature monkey was comparable to that of monkeys delivered at term in contrast with the marked acceleration of other functional steps (fig. 6).

The capacity of the newborn monkeys to concentrate bilirubin in bile was markedly restricted during the first 24 h of life, perhaps reflecting the defects in uptake and conjugation, or resulting from a specific defect in excretory transport of bilirubin by the hepatocyte (fig. 7).

This restriction in capacity to concentrate bilirubin indicates that the restriction in cellular excretion of bilirubin is not simply due to reduced bile volume. The postmature animal achieved a bilirubin concentration in bile equivalent to that of the adult animals (fig. 7).

Discussion

These studies have led us to conclude that physiologic jaundice of the newborn rhesus monkey and, probably that of the human newborn, results not simply from a single enzymatic defect in the conjugation of bilirubin, but rather from a complex interaction of many functional disabilities. Furthermore, physiologic jaundice of the rhesus monkey is characterized by two distinct phases. During the first phase the serum bilirubin concentrations rise rapidly to a peak of 4–5 mg/100 ml of serum by 24 h of age and rapidly decline to approximately 1.0 mg/100 ml by 36–48 h of age. Phase II begins at the end of the second day of life and persists until approximately the fifth day of life; it is characterized by a low grade hyperbilirubinemia of about 1.0 mg/100 ml of serum. Only after the fifth day of life does the serum bilirubin return to the normal value of approximately 0.1–0.2 mg/100 ml. Examination of daily serum bilirubin concentrations in full-term human neonates has revealed the same, and previously unrecognized pattern, but with each period of time lengthened approximately twofold.

From these studies in the newborn rhesus monkey we have drawn the tentative conclusions that the first phase of physiologic jaundice results primarily from a combination of markedly increased bilirubin load and severe deficiency of glucuronyl transferase activity with enzyme activity the rate-limiting step in the hepatic transport of bilirubin from blood to bile. The second phase is probably the result of the persisting exaggerated load of bilirubin superimposed on a *relative* deficiency in hepatic uptake of bilirubin. After the end of the first week of life, increased bilirubin production or re-absorbtion of bilirubin persists but the hepatic mechanisms have matured sufficiently to prevent further accumulation of bilirubin in the circulation.

The control of maturation of each of these functional steps remains to be investigated.

Of special note is the observation that hepatic uptake, conjugation and excretion were all at or near the adult level in a four-hour-old monkey delievered two weeks after expected delivery date, suggesting strongly that each of these mechanisms can mature while the fetus remains in the maternal-uterine environment. Thus, it is not the process of birth itself which leads to maturation of these hepatic functions, but rather some alteration of the biochemical environment. This, in turn, suggests that pharmacologic tools may be of great use in treatment or prevention of neonatal jaundice.

References

1 GARTNER, L. M. and ARIAS, I. M.: The transfer of bilirubin from blood to bile in the neonatal guinea pig. Pediat. Res. *3:* 171–180 (1969).

2 GARTNER, L. M.; LANE, D. L., and CORNELIUS, C. E.: Bilirubin transport by liver in adult *Macaca mulatta*. Amer. J. Physiol. *220:* 1528–1535 (1971).

3 GARTNER, L. M.; SNYDER, R. N.; CHABON, R. S., and BERNSTEIN, J.: Kernicterus: high incidence in premature infants with low serum bilirubin concentrations. Pediatrics *45:* 906–917 (1970).

4 LEVI, A. J.; GATMAITAN, Z., and ARIAS, I. M.: Deficiency of hepatic organic anion

binding protein, impaired organic anion uptake by liver and 'physiologic' jaundice in newborn monkey. New Engl. J. Med. *283:* 1136 (1970).

5 LUCEY, J. F.; BEHRMAN, R. E., and WARSHAW, A. L.: 'Physiologic' jaundice in newborn rhesus monkey. Amer. J. Dis. Child. *106:* 350 (1963).

6 SCHMID, R.: The identification of 'direct-reacting' bilirubin as bilirubin glucuronide. J. biol. Chem. *229:* 881–888 (1957).

Authors' address: Dr. LAWRENCE M. GARTNER, MD and Miss DONNA L. LANE, Department of Pediatrics and the Rose F. Kennedy Center for Research in Mental Retardation and Human Development, Albert Einstein College of Medicine, *Bronx, NY 10461* (USA)

Medical Primatology 1972. Proc. 3rd Conf. exp. Med. Surg. Primates, Lyon 1972, part I, pp. 248–254 (Karger, Basel 1972)

Lobular Architecture and Maternal Blood Supply of Primate Placentas

P. Gruenwald

Department of Pathology, Hahnemann Medical College and Hospital, Philadelphia, Pa.

The lobular architecture of hemochorial primate placentas and its relationship to uteroplacental vessels has been investigated in these four groups: *Tarsiidae,* the only prosimiae with hemochorial placentas; *Ceboidea,* the New World monkeys; *Cercopithecoidea,* the Old World monkeys exclusive of Hominoidea; and *Hominoidea,* the apes and man. Each of these groups is sufficiently uniform in placental structure and development to warrant a common description based on one species studied in detail, and confirmatory observations in others. The species examined are listed in table I.

The lobule is the structural unit of the placenta. It consists of an approximately spherical mass of densely arranged villi which contrasts with the loose interlobular areas. In *Cercopithecoidea* and *Hominoidea* the lobule may have a loose, even almost villus-free center. The lobule may or may not occupy an entire cotyledon as seen on the maternal surface of the separated placenta; several lobules may be partially fused. The lobular pattern is intimately related to the distribution of ostia of maternal arteries and veins. Venous drainage is generally at the periphery of the lobules. Uteroplacental arteries show characteristic differences between the four groups under consideration.

The relationship of the placenta to maternal vessels is intimately tied in with the manner in which the placental site expands after implantation [3]. In *Tarsiidae* no expansion of the placental site occurs. The attachment of the placenta remains narrow, and the growing placenta extends beyond it in the form of a button. In the other, multilobular forms the expanding placenta carries with it the superficial portions of the decidua. In *Ceboidea* the placental site expands shortly after implantation by additional sprouts of trophoblast invading the endometrium in an area surrounding the original site of implantation. The arteries tapped early in the enlarging area of implantation are carried along by the expanding periphery of the placenta; their number is about 10–15 in *Saimiri*. In both *Tarsiidae* and *Ceboidea* the trophoblast invades the endometrium without destroying it completely, and thus grows around maternal vessels. The arterial stems and some of their branches thus come to extend deep into the fetal placenta. Eventually the maternal tissue of the wall is destroyed while fibrin is deposited. This results in the formation of branched, fibrin-lined channels with innumerable openings into the intervillous space, extending from the artery in the decidua into the central parts of the lobules. No additional arteries are tapped, as shown by the continuation of all arteries into fibrin-lined channels.

Table I. List of species studied

Tarsiidae
Tarsius sp.
Ceboidea
Saimiri sciureus (squirrel monkey)
Cebus apella
macrocephalus
capucinus
gracilis
Saguinus nigrioullis
Ateles sp. (spider monkey)
Alouatta senicula (howler monkey)
palliata
Cercopithacoidea
Macaca mulatta (rhesus monkey)
fascicularis (irus)
nemestrina
arctoides (speciosa)
Cercopithecus aethiops (African green monkey)
Erythrocebus patas
Papio sp. (baboon)
Theropithecus gelada
Presbytis cristatus (Pithecus pyrrhus)
Colobus caudatus
Cercopithecus talapoin
Hominoidea
Hylobates sp. (gibbon)
Pongo pygmaeus (orangutan)
Pan troglodytes (chimpanzee)
Gorilla, gorilla
Homo sapiens (man)

In *Cercopithecoidea* the periphery of the expanding placental disc carries with it coiled arterioles which did not originally open into the intervillous space; these are tapped later on. Even though the mode of development differs, the result in both *Ceboidea* and *Cercopithecoidea* is that the peripheral portions of the disc are supplied by arteries coming from the vicinity of the original implantation site, and the numbers of arteries are also similar.

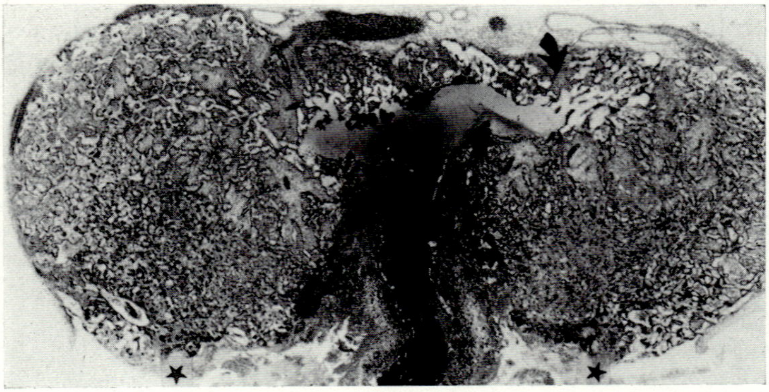

1

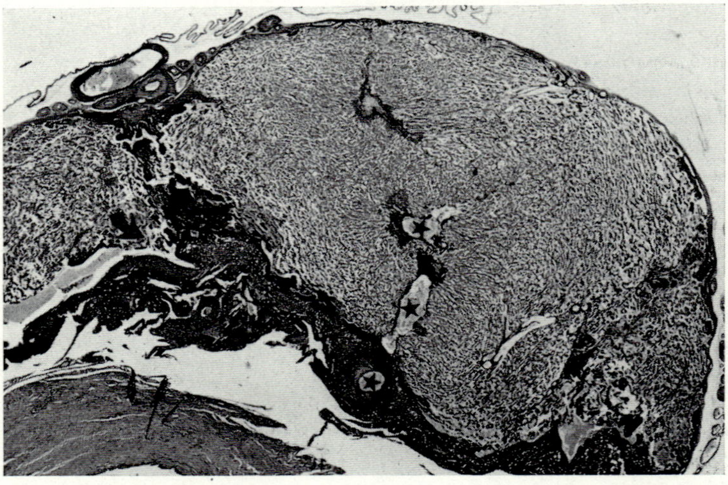

2

Fig. 1. Placental disc of an 8.1-mm *Tarsius* embryo. A large artery enters the disc from below; its branches open into the intervillous space, as is well seen near the arrow. The area of attachment to the uterus is between the asterisks. × 17. Reproduced from GRUENWALD [3] with the publisher's permission.

Fig. 2. Part of a placental disc of a near-term fetus of *Saimiri sciureus*. A maternal artery and the continuing channels entering the center of the lobule are marked*. A venous drainage area to the left of this lobule is marked by an arrow. × 8.

Hominoidea have, owing to deep implantation, a chorion frondosum and decidua completely surrounding the ovum. As the ovum grows, these layers apply themselves to the decidua parietalis, and fuse with it. Then, after reduction of the placenta to a circumscribed disc, deeper invasion and destruction of the fused decidua by trophoblast occurs until arteries are tapped in a large area extending far beyond the vicinity of the implantation site. Thus, 200–400 arteries from the areas supplied by both uterine arteries eventually supply the intervillous space [3]. This new feature of *Hominoidea* goes, phylogenetically speaking, beyond anything achieved in the development of the maternal arterial supply of either ceboid or cercopithecoid placentas.

In its mature form, the placental disc of *Tarsius* (fig. 1) consists of a single lobule with one maternal artery entering at the center, and continuing into fibrin-lined channels within the central portion of the lobule. Many small venous channels lined by fibrin, which may be derived from maternal veins, leave the placenta at the border of its attachment to the endometrium. In *Ceboidea* (fig. 2), each of the cotyledons contains several incompletely separated lobules. Arterial channels similar to those of *Tarsiidae* may extend within the placenta from one lobule to the next. Venous areas are distinct particularly at the borders of the cotyledons, and to a lesser extent between lobules within cotyledons. They consist of wider spaces in the intervillous space with few terminal villi. Uteroplacental veins originate at the base of these interlobular areas.

Cercopithecoidea have large lobules; in the rhesus monkey most of them occupy entire cotyledons. The uteroplacental arteries have a convolution below interlobular areas, and from there take a straight course to an opening below the center of a lobule. Some lobules have a loosely textured center in which the intervillous space is nearly free of villi. This is separated from the subjacent arterial ostium by a dense portion of the lobule. Venous drainage areas are again interlobular (fig. 3). The larger ones have wide spaces of free intervillous space traversed by big intervillous stems coated with fibrin. The anchoring portions of villous stems are more substantial than in *Ceboidea;* they are distributed nearly uniformly, with somewhat larger numbers in the interlobular areas, in both groups.

Hominoidea are represented by the human placenta which was previously studied [2]. There is great variation in the extent to which cotyledons are demarcated on the maternal surface of the separated placenta. The lobules are very numerous, about 200 according to CRAWFORD [1], and many are partially fused. They frequently have a loose center. Villous stems are distributed mostly along the circumference of lobules, and their anchoring points are accordingly at the base of interlobular areas, often associated with elevations of the decidua. Arterial coils are also below interlobular areas, and their openings are right there, frequently near large anchoring villi (fig. 4). They may be in close proximity of venous ostia. Information on apes is less complete, and agrees with findings in man.

The relationship of lobules to the ostia of uteroplacental vessels was confirmed by graphic reconstruction in specimens of all four groups.

Circulation in the intervillous space is clearly defined in *Tarsiidae* and *Ceboidea* where arterial channels distribute maternal blood to the central portions of the lobules, and veins drain the periphery. In *Cercopithecoidea* and *Hominoidea* the loose lobular centers seen more or less consistently, have been assumed to receive the bulk of maternal arterial blood, and distribute it to the lobules in a centrifugal manner. Injection studies have not substantiated this beyond doubt. There is no wide open communication from arterial ostia to the centers. In man, arterial blood appears to rise in the interlobular areas in the direction of the chorionic plate. Radiographic evidence in man and the rhesus monkey is inconclusive. The present findings in man agree best with the concept of LEMTIS [5] according to which maternal blood moves along the periphery of the lobules in the form of a shell, and from there enters the lobule along most of its circumference.

The following summary will serve to suggest the manner in which primates with hemochorial placentas may be related phylogenetically. *Tarsiidae* with their single-lobule placental disc, and no mechanism for expansion of its attachment or maternal blood supply, are obviously the simplest form. *Ceboidea* share with them the presence of arterial channels in the central portions of the lobules, but there is expansion of the placental site by means of additional trophoblastic invasion of the endometrium in early stages. *Cercopithecoidea,* on the other hand, resemble in their early development the tarsioid condition by having a sharply outlined placenta with a few arteries near the center [6]. This group expands the placental site by growth of the original disc, and recruitment of additional arteries which have stretched in the decidua to follow the growing peripheral parts of the disc from their base near the implantation site. There are no arterial channels within the placenta. In *Hominoidea* deep invasion of the decidua in an area far exceeding the immediate vicinity of the implantation site, occurs after the placenta has reached its final size relative to the uterus. This is a new development which makes possible a more extensive maternal blood supply, involving (in man) several hundred arterial openings from both uterine arteries. Thus, development and structure of the placenta suggest the form

3

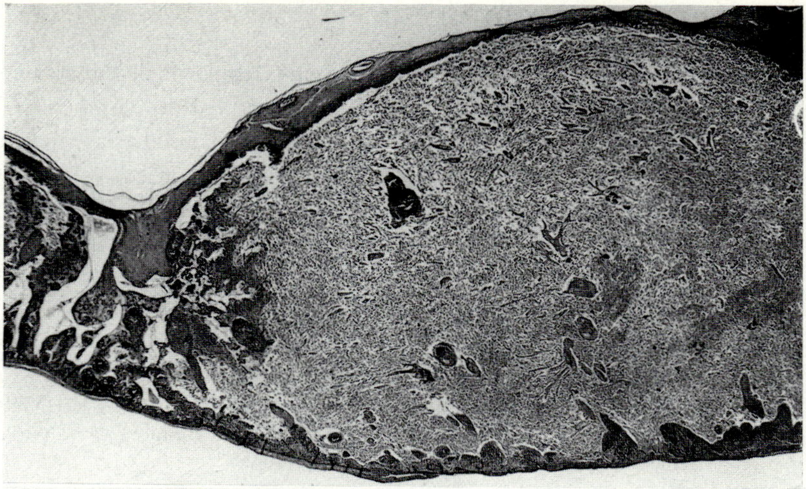

4

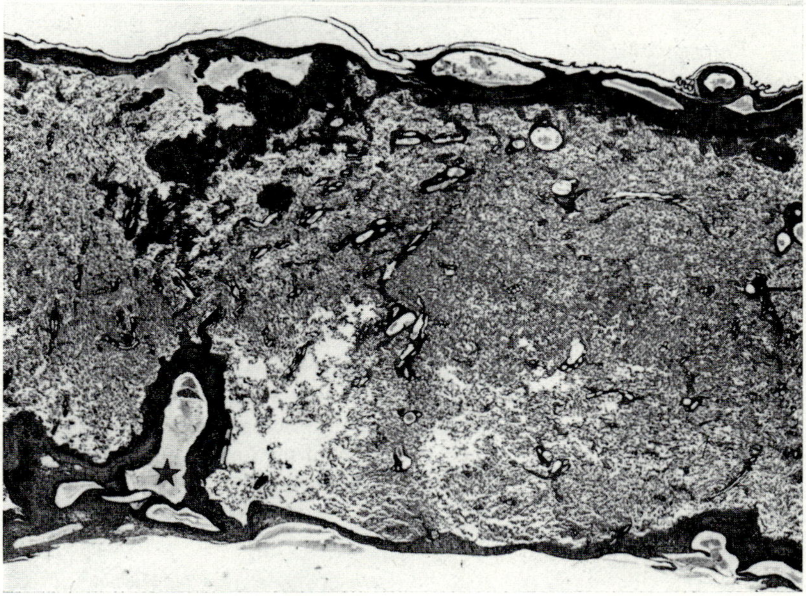

Fig. 3. The greater part of a lobule of the rhesus monkey is seen on the right, and a venous drainage area at a thin portion between lobules is shown on the left. Prominent anchoring villi are distributed over the entire base (bottom of figure) × 5. Reproduced with permission from P. GRUENWALD The Lobular Architecture of the Human and Rhesus Placenta; in Comparative Aspects of Reproductive Failure (Springer, New York 1967).

Fig. 4. A lobule of a human placenta at term is seen in the right half of the field. At the base of the adjacent interlobular area a convoluted artery (*) extends into an elevation of the decidua with anchoring villi. Adjacent sections show the opening of the artery in this interlobular area. × 8. Reproduced with permission from GRUENWALD [2].

now seen in *Tarsius* as the most primitive one, from which the ceboid and the cercopithecoid forms have originated by divergent development. *Hominoidea* appear to be related to the cercopithecoid form, and to have developed beyond it with the advantages mentioned above. This agrees with previous phylogenetic considerations based on placental structure [4, 7].

Acknowledgements

Many institutions and colleagues have made this work possible by supplying material. Some phases of the study were supported by grants from USPHS, NIH, NICHD (No. HD 00547), and the Association for the Aid of Crippled Children, New York.

References

1 CRAWFORD, J. M.: Vascular anatomy of the human placenta. Amer. J. Obstet. Gynec. *84:* 1543–1567 (1962).
2 GRUENWALD, P.: The lobular architecture of the human placenta. Bull. Johns Hopk. Hosp. *119:* 172–190 (1966).
3 GRUENWALD, P. Expansion of placental site and maternal blood supply of primate placentas. Anat. Rec. *173:* 189–204 (1972).
4 HILL, J. P.: On the developmental history of the Primates. Philos. Trans. Royal Soc. London B *221:* 45–178 (1932).
5 LEMTIS, H.: Physiologie der Plazenta; in Fortschr. Geburtsh. Gynäk. *41:* 1–53 (1970).
6 RAMSEY, E. M.: The vascular pattern of the endometrium of the pregnant rhesus monkey *(Macaca mulatta)*. Contrib. Embryol. Carneg. Inst. *33:* 113–147 (1949).
7 WISLOCKI, G. B. and STREETER, G. L.: On the placentation of the macaque *(Macaca mulatta)*, from the time of implantation until the formation of the definitive placenta. Contrib. Embryol. Carneg. Inst. *27:* 1–66 (1938).

Author's address: PETER GRUENWALD, MD, Department of Pathology, Hahnemann Medical College and Hospital, *Philadelphia, PA 19102* (USA)

Medical Primatology 1972. Proc. 3rd Conf. exp. Med. Surg. Primates, Lyon 1972, part I, pp. 255–262 (Karger, Basel 1972)

Histophysiology of Ciliated Cells in Oviduct and Cervix Uteri of Macaques[1]

E. S. E. HAFEZ and H. KANAGAWA[2]

Departments of Gynecology-Obstetrics and Physiology, Wayne State University School of Medicine, Detroit, Mich.

Introduction

The gross anatomy of the cervix uteri has been described for several species of primates [ZUCKERMAN, 1930; SANDYS and ZUCKERMAN, 1938; CUADROS, 1971]. The histological characteristics and ultrastructural features have been described for the mammalian oviduct [HAFEZ and BLACK, 1968] and cervix uteri [HAFEZ, in press; HAFEZ and JASZCZAK, in press]. The activity of the kinocilia [FAWCETT, 1961] and the production of luminal fluids [HAMNER and FOX, 1969] from the secretory granules play a major role in the transport and nutrition of gametes in the female reproductive tract. In the oviduct of the rhesus monkey, ciliogenesis and deciliogenesis are under the control of ovarian hormones. Various cellular processes occur during ciliogenesis such as lengthening of the microvilli, elongation of the mitochondria, hypertrophy of the Golgi material, enlargement of the endoplasmic reticulum, and formation of polysomes from the ribosomes [BRENNER, 1969]. Little is known about such cellular processes in the cervix.

The purpose of this investigation was to observe the surface structure and ultrastructure of the cervix and oviduct of macaque using scanning electron microscopy, phase contrast and Nomarski-interference microscopy.

1 This investigation was supported in part by The Ford Foundation grant No. 710-0287 and grant No. HD 06234-01 from the National Institute of Child Health and Human Development.

2 Senior Postdoctoral Fellow, International Training Program in Physiology of Reproduction.

Materials and Methods

Specimens of cervix and oviduct, taken from four rhesus monkeys *(Macaca mulatta)*, eight crab-eating macaques *(Macaca fascicularis)* and four stump-tailed macaques *(Macaca arctoides)*, were fixed in formalin or Bouin's solution. Transverse and sagittal sections, 5–7 μm thickness, were stained with HE, Mallory's triple stain and modified Masson's trichrome stain.

Tissues from the surface were removed to sterile Petri dishes with normal physiological saline. Pieces of longitudinal surface tissue were removed with scissors and transferred to a glass slide with a drop of culture medium (TC 199). The pieces of tissue were identified with regards to the vaginal and uterine ends in order to observe the direction of the ciliary beat. The tissue was covered with a coverslip and sealed to prevent the tissue from drying. The specimens were observed under phase contrast microscope or recorded by 16-mm cinematography under Nomarski-interference microscope.

Table I. Frequency of ciliated cells in the cervical epithelium of female macaques

Species	% of ciliated cells		
	external os	mid cervix	internal os
Stump-tailed macaque *(M. arctoides)*	3	11	8
Crab-eating macaque *(M. fascicularis)*	5	18	18

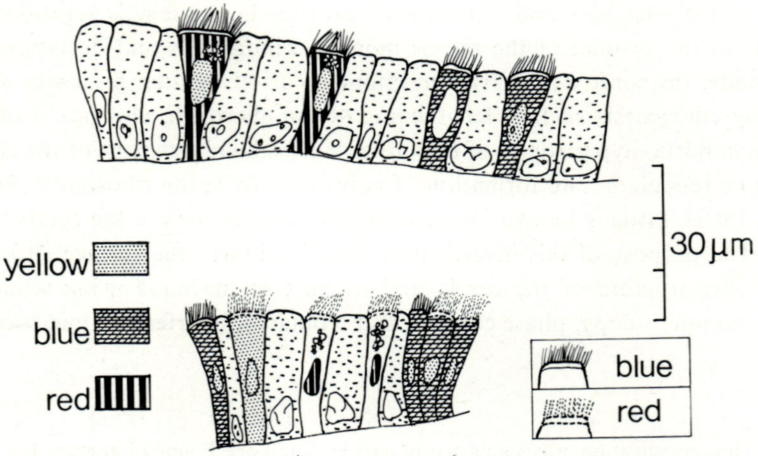

Fig. 1. Diagrammatic illustration of the cervical epithelium from crab-eating macaque. The ciliated cells in the cervix stained either blue or red with Mallory azan stain. It is not known whether these colors indicate several types of ciliated cells or different stages of one cell type and function.

For scanning electron microscopy, tissues from the cervix and oviduct were fixed in 6.5 % cacodylate-buffered glutaraldehyde for 24 h. The fixative was decanted and replaced with cacodylate buffered sucrose. After 24 h in sucrose, the specimens were postfixed for

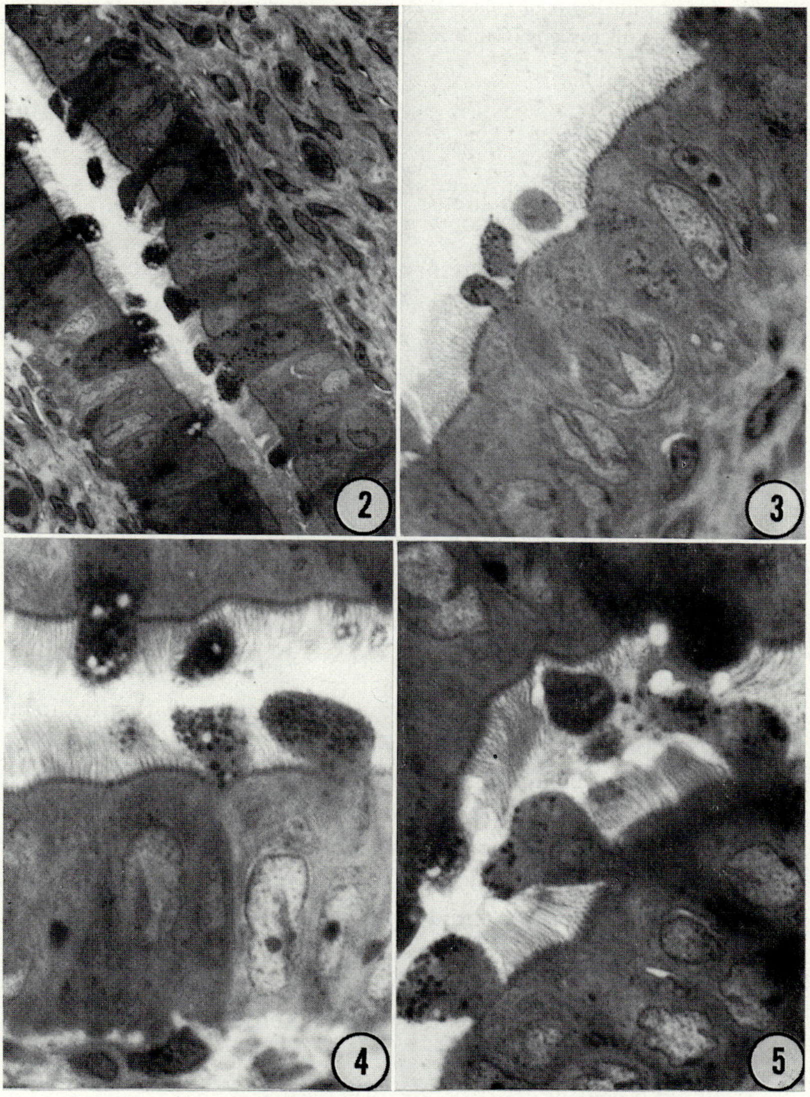

Fig. 2–5. Thin section preparations of oviductal epithelium of the macaque, stained with toluidine blue. Note the secretory granules released from the secretory cells. The ciliated cells, randomly distributed among secretory cells, bear kinocilia which may facilitate the release of secretory granules from adjacent secretory cells. *2* × 640; *3–5* × 1,600.

1 h in buffered 1.0% osmic acid. The tissues were dehydrated in graded concentrations of acetone from 30 to 100% and air dried [KANAGAWA *et al.*, in press], occasionally after fixation, the tissues were dehydrated in graded concentrations of alcohol and propylene oxide, and treated with maraglas mixture (Cleveland and Schneider, 1969). The samples were shadowed with 300 Å of gold-palladium *in vacuo* by rotation and then examined in scanning electron microscope (Cambridge Stereoscan Mark II).

Results

Histology of the Epithelium

The percentage of the secretory and ciliated cells varied within different species and different sections of the cervix (table I) and oviduct. The secretory cells were more abundant and contained many granules which crowded the nuclei toward the basement membrane. Some cells showed ruptured cell membrane allowing the secretory granules to pass into the lumen (fig. 2–5). Different species varied in the abundance of secretory granules in the cervical mucosa and cyclical changes in storage and secretory phase of epithelium. The cilia in the cervix and oviduct stained either blue or red with Mallory azan stain (fig. 1). Whether these colors indicate two types of ciliated cells or different stages of one cell type was not known.

Cilia Beat

The cilia beat rhythmically toward the vagina creating a directional flow of secretory material on the surface epithelium of the cervix and oviduct. Two types of cilia motility were recognized: an effective stroke and a recovery stroke. In the effective stroke, the cilia sweep rapidly in a stiff and slightly wavy motion. In the recovery stroke, the cilia bend near the basal body and degree of bending proceeds slowly as a wave toward the tip. Most motile sperm in the lumen of midcervix were swimming against the effective stroke of cilia beat. Some inactive sperm heads were attached to the cilia. In cultured tissue, cilia from the cervix and oviduct continued to be active for about ten days; however, on the fifth day fibroblasts started to grow out.

Scanning Electron Microscopic Structure

Ciliated cells were found singly or in groups. Kinocilia were overlapping over the surface of secretory cells. The nonciliated cells had a dome-shaped surface and were covered with microvilli. Secretory granules of varying size were observed on the surface of microvilli and kinocilia. The

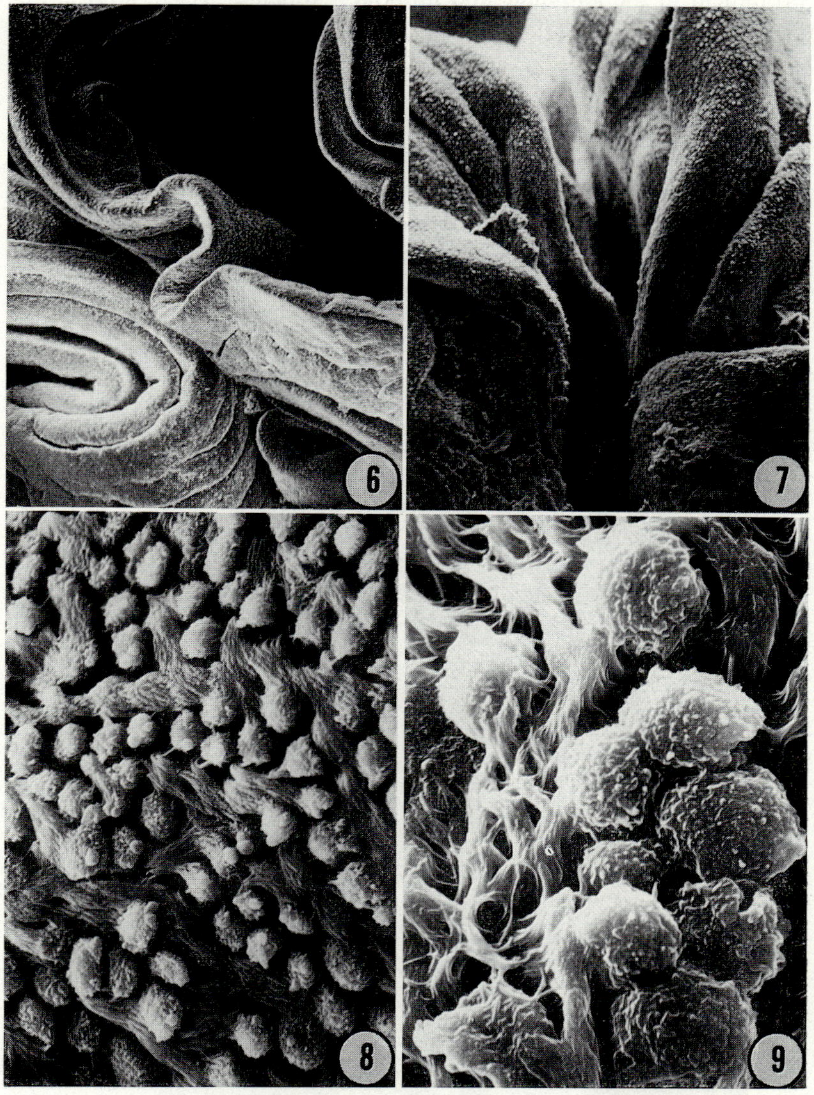

Fig. 6–9. Scanning electron micrographs of the fimbriae in the female macaque.
6 Bonnet macaque *(M. sinica)*. Note the complexity of mucosal folds. × 53.
7 Crab-eating macaque *(M. fascicularis)*. Note the fan-shaped folds. × 117.
8 Crab-eating macaque *(M. fascicularis)*. Note distribution of ciliated cells, × 1,175.
9 Bonnet macaque *(M. sinica)*. Note microvilli covering nonciliated cells. × 2,200.

luminal surface of the isthmus contained clusters of ciliated and nonciliated cells in approximately equal numbers, but the ciliated cells were more prominent in the ampullae and fimbriae. Nonciliated cells were prominent in the cervix (fig. 6–11).

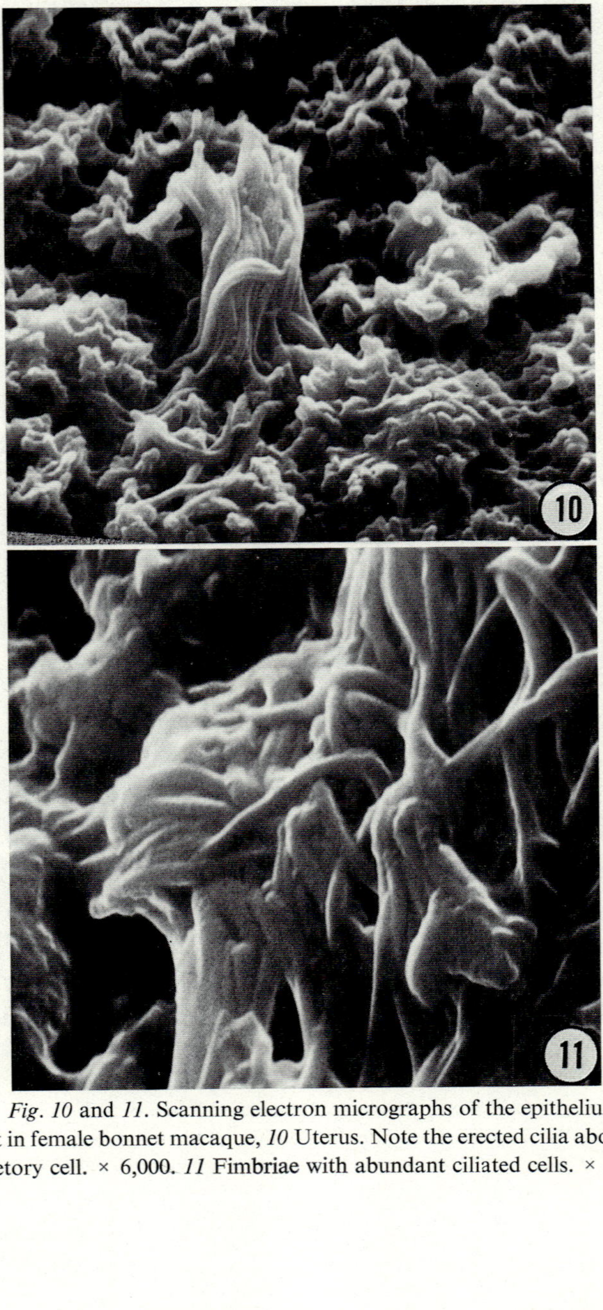

Fig. 10 and *11*. Scanning electron micrographs of the epithelium of the reproductive tract in female bonnet macaque, *10* Uterus. Note the erected cilia above the convex-shaped secretory cell. × 6,000. *11* Fimbriae with abundant ciliated cells. × 10,000.

Discussion

The percentage of ciliated and secretory cells varies in different species and in different segments of the female reproductive tract. The percentage of ciliated cells in the cervix is about three in man, 12 in rhesus, 15 in crab-eating and nine in stump-tailed macaque [HAFEZ and JASZCZAK, in press]. The physiological significance of these differences are not known.

Cilia may play an important role in the transport of particles and/or gametes and directing the flow of luminal fluids. The epithelial cilia of the fimbriae, oviduct and cervix as observed *in vitro,* beat toward the vagina. Motile spermatozoa in the cervical lumen swim against the direction of cilia beat. The kinocilia of the cervix may facilitate the release of the secretory granules adhering to the surface of adjacent secretory cells [HAFEZ and KANAGAWA, 1972]. Some relationship may exist between the amount of cervical mucus and the percentage of ciliated cells observed in different species. The smaller peg cells in the oviduct, found by transmission electron microscopy to be compressed between the larger ciliated and secretory cells [HASHIMOTO *et al.,* 1964] have been identified with a reserve precursor function [FERENEZY *et al.,* 1972]. It is possible that this is not so and that the peg cells represent exhausted secretory cells.

References

BRENNER, R. M.: The biology of oviduct cilia; in HAFEZ and BLANDAU The mammalian oviduct, pp. 203–229 (University of Chicago Press, Chigaco 1969).

CUADROS, A.: New findings relating to the gross and microscopic morphology of the uterine cervix in the rhesus monkey. Fertil. Steril. *22:* 138 (1971).

FAWCETT, D. W.: Cilia and flagella; in BRACHET and MIRSKEY The cell, vol. 2, pp. 217–297 (Academic Press, New York 1961).

FERENEZY, A.; RICHART, R. M.; AGATE, F. J. jr.; PURKERSON, M. L., and DEMPSEY, E. W.: Scanning electron microscopy of the human fallopian tube. Science *175:* 783–784 (1972).

HAFEZ, E. S. E.: Comparative anatomy of the cervix; in BLANDAU and MOGHISSI Biology of the cervix (University of Chicago Press, Chigaco, in press).

HAFEZ, E. S. E. and BLACK, D. L.: The mammalian uterotubal junction; in HAFEZ and BLANDAU The mammalian oviduct, pp. *85*–126 (University of Chicago Press, Chicago 1968).

HAFEZ, E. S. E. and JASZCZAK, S.: Comparative anatomy and histology of the cervix uteri in non-human primates. Primates (in press).

HAFEZ, E. S. E. and KANAGAWA, H.: Ciliated epithelium in the uteri cervix of the macaque and rabbit. J. Reprod. Fertil. *28:* 91–94 (1972).

Hamner, C. E. and Fox, S. B.: Biochemistry of oviductal secretion: in Hafez and Blandau The mammalian oviduct (University of Chicago Press, Chicago 1969).

Hashimoto, M.; Shimoyama, T.; Kosaka, M.; Komori, A.; Hirasawa, T.; Yokoyama, Y.; Kawase, N., and Nakamura, T.: Electron microscopic studies on the epithelial cells of the human fallopian tube (report II). J. Jap. Obstet. Gynaec. Soc. *11:* 92–100 (1964).

Kanagawa, H.; Hafez, E. S. E.; Beachler, C. A.; Pitchford, W. C., and Barnhart, M. I.: Improved methodology for scanning electron microscopy of the female reproductive tract. Int. J. Fert. (in press).

Sandys, D. C. and Zuckerman, S.: Observations on the cervix uteri and the urethra of monkeys. J. Anat., Lond. *72:* 352 (1938).

Zuckerman, S.: The menstrual cycle of the primates. I. General nature and homology. Proc. zool. Soc., Lond. *391:* 754 (1930).

Authors' address: Dr. E. S. E. Hafez and Dr. H. Kanagawa, Departments of Gynecology-Obstetrics and Physiology, Wayne State University School of Medicine, *Detroit, MI 48201* (USA)

Medical Primatology 1972. Proc. 3rd Conf. exp. Med. Surg. Primates, Lyon 1972, part I, pp. 263–270 (Karger, Basel 1972)

The Cervix Uteri and Sperm Transport in Female Macaques[1]

S. Jaszczak[2] and E.S.E. Hafez

Departments of Gynecology-Obstetrics and Physiology, Wayne State University School of Medicine, Detroit, Mich.

The uterine cervix serves a variety of functions in the reproductive process. Failures in the function of the cervix will result in sterility or lowered fertility [Moghissi, 1972]. The cervix is a dynamic organ with cyclic changes in its anatomy, biochemistry and physiology as a result of ovarian hormones. Throughout most of the reproductive cycle the cervix and its secretion serve as a biologic barrier against microorganism. Near ovulation the cervix facilitates sperm migration to the uterus. Basic concepts on the mechanisms of sperm migration through the human cervix are emerging from animal studies and are not confirmed in man. Quantitative and qualitative differences exist among various species of nonhuman primates in the length, anatomy and diameter of the cervical canal, morphological structure of the cervical mucosa and relative percentage of the secretory and ciliated cells [El banna and Hafez, 1972; Hafez and Jaszcak, in press; Hafez, in press].

The present study was designated to show similarities and dissimilarities in the gross anatomy of the cervix, histological characteristics of cervical epithelium and sperm migration in macaques.

Material and Methods

Four adult, female rhesus monkeys *(Macaca mulatta)*, nine crab-eating macaques *(M. fascicularis)*, three pig-tailed macaques *(M. nemestrina)* and four stump-tailed maca-

1 This investigation was supported in part by The Ford Foundation grant No. 710–0287 and grant No. HD 06234 from the National Institute of Child Health and Human Development.

2 Senior Postdoctoral Fellow, International Training Program in Physiology of Reproduction.

ques *(M. arctoides)* were used to study the gross anatomy of the cervix and histological characteristics of cervical epithelium.

Four adult crab-eating macaques and one rhesus monkey were used to study the patterns of sperm migration in the cervix uteri. The onset of menstrual bleeding was considered as day 1 of the cycle. The menstrual cycle was also checked by examination of vaginal and cervical smears and from quantity, arborization and spinnbarkeit of cervical mucus. The monkeys were bred at midcycle as judged by maximal ferning and spinnbarkeit and quantity of mucus. Vaginal fluid was examined for the presence and motility of spermatozoa. Autopsy was performed at 1, 3, 6, 16 and 24 h *post coitum (p.c.)*. Sagittal and transverse sections at 10 μ were made from the cervix. Spermatozoal heads and nuclei of leukocytes were counted in the upper vagina, ectocervix, midcervix and endocervix. A microscopic counting square (0.144 mm^2 at 480 magnification) was used. Mean values from 75 to 150 sites taken at random in each section were counted.

Results

Differences exist in the gross anatomy and histological structure of the cervix uteri in the four species of macaques (fig. 1). In the rhesus monkey, crab-eating and pig-tailed macaques the cervical canal is thrown into several folds by colliculi extending from the cervical walls into the lumen. The main ventral cervical colliculus protrudes dorsally from the ventral wall of the cervix between the external and internal ora. The two colliculi which protrude from the dorsal wall of the cervix are much smaller than ventral colliculus, and in some females they are rudimentary or even absent. On the other hand, the dorsal cephalic colliculus in some specimens may be quite large. The dorsal caudal colliculus is often superimposed on the dorsal lip of the cervix. The ventral lip of the ectocervix is relatively thin, the dorsal lip is much thicker in consequence of superimposition of the caudal dorsal colliculus. The external os, which lies nearer the ventral fornix, is oval. The area surrounding the external os (2–5 mm) is covered by columnar epithelium resembling congenital erosion in the human cervix.

The sagittal sections reveal a remarkable thickness of the cervical mucosa which in some specimens of crab-eating macaque constitutes the majority of the entire wall of the cervix. There are remarkable differences in the thickness and degree of branching of cervical mucosa of different species as well as within specimens of the same species. Cervical mucosa contains large amounts of complex crypts filled with mucin (fig. 2). In the ectocervix, crypts with secondary and tertiary branches are directed toward the external os. The endocervix contains both short crypts and simple clefts and tunnels.

The histology of epithelium lining the cervical mucosa vary with the species and stage of the ovarian cycle. The percentage of ciliated cells in cervical columnar epithelium was about 12 in rhesus monkey, 15 in crab-eating macaque and 5–9 in stump-tailed macaque. The columnar epithelium cells rest directly on the fibrous stroma without distinctive submucosa. Ciliated cells, found singly or in groups, contain few cytoplasmic granules and well-developed kinocilia. In the rhesus monkey, pig-tailed and crab-eating macaques, the location of squamo-columnar junction varied among specimens; in some it was in the external os and in others it extended even to the vaginal fornix. Extensive variations in the length of transitional zone was observed between squamous and columnar epithelium.

In stump-tailed macaque, the cervix is relatively simple, there is no cervical colliculi and the cervical canal is relatively unobstructed, straight or slightly bent. The vaginal fornices and portio vaginalis are poorly developed. Mucosa of ectocervix and lower part of midcervix consists of deep, relatively simple crypts. The surface of these crypts are covered with columnar epithelium at the base, squamous or mixed squamo-columnar epithelium in the top of the crypts. The cervical mucosa of upper midcervix and endocervix con-

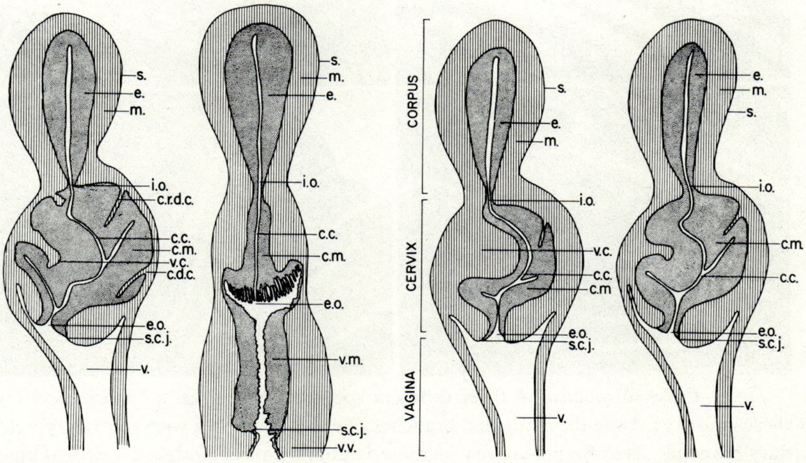

Fig. 1. Schematic drawings of the uterus of four species of macaques. Note similarities in the gross anatomical structure of the cervix in rhesus monkey *(M. mulatta)*, pig-tailed macaque *(M. nemestrina)*, crab-eating macaque *(M. fascicularis)* and different anatomical structure of the cervix and vagina in stump-tailed macaque *(M. arctoides)*. c.c. = Cervical canal; c.d.c. = caudal dorsal colliculus; c.m. = cervical mucosa; c.r.d.c. = cranial dorsal colliculus; e. = endometrium; e.o. = external os; i.o. = internal os; m. = myometrium; s. = serosa; s.c.j. = squamo-columnar junction; v. = vagina; v.m. = vaginal mucosa; v.v. = vaginal vestibule; v.c. = vaginal colliculus.

tains clefts and tubular tunnels of variable structure, length, width, direction and degree of branching.

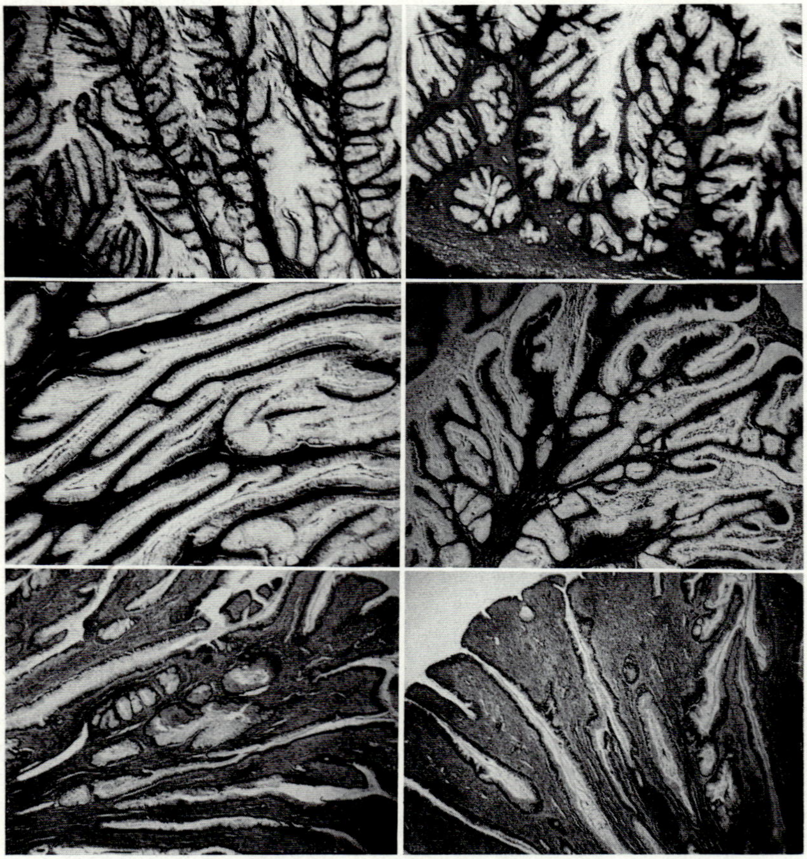

a b c d e f

Fig. 2. Cervical mucosa of three different species of macaque. *a* Ectocervical crypts of rhesus monkey. Note the elongated branches of cervical mucosa with secondary, seldom tertiary branches. Thin branches are usually separated from each other. *b* Cervical mucosa of rhesus monkey in the midcervix. Branches of cervical mucosa are usually thicker than in ectocervix. Secondary and tertiary branches are connected with each other forming multiple and complicated clefts and tunnels. *c* Ectocervical crypts of crab-eating macaque with thin, elongated branches. *d* Vaginal mucosa of stump-tailed macaque, between squamocolumnar junction at the vaginal vestibule and external os of cervix. *e* Cervical mucosa of stump-tailed macaque near the external os of the cervix, with complex crypts. *f* Ectocervical mucosa between external os and vaginal fornix. Note the deep crypts, high columnar epithelium at bottom of crypts, and the mixture of squamous and columnar cells. Mallory triple stain. × 45.

In stump-tailed macaque, the zone of transition between the stratified squamous epithelium and columnar epithelium is located near the vaginal vestibule about 1–3 cm from the external os of the cervix. From the squamo-columnar junction toward the cervix, the vaginal wall is covered with thick mucosa containing multiple crypts. The crypts extend from the squamo-columnar junction to midcervix, are filled with mucus. In this area the surface of the crypts is covered with columnar epithelium with a few ciliated cells. A mixture of columnar and squamous epithelium appear in the vaginal crypts. The transition from the predominantly fibrous tissue of the cervix to the muscular tissue of the corpus is usually abrupt, but in some specimen it is gradual. Near the internal os large numbers of muscle fibers extend transversally or obliquely to form a sphincter-like structure.

Sperm Migration in the Cervix

In female crab-eating macaques bred at midcycle and autopsied at 1, 3 and 6 h p.c., density of spermatozoa was maximal in the vagina near the external os and within the cervical canal of the ectocervix (fig. 3). The density

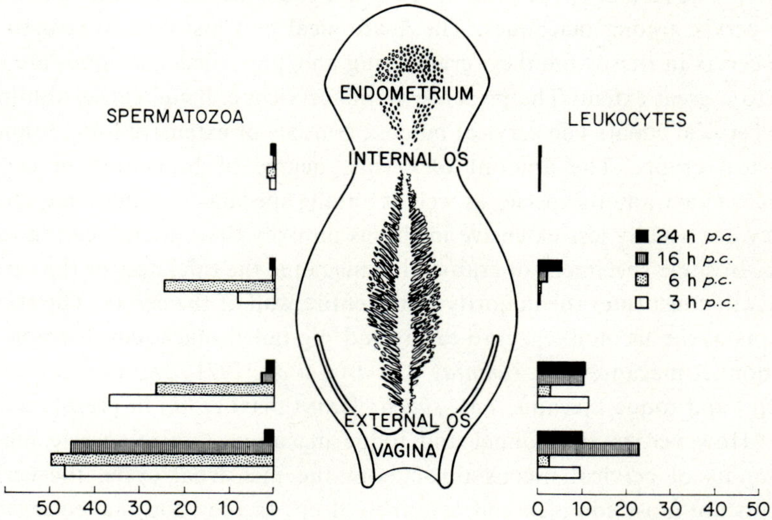

Fig. 3. Diagramatic presentation of the distribution of spermatozoa in different portions of the cervix 3, 6, 16 and 24 h *post coitum* in monkeys mated at the time of ovulation. Values represent the mean number of spermatozoa per one microscopi square (0.144 mm^2) of 10-μ histological sections.

of spermatozoa in the cervical canal decreased gradually toward the uterus. The number of spermatozoa in the endocervical canal was relatively small and did not vary from 1 to 6 h *p.c.*

In the ectocervical canal the number of spermatozoa decreased with the time after copulation. At 16 and 24 h *p.c.* the difference in the number of sperm between ectocervix and endocervix was minimal due to a decline of sperm number in the exocervical canal and not due to the increase of sperm number in the endocervix. Although spermatozoa appear to migrate at random, they were often densely crowded in strands of mucus in the midcervix and especially in the endocervix. Some strands of mucus containing spermatozoa disappeared into the crypts of cervical mucosa, when others continued through the internal os to the uterus. Some cervical crypts contained a large number of spermatozoa.

Discussion

Extensive differences exist in the anatomical and histological structure of the uterine cervix among nonhuman primates [HAFEZ and JASZCZAK, in press]. The results of the present study indicate variations in the structure of the cervix among macaques. The anatomical and histological structure of the cervix in rhesus monkey, crab-eating and pig-tailed macaques are similar to a great extent. The presence of the cervical colliculi causes folding of the cervical canal. The cervical mucosa consists of extensive long, branched cervical crypts. The amount of crypts, degree of branching of cervical mucosa vary among species as well as among specimens of the same species. They are usually less extensive in rhesus monkey than in crab-eating macaque. In some specimens of crab-eating macaque the thickness of the cervical mucosa constitutes the majority of the entire wall of the cervix. The cervical crypts of rhesus monkey, crab-eating and pig-tailed macaques resemble that of bonnet macaque *(M. radiata)* [OVADIA *et al.*, 1971; JAINUDEEN *et al.*, in prep.] and toque macaque *(M. sinica)* [JAINUDEEN *et al.*, in prep.].

However, in the bonnet and toque macaques, which secrete massive amounts of cervical mucus throughout the menstrual cycle, the cervical crypts are most complex and extremely deep. In the stump-tailed macaque the cervix is relatively simple, cervical colliculi are absent and cervical canal is unobstructed. Squamo-columnar junction is located near the vaginal vestibule, and the vaginal wall between the squamo-columnar junction and the cervix consists of large, branching crypts, covered with columnar epithelium. This also results in a relatively large surface area of mucus secreting epithelium in upper vagina. Cervical secretion at the time of ovulation is particularly favorable for sperm penetration [ODEBLAD, in press; JASZCZAK and HAFEZ, in press].

The complex structure of the uterine cervix in rhesus monkey and crab-eating macaque and cyclic changes in the quantity and physicochemical properties of the cervical mucus may indicate that the cervix plays an active role in sperm migration in these monkeys. Remarkable concentration gradients of spermatozoa between ectocervix and endocervix were evident. The ectocervix is the first and important barrier which prevents the migration of vast majority of spermatozoa toward the upper reproductive tract. Only a relatively small number of spermatozoa migrate through the ectocervix, even at the time of ovulation when the penetrability of cervical mucus is maximal. The role of the cervix in preventing the transport of abnormal spermatozoa to the site of fertilization has been suggested [BERGMAN, 1955; BOTELLA-LLUSIA, 1967; BEDFORD, 1970].

Spermatozoa are not uniformly distributed within the lumen of the cervical canal. Large numbers of spermatozoa accumulate in the cervical crypts. The role of the cervix as a reservoir, where spermatozoa are stored and released at a constant rate into the uterus was suggested in ruminants [GIBBONS and MATTNER, 1971; MATTNER, 1972; EL BANNA and HAFEZ, 1970] and rabbits [EL BANNA and HAFEZ, 1970; DOBROWOLSKI and HAFEZ, 1970]. Probably similar phenomenon occurs in primates.

Further studies are needed to relate possible species differences in physiological mechanisms of sperm transport to phenomena such as anatomical differences in the uterine cervix, copulatory behavior, coagulation and liquefaction of semen in the vaginal pool, and degree of leukocytosis in the uterus in response to copulation.

References

BEDFORD, J. M.: The saga of mammalian sperm from ejaculation to syngamy; in GIBIAN and PLOTZ Mammalian reproduction, pp. 124–182 (Springer, Heidelberg 1970).

BERGMAN, P.: Sperm migration and its relation to the morphology and motility of spermatozoa. Int. J. Fert. *1:* 45–54 (1955).

BOTELLA-LLUSIA, J.: Methods for determining the type and degree of spermatic motility; in WESTIN and WIQVIST Proc. 5th World Congr. Fert. Steril., Stockholm 1966. Int. Congress Series, vol. 133, pp. 636–644 (Excerpta Medica Foundation, Amsterdam 1967).

DOBROWOLSKI, W. and HAFEZ, E.S.E.: Transport and distribution of spermatozoa in the pseudopregnant rabbit. Amer. J. vet. Res. *31:* 2243–2246 (1970).

EL BANNA, A.A. and HAFEZ, E.S.E.: Sperm transport and distribution in rabbit and cattle female tract. Fertil. Steril. *21:* 534–540 (1970).

EL BANNA, A.A. and HAFEZ, E.S.E.: The uterine cervix in mammals. Amer. J. Obstet. Gynec. *112:* 145–164 (1972).

GIBBONS, R.A. and MATTNER, P.E.: The chemical and physical characteristics of the cervical secretion and its role in reproductive physiology; in SHERMAN Pathways to conception, 143–155 (Thomas, Springfield 1971).

HAFEZ, E.S.E.: Comparative anatomy of the cervix; in BLANDAU and MOGHISSI Biology of the cervix (University of Chicago Press, Chicago, in press).

HAFEZ, E.S.E. and JASZCZAK, S.: Histophysiology of the cervix in non-human primates. Primates (in press).

HAFEZ, E.S.E. and KANAGAWA, H.: Ciliated epithelium in the uterine cervix of the macaque and rabbit. J. Reprod. Fertil. *28:* 91–94 (1972).

JAINUDEEN, M.R.; HAFEZ, E.S.E., and KOTHARI, L.: Histomorphology of the cervix uteri of macaques (in preparation).

JASZCZAK, S. and HAFEZ, E.S.E.: Endocrine control of spermatozoal migration through the uterine cervix in macaque. Amer. J. Obstet. Gynec. (in press).

MATTNER, P.E.: The cervix and its secretion in relation to fertility in ruminants; in BLANDAU and MOGHISSI Biology of the cervix (University of Chicago Press, Chicago, in press).

MOGHISSI, K.S.: The function of the cervix in fertility. Fertil. Steril. *23:* 295–306 (1972).

ODEBLAD, E.: Types of cervical secretions, biophysical characteristics; in BLANDAU and MOGHISSI Biology of the cervix (University of Chicago Press, Chicago, in press).

OVADIA, J.; MCARTHUR, J.W.; KOPITO, L., and ULFELDER, H.: The cervical mucus secretion of the bonnet monkey *(M. radiata)*. Anatomical basis and physiological regulation. Biol. Reprod. *5:* 127–145 (1971).

Author's address: DR. S. JASZCZAK, Department of Gynecology-Obstetrics and Physiology, Wayne State University School of Medicine, *Detroit MI 48201* (USA)

Medical Primatology 1972. Proc. 3rd Conf. exp. Med. Surg. Primates, Lyon 1972, part I, pp. 271–287 (Karger, Basel 1972)

Ultrastructural Observations on Protein Transport in the Placenta of the Marmoset, *Saguinus oedipus*

W. P. JOLLIE

Department of Anatomy, Medical College of Virginia, Health Sciences Division of Virginia Commonwealth University, Richmond, Va.

Introduction

In contradistinction to the villous hemochorial organization of the hominid placenta, the placentas of lower primates show a progression from this condition, through intervillous fusions in Old World simians (e.g., the 'bonnet monkey' *Macaca* [WYNN, 1964]), to a pattern where syncytical connections between adjacent villi convert most of the placental mass to a pseudolabyrinth (e.g., the New World 'squirrel monkey', *Saimiri* [WYNN, 1964]) and, finally, to a condition where the placental mass is typically labyrinthine (e.g., the New World 'howling monkey', *Alouatta* [WISLOCKI, 1929], 'spider monkey', *Ateles* [WISLOCKI, 1930], and the marmoset, *Oedipomidas* – now called *Saguinus* – [WISLOCKI, 1932]). Consequently, a labyrinthine architecture can be considered to represent the most primitive anthropoid placental condition. It also characterizes the chorioallantoic placentas of lagomorphs, most rodents, many chiropterans, the hydrachoids, sirenians, and some insectivores [WIMSATT, 1962].

In his classic descriptions of the microscopic anatomy of platyrrhine placentas, WISLOCKI characterized the labyrinth as consisting of anastomosing trabeculae enclosing intertrabecular maternal blood sinuses; in the marmoset during midgestation, such trabeculae were described as 'stout, containing much (cellular) fetal mesoderm and clothed by a continuous layer of plasmoditrophoblast'; cytotrophoblast was said to be absent [WISLOCKI, 1932]. The major purpose in this early report, as well as in WISLOCKI's [1939, 1943] subsequent papers on marmoset fetal membranes, was not so much to describe the nature of the placental membrane, or barrier, within the labyrinth, as to determine the mode of origin of twinning in this genus and to characterize the unique phenomenon of erythropoiesis in the chorioallantoic placental mass, the discovery of which is credited to HILL in 1929 and later reported by him [HILL, 1932].

Neither twinning nor hemopoiesis was the concern of the present investigation. Our purpose has been to characterize the normal fine structure of the chorioallantoic placental membrane (i.e., the layers across which maternofetal exchange occurs) in the labyrinth of the cotton-top marmoset, *Saguinus oedipus*. Further, by examining placentas both early in pregnancy and near term, fine structural changes in the membrane with increasing gestational age in this species are reported. And finally, following an i.v. injection of an exogenous enzyme, viz., peroxidase, into pregnant marmosets and by subsequent cytochemical localization of an electron opaque reaction product, an attempt is made to correlate variations in the uptake and transport of this low molecular weight protein by the placental membrane with gestational age. It has been shown physiologically in several experimental animals [ANDERSON, 1959; reviewed by BRAMBELL, 1958 and by SCHULTZ, 1970], anatomically in some laboratory rodents [JOLLIE, 1968; KING and ENDERS, 1970, 1971; LAMBSON, 1966], and both clinically [DANCIS *et al.*, 1961] and anatomically [SATO, 1966] in humans that placental transport of protein is confined to a period of gestation immediately before term.

Materials and Methods

As a source of tissues, 16 marmosets, eight of each sex, were maintained over an 18-month period in air-conditioned quarters under a daily regimen of 10 h of darkness and 14 h of light. Animals were caged as breeding pairs in plastic dog cages (1 m^3), each cage with a separate nesting box (35 × 20 × 20 cm). They were provided high-protein Purina monkey chow and water *ad libitum*.

During the course of their maintenance, three females delivered spontaneously, in each case twins; none of the young survived. Because of the lack of success in raising young, a decision was made to undertake the present investigation by surgically terminating pregnancies.

The animals were restrained manually and palpated for uterine enlargement. Six females were autopsied. Of these, one was not pregnant and one was found to be in an implantation (i.e., a 'preplacental') stage. Two pregnant marmosets served as experimental animals; they were anesthetized with an intraperitoneal injection of 5 mg/g body weight sodium pentobarbital (Nembutal: Abbot Laboratories). Over a period of 1 min, 5 mg horseradish peroxidase (Type II, Sigma Chem. Co.) per 100 g body weight, dissolved in Locke's solution (20 mg/ml) was injected into the femoral vein. The uterus was exposed by a midventral abdominal incision; and the uterine wall was incised in the least vascular region to expose the fetal sacs. In each case (for these experimental animals, after a 5-min postinjection interval), both placentas (i.e. primary and secondary) were removed and cut into 1-cm^3 blocks. The blocks were transferred directly to fixative (for light microscopy: formol-sublimate-dichromate; for electron microscopy; 3% glutaraldehyde in 0.1 M phosphate buffer at pH 7.4). The crown-rump lengths (CRL) of the twin fetuses were recorded (for the two peroxidase-injected animals, 10 and 12 mm and 75 and 80 mm, respectively.

As normal controls (and to determine if endogenous peroxidatic activity was present in placental tissues) two pregnant animals were autopsied, and the tissues were processed without prior injection of peroxidase. The CRL of their twin fetuses were 22 and 24 mm and 38.5 and 41 mm, respectively.

For light microscopy, tissue blocks were fixed for 24 h, embedded in paraffin (Tissuemat), cut at 5 μm, and stained with Harris' hematoxylin and triple eosin, modified Masson's trichrome or Harris' hematoxylin-Alcian blue-PAS.

For electron microscopy, tissue blocks were minced in the fixative into 2-mm^3 blocks, fixed for 3 h at 4 °C, then washed in 0.1 M phosphate buffer. Half of the blocks were postfixed directly and subsequently processed as described below. From buffer the remaining tissue blocks were transferred to a Smith-Farquhar Tissue Sectioner, were chopped into 50-μm sections, and were incubated first in a medium containing a tris-HCl buffer at pH 7.6, plus 3,3'-diaminobenzidine, then in this solution containing 0.01% hydrogen peroxide after a modification [KRAEHENBUHL and CAMPICHE, 1969] of the cytochemical localization method of GRAHAM and KARNOVSKY [1966]. From the control series, tissues were similarly incubated to ascertain the possibility of endogenous peroxidatic activity. Subsequently, the incubated sections were washed, postfixed in 2% osmium tetroxide in 0.1 M phosphate buffer for 1½ h, rapidly dehydrated in ethanol and embedded in durcupan (Fluka). Thin durcupan sections were cut with a diamond knife in the range of 300–500 Å and mounted on uncovered 200 mesh copper grids. Some sections were viewed unstained; others were stained with lead citrate [REYNOLDS, 1963]. All sections were examined and photographed with an RCA EMU-3G electron microscope.

Adjacent 1- to 2-μm durcupan sections were cut, mounted on glass slides, stained with toluidine blue [LYNN *et al.*, 1966], and examined with a light microscope for orientation and for trimming the tissue sections before thin sections were cut.

Results

The Normal Placental Membrane at Increasing Gestational Age

In the earliest placental labyrinthine tissue which was examined (twin fetal CRL of 10 and 12 mm), the trabecular wall, i.e. the barrier across which maternofetal exchange takes place, was seen to consist of three cellular layers: two of trophoblast, and a layer of allantoic capillary endothelium (fig. 1–4). The nature of this membrane was discernible at the light microscopic level when 1- to 2-μm durcupan sections were so examined (fig. 1).

The outer layer of trophoblast (i.e. that which faced the maternal blood sinus) was syncytial and stained more deeply with toluidin blue than the inner

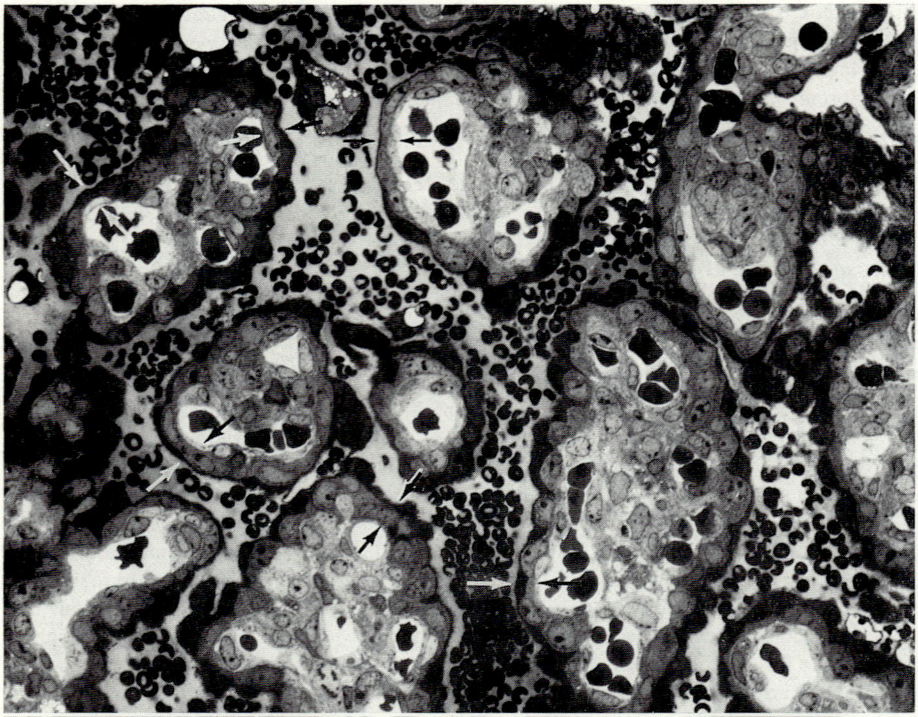

1

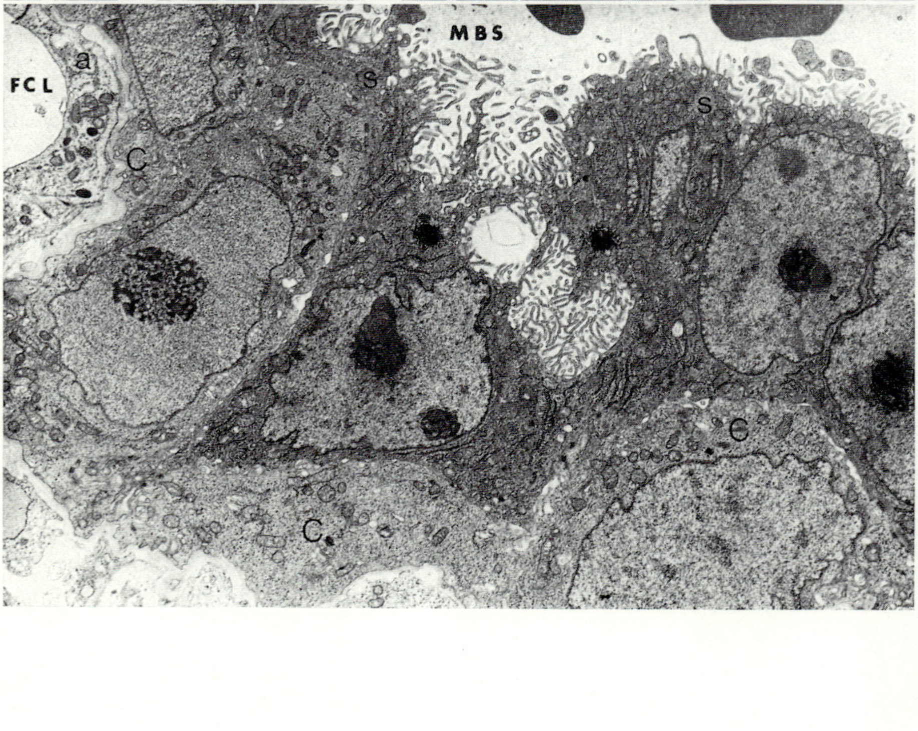

2

layer (fig. 1). With the electron microscope, such basophilia was seen to result from a greater abundance of granular reticulum in the outer syncytial layer (fig. 2 and 3). Often this outer layer was attenuated, so that the syncytial nuclei bulged into the maternal blood sinuses. These nuclei of the syncytiotrophoblast appeared somewhat smaller and often were more irregular in outline than those of the inner, cellular trophoblast; further, they showed both a greater amount of heterochromatin and a different nucleolar morphology (fig. 2). Lipid droplets, ranging from 800 to 900 nm in diameter, were prevalent (fig. 2). The outer plasma membrane of the syncytium showed densely arranged, slender, irregular microvilli and abundant evidence of micropinocytotic activity. Micropinocytotic vesicles frequently were identified throughout the thickness of this layer (fig. 2). The inner plasma membrane of the syncytium closely adhered to the cell membranes of the deeper individual cytotrophoblastic elements. It did not reach the adepithelial basement membrane. Frequently, desmosomes were identified, attaching the two trophoblastic layers together (fig. 3).

Cytotrophoblastic cells were large, relatively undifferentiated elements with comparatively little granular endoplasmic reticulum, but with a marked abundance of polyribosomes (fig. 2 and 3). Glycogen granules were dispersed throughout the cytoplasm (fig. 3). Lipid droplets were scarce, or absent; micropinocytotic vesicles were present (fig. 2 and 3). In these cells, the nucleus to cytoplasm volume ratio was consistently high; and the nuclei were vesicular, i.e. relatively electron lucid, with very little or no heterochromatin, and with fairly smooth contours (fig. 2–4). Cytotrophoblast cells were attached to each other by desmosomes like those which were seen between syncytio- and cytotrophoblasts (fig. 3). In all tissues which were examined, there was a well-developed basement membrane underlying the basal aspect of the cytotrophoblast cells. It followed the contours of these cells

---

Fig. 1. Photomicrograph of a 2-μm durcupan-embedded section of early labyrinthin. The trabeculae are cut in cross-section. Note the dark (basophilic) layer, i.e. syncytiotrophoblast, which covers the trabecular wall. At the earliest stage of pregnancy which was examined, this is consistently underlain by cytotrophoblast. The layers between the paired arrows constitute the placental membrane. × 500.

Fig. 2. A portion of the trabecular wall at the earliest stage of pregnancy. From maternal blood sinus (MBS) to fetal capillary lumen (FCL), three cytoplasmic layers constitute the placental membrane, viz., trophoblastic syncytium (s), cytotrophoblast (c), and allantoic capillary endothelium (a). Note the differences in cytoplasmic density and nuclear morphology between synctial trophoblast and the four cytotrophoblast cells which are illustrated. × 6,500.

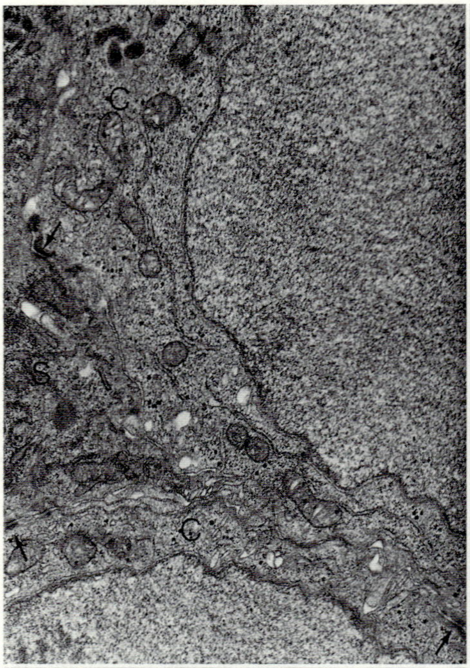

3

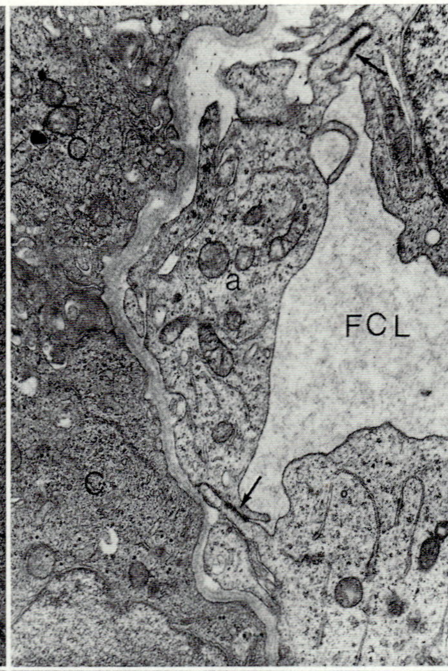

4

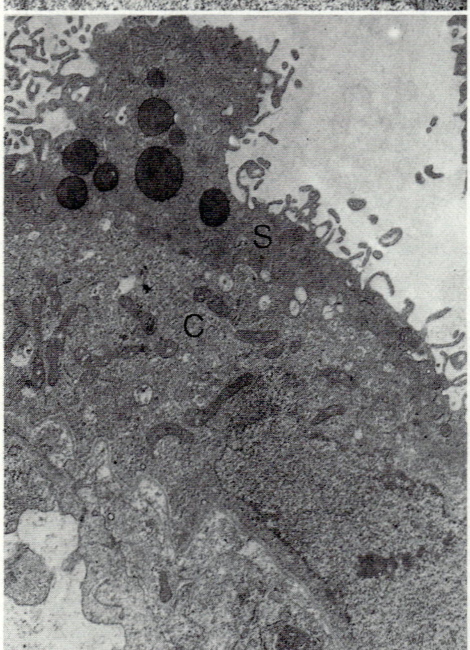

5

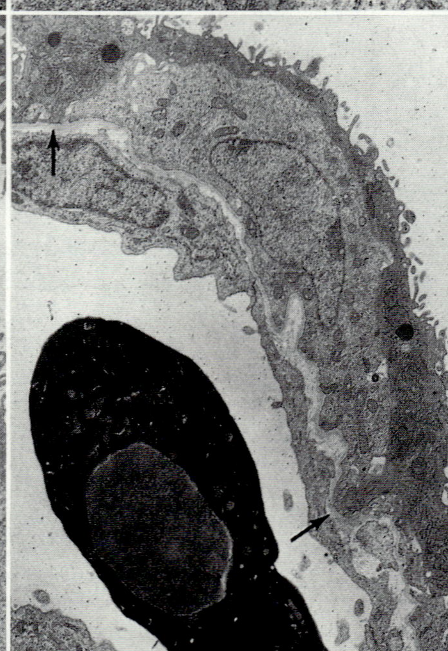

6

closely and was composed of both a wide lamina densa which measured 130 nm in thickness and a 40-nm lamina lucida, the latter separating the dense layer from the plasmalemmas of the cytotrophoblast cells (fig. 4). Occasionally two parallel dense layers were seen, situated between the cytotrophoblast and the endothelium; where these cellular layers separated, however, the basement membrane always followed the trophoblast (fig. 4).

The endothelium which delimited the allantoic capillaries was similar to unfenestrated endothelia which have been described in a great variety of tissues and organs elsewhere. Consistently it was considerably thinner than the cytotrophoblastic layer (fig. 2 and 4). Interendothelial junctional complexes were typical (i.e. gap-type junctions); and evidence for micropinocytosis was rare (fig. 4). At this early gestational age, an adendothelial basement membrane was very poorly developed or absent (fig. 2 and 4). In many places, however, the basement membrane of the chorionic epithelium was 'shared' with the endothelium (fig. 4).

At this stage, there was no evidence of hematopoietic foci in either primary or secondary placentas; furthermore, neither in general organization, nor in fine structure, were there discernible differences between the two placental masses from this single pregnancy.

In placental tissues which were autopsied later in gestation (two pregnancies, with twin fetal CRL of 22 and 24 mm and of 38.5 and 40 mm, respectively), hemopoietic foci were seen; for the present investigation, however, they were not studied. Again, primary and secondary placentas

---

Fig. 3. The interrelationship between two cytotrophoblast cells (c) and between these cells and the syncytiotrophoblast (s) at the earliest stage of pregnancy are shown. Note the desmosomes (arrows) by which these elements are attached. Also note that dispersed glycogen granules are confined to the cytotrophoblast. × 17,000.

Fig. 4. At the earliest stage of pregnancy, a well-developed basement membrane intervenes between the cytotrophoblast cells (c) and the endothelium of the allantoic capillary (a). Note that where these constituent layers separate, the basal lamina follows the trophoblast. Also note that gap-type junctions (arrows) seal off the fetal capillary lumen (FCL). × 17,000.

Fig. 5. A portion of the placental membrane at a stage near midpregnancy. Note that cyotrophoblast (c) is still present, and that the syncytium (s) is still lipid-rich, but appears attenuated. The microvilli at its free surface appear less densely arrayed than previously. × 6,800.

Fig. 6. The placental membrane from another midpregnancy. In places (arrows), the lipid-rich syncytiotrophoblast reaches the corionic epithelial basement membrane. In this section, which was incubated on substrate as a control (i.e. without prior injection of peroxidase), only the fetal erythroblast shows peroxidatic activity. × 5,000.

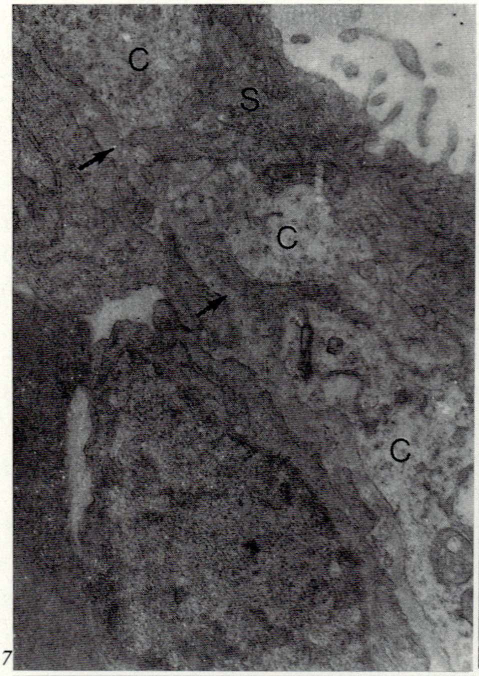

7

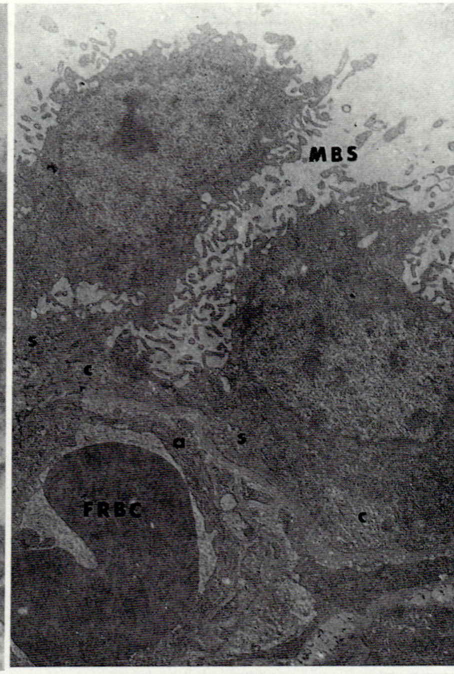

8

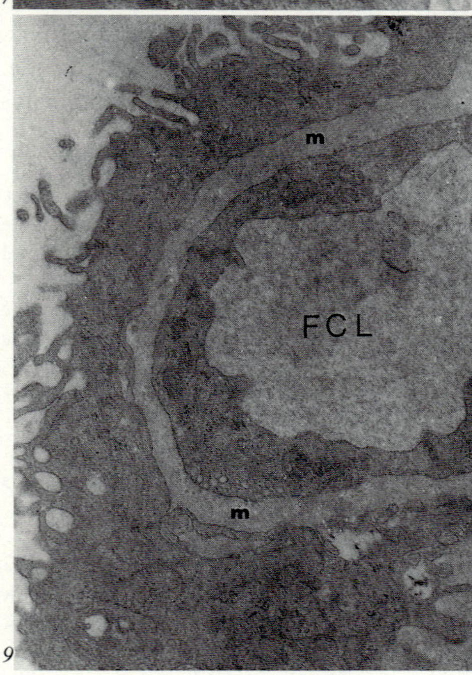

9

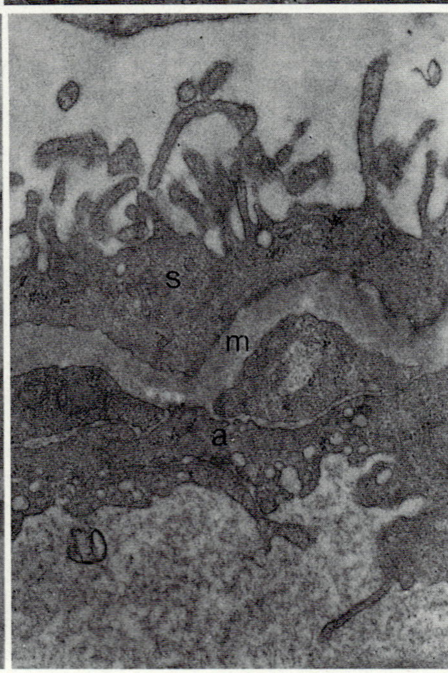

10

were identical. Within the placental membrane itself, the distinction between the cytoplasms of syncytio- and cytotrophoblastic layers which is described above, though often discernible, was much less pronounced than in the earlier stage (fig. 5–7). Granular reticulum was no more elaborately developed in the cytotrophoblast than during the previous stage. Both trophoblastic layers appeared more attenuated than previously; and microvilli at the free surface of the syncytium were fewer and shorter than at the early stage (fig. 5–7). Lipid droplets, though seemingly reduced in numbers, were still confined to the outer syncytial layer (fig. 5 and 6). Micropinocytotic vesicles were evident in all three cytoplasmic layers of the trabecular wall; they were especially numerous in the syncytiotrophoblast (fig. 5 and 7). Occasionally, the outer layer of syncytiotrophoblast was seen to reach the adepithelial basement membrane (fig. 6 and 7). This basement membrane, though more closely applied to the trophoblast (i.e. chorionic epithelium), also underlay the allantoic endothelium of the fetal capillaries; in many cases, it appeared equidistant between epithelium and endothelium (fig. 7).

In tissues from the latest pregnancy which was examined (twin fetal CRL of 75 and 80 mm), again hemopoietic foci (not pictured) were prevalent in the structurally identical placental masses. Within the placental membrane, cytotrophoblast cells very often were absent; consequently, the inner plasmalemma of the syncytiotrophoblast most often rested directly on the basement membrane (fig. 8–10). This membrane usually was situated equidistant between the syncytium and the allantoic endothelium (fig. 9 and 10). The syncytiotrophoblast layer was markedly thinner than previously, often barely exceeding the thickness of the endothelium (fig. 9 and 10).

Fig. 7. A small portion of placental membrane at a stage near midgestation in which tongues of syncytium (s) insinuate themselves between cytotrophoblast cells (c) to reach the basement membrane (arrows). × 16,500.

Fig. 8. The placental membrane from the latest stage of pregnancy which was examined (believed to be near term). Though sometimes present, cytotrophoblast cells (c) are not a constant constituent of the membrane. The syncytium (s) often appears as thin as the allantoic capillary endothelium (a). MBS = maternal blood space; FRBC = fetal erythroblast. × 6,800.

Fig. 9. A portion of the near-term placental membrane in which cytotrophoblast is absent. Note that gap-type junctions seal off the capillary lumen (FCL) and that the basement membrane (m) is 'shared' by syncytiotrophoblast and endothelium. × 16,700.

Fig. 10. Another portion of near-term placental membrane. The syncytiotrophoblast (s) is reduced in thickness to approximate that of the capillary endothelium (a). Note the presence of micropinocytolic vesicles in both of these cytoplasmic layers. These two layers and the single basement membrane (m) constitute the placental barrier near term. × 23,700.

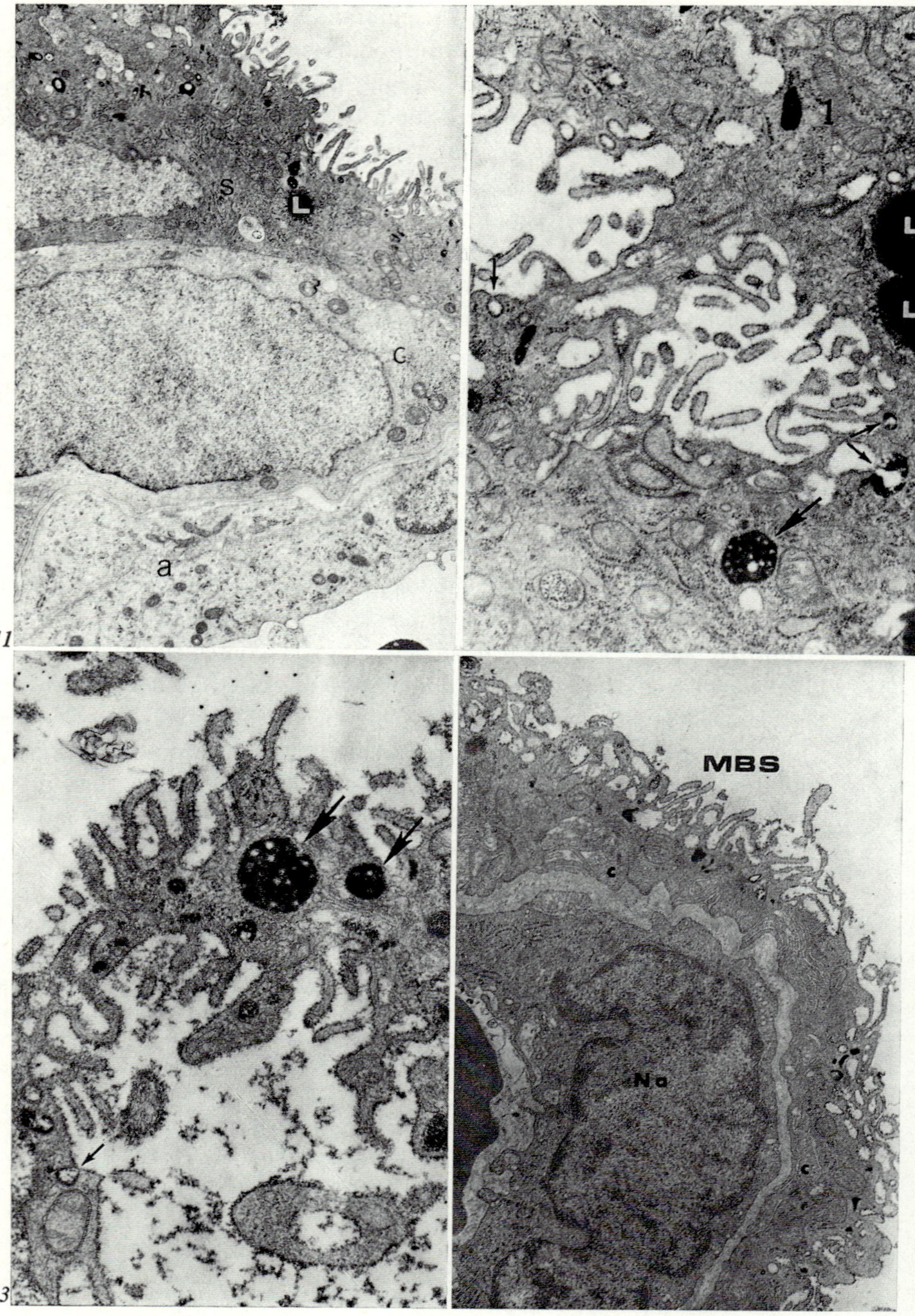

11

13

1

Occasionally sections of attenuated processes were seen which were interposed between the syncytium and the basement membrane and which were attached to the former by desmosomes (fig. 14). On rare occasions, more of the cytoplasm and even the nucleus of persistent cytotrophoblast cells were identified. Cytologically such residual cytotrophoblast cells resembled the syncytium. Both cytoplasmic constituents of the minimal placental barrier, viz., syncytiotrophoblast and allantoic capillary endothelium, showed abundant evidence of micropinocytotic activity; micropinocytotic vesicles were especially numerous in the endothelium (fig. 10).

Localization of Peroxidatic Reaction Product within Constituents of the Placental Membrane

Horseradish peroxidase was injected 5 min before placental tissues were taken from two pregnant animals, viz., from the earliest pregnancy which was autopsied (fetal CRL of 10 and 12 mm) and from the latest (fetal CRL of 75 and 80 mm). In the earlier tissues, the cytochemical reaction product of peroxidase localization was scattered diffusely within the maternal blood sinuses, but more especially along the outer plasmalemma of the syncytiotrophoblast; often it appeared particularly concentrated within invaginated recesses of this plasma membrane between the microvilli at the free surface

---

Fig. 11. A portion of the placental membrane from the earliest pregnancy which was autopsied 5 min after injection of peroxidase and which subsequently was processed cytochemically for localization of reaction product. Note that the electron opacities are confined to the syncytiotrophoblast (s). Neither cytotrophoblast (c) nor endothelium (a) contains the product of peroxidase localization. L = lipid droplet. × 5,900.

Fig. 12. A tangential section through the free surface of the syncytium from the earliest pregnancy after cytochemical localization of the reaction product of injected peroxidase. Note that at several points, the product is localized in vesicles which have invaginated from the free surface (small arrows). At this early stage, it also appears to be concentrated into larger vacuoles (large arrow). L = lipid droplets. × 17,000.

Fig. 13. Another tangential section of the syncytial free surface in an early placenta which has been processed for localization of peroxidase. Reaction product is seen free in the maternal blood sinus, dispersed along the outer leaflet of the syncytial plasmalemma, and concentrated into invaginations of this membrane (small arrow); within the syncytial cytoplasm, it is also being concentrated into larger vacuoles (large arrows). × 33,000.

Fig. 14. A portion of placental membrane near term. After cytochemical localization of previously-injected peroxidase, the reaction product again is seen along the syncytial plasmalemma and enclosed into vesicles within its cytoplasm. Note that a few cytotrophoblastic processes are present (c), attached to the syncytium by desmosomes. MBS = maternal blood sinus; Na = nucleus of allantoic capillary endothelium. × 17,500.

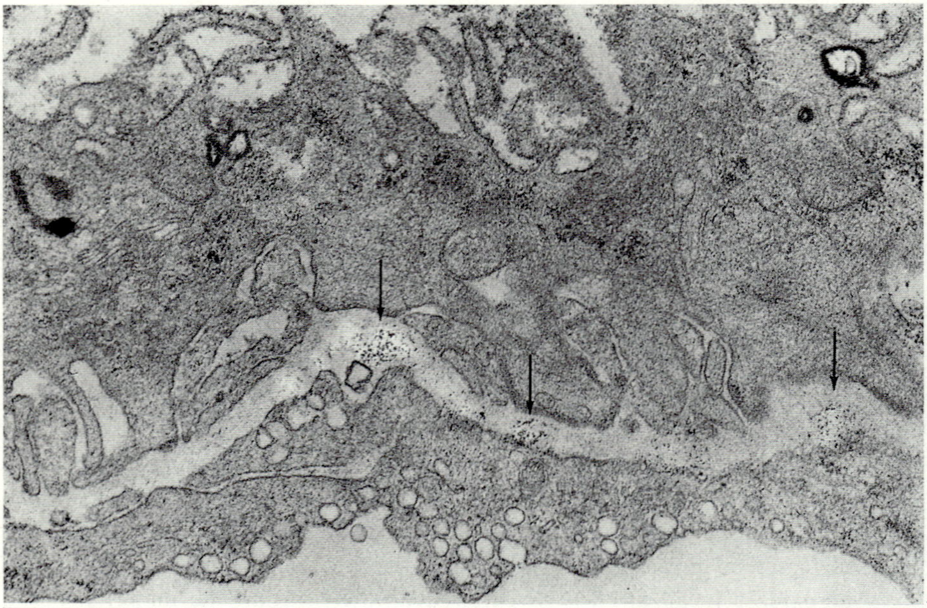

15

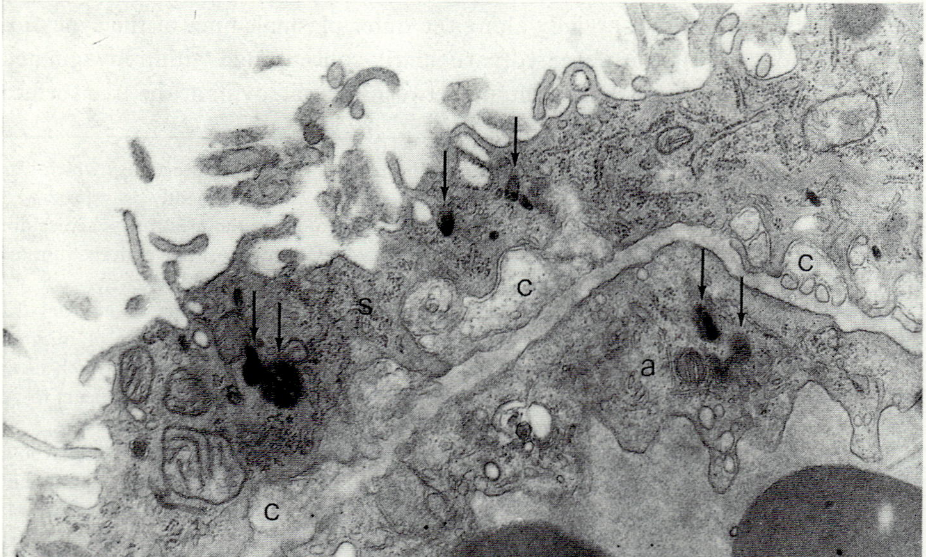

16

Fig. 15. At higher magnification of the near-term placental membrane after peroxidase localization, occasionally reaction product is seen dispersed in the basement membrane between the syncytium and allantoic endothelium (arrows). Note also the abundance of micropinocytotic vesicles in the latter cytoplasmic layer. × 43,000.

of the trabecular wall (fig. 12). Reaction product was also localized inside membranous profiles within the cytoplasm (fig. 11 and 13), but whether such profiles always represented closed vesicles or represented arbitrary sections through patent invaginations from the free surface could not be determined from the methods which were used (i.e., serial sections of the tissues were not cut for evaluation). In each case, however, the electron dense reaction product was associated with the inner, i.e. noncytoplasmic, surface of the profile membrane. Deeper in the syncytial cytoplasm, reaction product was concentrated into large (i.e. 850 nm to 1 μm) vesicles (fig. 12 and 13). The electron dense product of peroxidase localization was confined to such profiles in the syncytiotrophoblast, to its outer surface membrane, and to the maternal blood sinus. In no instance was reaction product identified in the cytotrophoblast cells which were numerous at this stage, in the adepithelial basement membrane, in the interstitium of the trabeculae, or in the allantoic endothelium. Aside from the presence of reaction product in this tissue, the electron microscopic appearance of the placental membrane was the same as in control tissues.

In placental tissues from the latest pregnancy, which was also an experimental animal, the electron dense product of cytochemical localization of peroxidase again was identified with maternal blood and concentrated along the free-surface plasma membrane of the syncytiotrophoblast, especially in association with intermicrovillous recesses; as seen at the earlier stage, the product was localized inside membrane-delimited vesicular profiles within the syncytiotrophoblastic cytoplasm (fig. 14). Unlike the previous stage, however, collections of electron dense particles, presumably representing peroxidase reaction product, occasionally were seen within the substance of the basement membrane (fig. 15). Furthermore, electron opacities were often seen enclosed within vesicular profiles in the cytoplasm of the allantoic endothelium (fig. 16).

Discussion

There are profound differences between the placentas of marmosets and

---

Fig. 16. Following peroxidase injection and cytochemical localization, in certain portions of the placental membrane near term, reaction product (arrows) can be identified not only in the syncytium (s), but also in the allantoic capillary endothelium (a). Note also the cross-sections of processes (c), on the outer surface of the basement membrane; these represent persistent cytotrophoblast cells. × 33,000.

those of higher primates. One very obvious difference at the histologic level is in the basic organization of fetal tissues (i.e. villous, as compared to labyrinthine patterns). Another is in the hemodynamics of maternal blood flow; in the villous placentas of the human [RAMSEY *et al.*, 1967] and rhesus monkey [DONNER *et al.*, 1963] maternal blood has been shown to gain access to intervillous sinuses by periodic spurts whereas in labyrinths presumably it is more steady and countercurrent to fetal blood flow [MOSSMAN, 1926; WYNN, 1964]. However, despite these differences and despite the presence of hemopoietic foci (and presumably function) in the placental mass of the marmoset chorioallantois, the actual sites of maternofetal exchange in the marmoset, a primitive anthropoid, and in hominoids, the highest primates, are remarkably similar. They are similar both in the organization of the constituent layers of the placental membrane and in the changes in these layers with increasing gestational age: early in gestation the placental membranes of both are composed of two layers of trophoblast (a situation which has been called hemodichorial [ENDERS, 1965a], and both membranes are reduced to a single trophoblastic layer (i.e., the membrane becomes hemomonochorial) near term. Further, in both cases the outer trophoblastic layer is lipid rich and syncytial and the inner is cellular, and this cellular layer (called Langhans cells in humans) is more undifferentiated in its ultrastructure and is attached both to other cytotrophoblast cells and to the more superficial syncytium by desmosomes. In the human placenta, cytotrophoblast cells have been shown to be the source of the syncytium, and intermediate cell types have been described [ENDERS, 1965b; BOYD and HAMILTON, 1966]; whether this situation pertains in the marmoset placenta has not been demonstrated. However, in their fine structural appearances the syncytial and cellular trophoblastic components of the marmoset placental membrane are strikingly similar to those of the human [ENDERS, 1965a and b; BOYD and HAMILTON, 1966] and rhesus monkey [PIERCE *et al.*, 1964], and are quite different from the trophoblastic components of other labyrinthine placentas, viz., those of rodents, i.e., rat [JOLLIE, 1964; ENDERS, 1965a; DAVIES and GLASSER, 1968], guinea pig [KING and ENDERS, 1971], hamster [BUTLER, 1970] and chiropterans [WIMSATT, 1962; ENDERS and WIMSATT, 1968]. Further, the hominoid placenta has been shown to have unfenestrated allantoic capillary endothelium, whereas the allantoic endothelium of rodents exhibits pores which are closed by diaphragmata [JOLLIE, 1964]. In this respect, then, the marmoset placentas, as observed in the present study, are also more like those of higher primates.

The chorioallantoic placentas of rodents in general have been demonstrated to be impermeable to the bulk transport of proteins [BRAMBELL and HALLIDAY, 1956; JOLLIE, 1965]. Although physiologically it has been demonstrated that maternal serum proteins are transferred from mother to fetus very late in pregnancy [BRAMBELL, 1958; MAYERSBACH, 1958] (a mechanism by which, presumably, γ-globulins are transported near term to give the late fetus passive immunity) anatomically it has been shown (except in the guinea pig [KING and ENDERS, 1971]) that such protein transport occurs solely across a functional visceral yolk-sac placenta [LAMBSON, 1966; SEIBEL, 1972]. Such materials do not cross the chorioallantoic placental barrier. In the chorioallantois of the rat where three layers of trophoblast are persistent until term [JOLLIE, 1964; DAVIES and GLASSER, 1968] electron opaque colloids effectively are blocked at the layer which has been designated trophoblast II [JOLLIE, 1965]. In man who lacks a functional yolk-sac placenta late in gestation, transport of protein has been demonstrated to occur only near term [DANCIS *et al.*, 1961]. More recently, at the electron microscopic level, it has been shown that human placenta in tissue culture absorbs both ferritin and ferritin-tagged γ-globulins, but only into the syncytiotrophoblast [SATO, 1966].

In the present study, an exogenous protein has been shown readily to be absorbed by the marmoset trophoblastic syncytium, both early in gestation and near term. At early stages of pregnancy, however, it is not transferred more deeply through the membrane, i.e. into the underlying cytotrophoblast cells. Late in gestation – indeed, at a stage of pregnancy believed to be very near term – by virtue of the micropinocytotic activity of the syncytium, peroxidase has been demonstrated to have penetrated through the syncytiotrophoblast and, in the absence of a persistent layer of cytotrophoblast at this stage, to have penetrated both into the fused basement membrane of the chorionic epithelium and capillary endothelium and through it into vesicles in a now micropinocytotically active allantoic endothelium. It is of interest to note that in culturing fragments of human placental villi in tissue culture containing electron densely tagged protein, such protein was not identified in the Langhans' cells [SATO, 1966]. From this *in vitro* study and from the present *in vivo* observations, it is tempting to conclude that the cytotrophoblast cells are impermeable to protein transport. Their presence as a discretely layered constituent of the placental membrane during early pregnancy would serve to block protein transport at this stage. Their reduction in numbers later in pregnancy and their ultimate disappearance from large portions of the placental membrane near term effectively would permit passage of proteins by endocytosis and exocytosis across the persistent syncytiotrophoblast, diffusion across the fused basement membrane and endocytosis into the allantoic endothelium. Although exocytosis from this latter cellular layer has not been observed, its occurrence in the endothelium would deliver the unaltered protein directly into the fetal circulation.

References

ANDERSON, J.W.: The placental barrier to gamma globulins in the rat. Amer. J. Anat. *104:* 403–429 (1959).

BOYD, J.D. and HAMILTON, W.J.: Electron microscopic observations on the cytotrophoblast contribution to the syncytium in the human placenta. J. Anat., Lond. *100:* 535–548 (1966).

BRAMBELL, F.W.R.: The passive immunity of the young mammal. Biol. Rev. *33:* 488–531 (1958).

BRAMBELL, F.W.R. and HALLIDAY, R.: The route by which passive immunity is transmitted from mother to fetus in the rat. Proc. roy. Soc. B *145:* 170–178 (1956).

BUTLER, H.: The formation of the allantoic placenta in the golden hamster *(Mesocricetus auratus)*. Z. Anat. Entw. Gesch. *131:* 68–85 (1970).

DANCIS, J.; LIND, J.; GRATZ, M.; SMOLENS, J., and VARA, P.: Placental transfer of proteins in human gestation. Amer. J. Obstet. Gynec. *81:* 167–171 (1961).

DAVIES, J. and GLASSER, S.R.: Histological and fine structural observations on the placenta of the rat. Acta Anat. *69:* 542–608 (1968).

DONNER, M.W.; RAMSEY, E.M., and CORNER, G.W.: Maternal circulation in the placenta of the rhesus monkey. Amer. J. Radiol. Roentgenol. Ther. *90:* 638–649 (1963).

ENDERS, A.C.: A comparative study of the fine structure of the trophoblast in several hemochorial placentas. Amer. J. Anat. *116:* 29–68 (1965a).

ENDERS, A.C.: Formation of the syncytium from cytotrophoblast in the human placenta. Obstet. Gynec. *25:* 378–386 (1965b).

ENDERS, A.C. and WIMSATT, W.A.: Formation and structure of the hemodichorial chorio-allantoic placenta of the bat *(Myotis lucifugus lucifugus)*. Amer. J. Anat. *122:* 453–490 (1968).

GRAHAM, R.C. and KARNOVSKY, J.M.: The early stages of absorption of injected horseradish peroxidase in the proximal tubules of the mouse kidney: ultrastructural cytochemistry by a new technique. J. Histochem. Cytochem. *14:* 291–302 (1966).

HILL, J.P.: The developmental history of the primates. Croonian Lecture. Philos. Trans. roy. Soc., Lond. *221:* 45–175 (1932).

JOLLIE, W.P.: Fine structural changes in placental labyrinth of the rat with increasing gestational age. J. Ultrastruct. Res. *10:* 27–47 (1964).

JOLLIE, W.P.: Visualization of a block to placental transport of protein and dextrin in the rat. Bull. Tulane med. Fac. *24:* 213–224 (1965).

JOLLIE, W.P.: Changes in the fine structure of the parietal yolk sac of the rat placenta with increasing gestational age. Amer. J. Anat. *122:* 513–532 (1968).

KING, B.F. and ENDERS, A.C.: Protein absorption and transport by the guinea pig visceral yolk sac placenta. Amer. J. Anat. *129:* 261–288 (1970).

KING, B.F. and ENDERS, A.C.: Protein absorption by the guinea pig chorioallantoic placenta. Amer. J. Anat. *130:* 409–430 (1971).

KRAEHENBUHL, J.P. and CAMPICHE, M.A.: Early stages of intestinal absorption of specific antibodies in the newborn. J. Cell Biol. *42:* 345–365 (1969).

LAMBSON, R.O.: An electron microscopic visualization of transport across rat visceral yolk sac. Amer. J. Anat. *118:* 21–32 (1966).

LYNN, J.A.; MARTIN, J.H., and RACE, G.J.: Recent improvements of histologic technics for the combined light and electron microscopic examination of surgical specimens. Amer. J. clin. Path. *45:* 404–713 (1966).

MAYERSBACH, H.: Zur Frage des Proteinüberganges von der Mutter zum Föten. I. Befunde an Ratten am Ende der Schwangerschaft. Z. Zellforsch. *48* 479–504 (1958).

MOSSMAN, H.W.: The rabbit placenta and the problem of placental transmission. Amer. J. Anat. *37:* 433–497 (1926).

PIERCE, G.B.; MIDGLEY, A.R., and BEALS, T.F.: An ultrastructural study of differentiation and maturation of trophoblast of the monkey. Lab. Invest. *13:* 451–464 (1964).

RAMSEY, E.M.; MARTIN, C.B., and DONNER, M.W.: Fetal and maternal placental circulations. Amer. J. Obstet. Gynec. *98:* 419–423 (1967).

REYNOLDS, E.S.: The use of lead citrate at high pH as an electronopaque stain in electron microscopy. J. Cell Biol. *17:* 208–212 (1963).

SATO, M.: An electron microscopic observation on the transfer of gamma-globulin and of ferritin protein through the human placenta. J. jap. obstet. gynaec. Soc. *13:* 227–234 (1966).

SCHULTZ, R.L.: Placental transport. A review. Obstet. Gynec. *25:* 979–1020 (1970).

SEIBEL, W.: An electron microscopic investigation of the uptake, utilization, and transport of an exogenous protein by the visceral yolk-sac placenta following intravenous injection in pregnant rats; Ph. D. diss., Virginia Commonwealth University, Richmond, Va. (1972).

WIMSATT, W.A.: Some aspects of the comparative anatomy of the mammalian placenta. Amer. J. Obstet. Gynec. *84:* 1568–1594 (1962).

WISLOCKI, G.B.: On the placentation of primates, with a consideration of the phylogeny of the placenta. Contrib. Embryol. Carneg. Inst. *20:* 51–80 (1929).

WISLOCKI, G.B.: On a series of placental stages of a platyrrhine monkey *(Ateles geoffroyi)*, with some remarks upon age, sex and breeding period in platyrrhines. Contrib. Embryol. Carneg. Inst. *22:* 173–192 (1930).

WISLOCKI, G.B.: Placentation in the marmoset *(Oedipomidas geoffroyi)*, with remarks on twinning in monkeys. Anat. Rec. *52:* 381–399 (1932).

WISLOCKI, G.B.: Observations on twinning in marmosets. Amer. J. Anat. *64:* 445–483 (1939).

WISLOCKI, G.B.: Hemopoiesis in the chorionic villi of the placenta of platyrrhine monkeys. Anat. Rec. *85:* 349–363 (1943).

WYNN, R.M.: Comparative morphogenesis and vascular relationships of the villous hemochorial placenta. Amer. J. Obstet. Gynec. *90:* 758–768 (1964).

Author's address: Prof. W.P. JOLLIE, Department of Anatomy, Virginia Commonwealth University, MCV Station, *Richmond, VA 23219* (USA)

Medical Primatology 1972. Proc. 3rd Conf. exp. Med. Surg. Primates, Lyon 1972, part I, pp. 288–297 (Karger, Basel 1972)

Endometriosis in Primates

W. F. MacKenzie, G. A. Splitter and Marion G. Valerio[1]

USAF School of Aerospace Medicine (AFSC), Brooks Air Force Base, Tex. and Department of Pathology, Litton Bionetics Research Laboratories, Kensington, Md.

Endometriosis in Primates

Endometriosis is a disease that affects up to 20–30% and maybe more of women and is becoming a well-known entity in breeding colonies of primates. However, very little has been published on this syndrome in simians. A description of the clinical appearance of the disease and the gross and microscopic pathology will be presented based on experience of the authors and a review of the literature; possible etiologies and the use of primates as models for human disease will also be briefly discussed.

Endometriosis is characterized by the presence of viable endometrial tissue in ectopic sites. It is usually divided into two types [18]: internal endometriosis or adenomyosis, where endometrial tissue invades the myometrium; and, external or true endometriosis, where endometrial tissue is found outside the uterus on the serosa of the ovaries, uterine ligaments and other pelvic organs. Adenomyosis has been described in dogs, cattle, cats, and rabbits [7]. External or true endometriosis has been found only in the menstruating primates. It has been suggested that this is because primates do not have a utero-tubal valve, thereby allowing menstrual detritus to pass to the abdominal cavity [9, 10]. They are also the only mammals that cyclically regenerate and slough the endometrium. This combination of reflux menstruation implanting functional uterine fragments on the serosa and the subsequent reaction of these fragments to the hormonal stimulation for growth and then destruction, produce the widespread adhesions and hemorrhage that cause the symptoms of true endometriosis [18]. The menstrual cycle is probably also related to the frequency with

1 Conducted under USPHS contract NIH-NCI 71-2025.

which adenomyosis is seen in women and monkeys since adenomyosis is rare in nonmenstruating animals.

Endometriosis can be a progressive and fatal disease with a varied clinical picture. Although adenomyosis and true endometriosis occur separately, both are often found concurrently with one dominating the picture. Adenomyosis is characterized by infertility, dysmenorrhea, menorrhagia and irregular cycles. In severe cases palpable abdominal masses and abdominal enlargement resembling pregnancy occur. The enlarged uterus can obstruct the ureters causing secondary hydronephrosis. Radiography using simultaneous pyelogram, pneumoperitoneum, and pneumocystogram is useful in diagnosis (fig. 1). True endometriosis produces vague but often severe signs of dysmenorrhea. Animals can evidence signs of abdominal discomfort, even prostration or death, during menstrual periods. True endometriosis can be progressive and produce serious bowel obstruction [8]. Radiograms after barium feedings are helpful in diagnosis. The age range when the disease occurs cannot be given since the age of few monkeys is known and the disease is often present for many years before it is diagnosed.

The gross and microscopic findings are very similar to those seen in women. The appearance of adenomyosis in simians varies from small

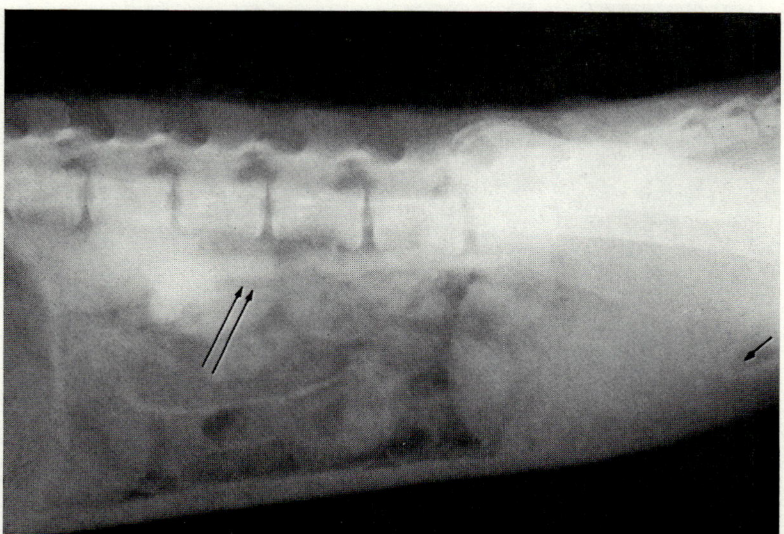

Fig. 1. Lateral simultaneous pneumoperitoneum, pneumocystogram and intervenous pyelogram showing enlarged uterus. Single arrow points to urinary bladder, double arrow to enlarged ureter.

discrete nodules in the myometrium to massive erosion and distortion of the uterine musculature (fig. 2 and 3) in which cases the serosal surface is uneven and bossalated, the wall is fibrous with cystic spaces, and the lumen is filled with as much as 50 ml of brown mucoid fluid and necrotic debris (fig. 3). These uteri can be as much as 7 cm in diameter. Mild cases of true endometriosis are described [4, 6] as hemorrhagic or 'chocolate' cysts associated with adhesions that cause puckering of the serosa. Wider involvement is more frequent [2, 4–6, 15] with extensive spread over serosal surfaces (fig. 4 and 5). Large pseudocysts can be formed filled with mucoid to hemorrhagic 'chocolate' material (fig. 5). Extensive adhesions often cause secondary involvement of the small and large intestine. Small cysts must be distinguished from lesions due to parasites (*Esophagostomum* sp.), and more severe lesions may resemble nonspecific peritonitis and abscesses. Widely disseminated endometriosis with metastases to lymph nodes, liver, and lungs has been described [5].

The microscopic picture is identical to that seen in women. The endometrium consists of a single layer of cuboidal to columnar epithelial cells supported by a richly vascular stroma identical to normal endometrium (fig. 6 and 7). Complex papillary folds are formed and cystic spaces are

2

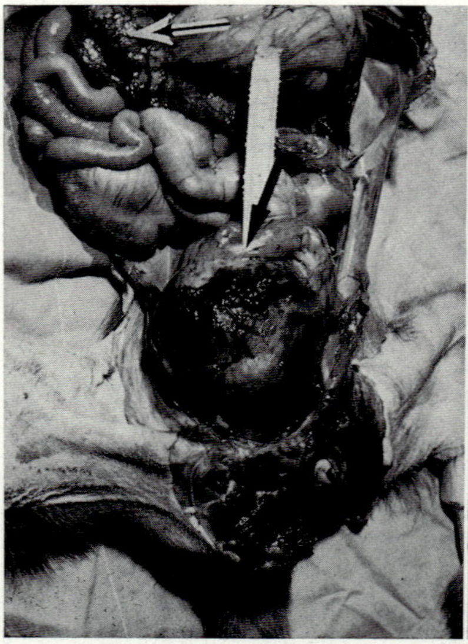

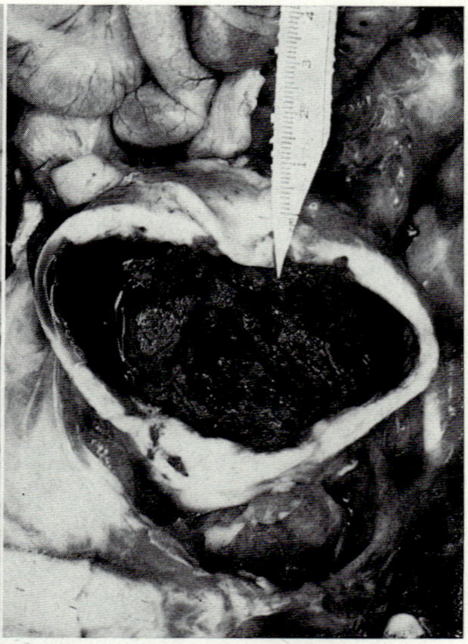

3

Fig. 2. Uterus (large arrow) of female *M. mulatta* shown in figure 1. Endometriosis and adhesions also involve the colon (small arrow).

Fig. 3. Uterus of female *M. mulatta* shown in figure 1 showing lumen containing hemorrhagic necrotic debris.

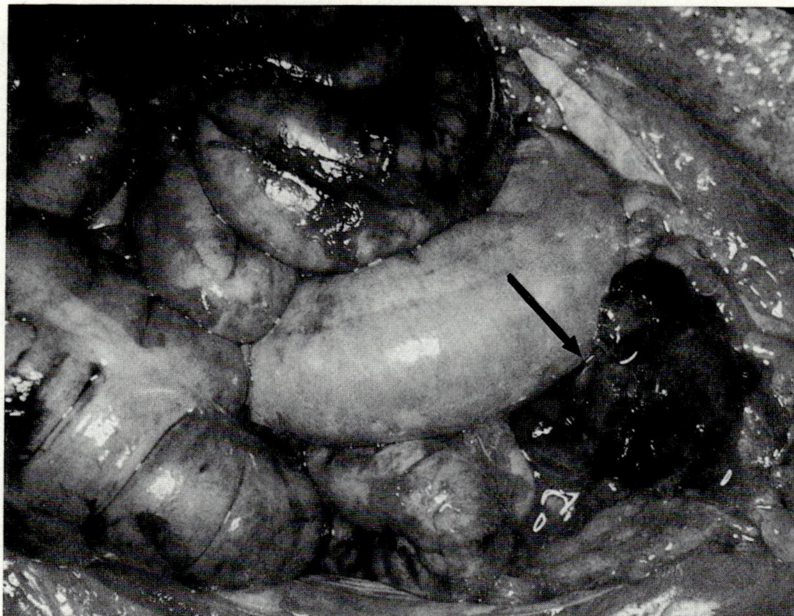

4

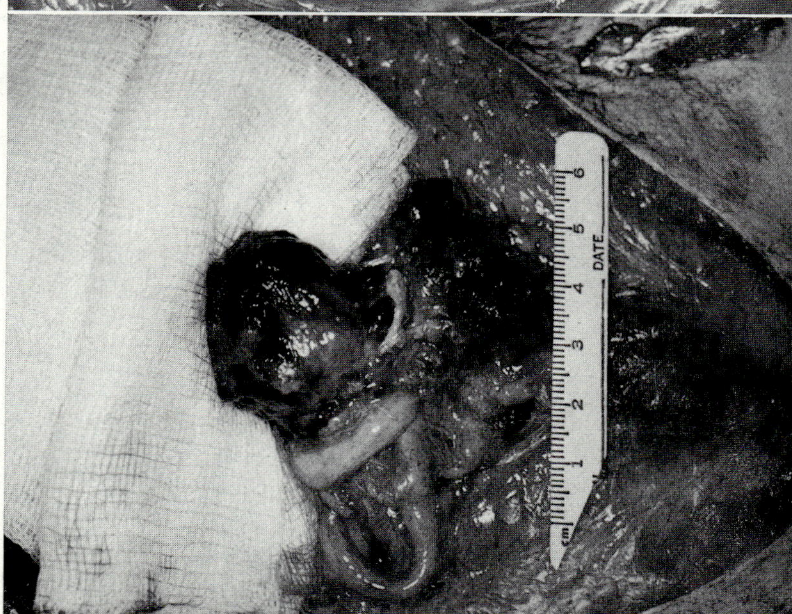

5

Fig. 4. Extensive fatal endometriosis in a female *M. mulatta.* Arrow points to large endometrial cyst. Note distention of bowel due to partial obstruction.

Fig. 5. Endometrial cyst from figure 4 dissected free showing extensive involvement of small intestines. This cyst was not attached to the uterus.

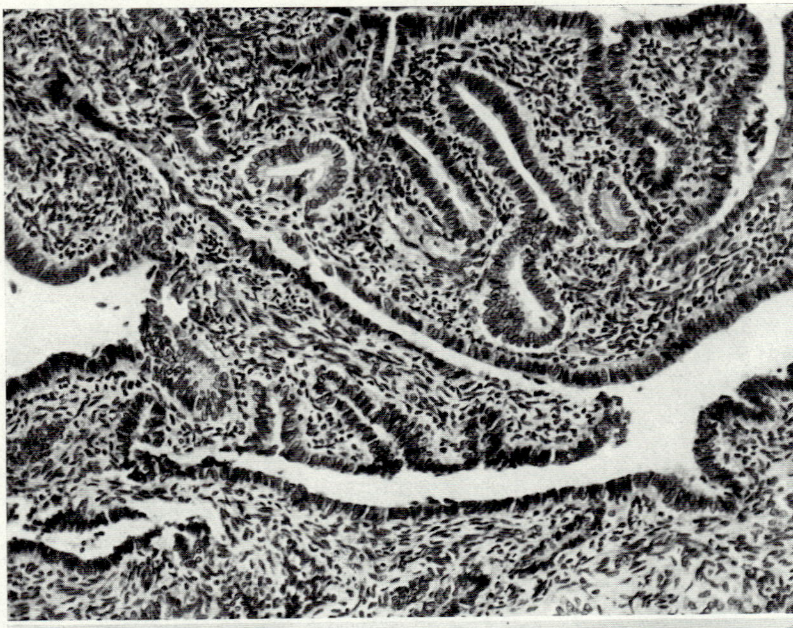

6

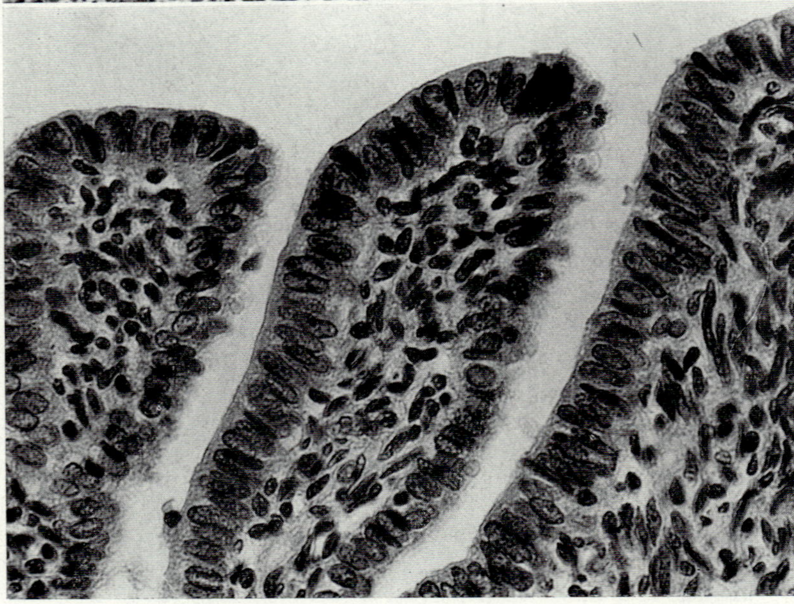

7

Fig. 6. Photomicrograph of endometriosis showing typical mucosa and stroma undistinguishable from normal endometrium. HE. × 170.

Fig. 7. Photomicrograph of mucosal cells of endometriosis showing appearance similar to normal endometrium. HE. × 500.

frequent (fig. 8). Necrosis and hemorrhage are commonly widespread and associated with hemosiderin filled macrophages and extensive fibrosis (fig. 9). Usually aberrant endometrium spreads over the serosal surface but it can penetrate and invade deeper structures (fig. 10). Glands and stroma may be present in these aberrant foci.

A relationship between hysterotomy and endometriosis is apparent [1, 4, 9]. It has also been suggested that radiation might induce the disease [1, 5, 15]. As noted in table 1, 17 of the reported monkeys have been irradiated, which amounts to 29.3 % of the total reported cases not having hysterotomies. This might represent a biased population of monkeys. However, EISENFELD *et al.* [1] described endometriosis in eight of 13 irradiated female monkeys, an incidence much larger than can be expected in normal animals. 35 % of all reported cases have been associated with hysterotomy. Incidences as high as 51 % of animals following cesarean sections have been reported [4]. However, WILSON [17] reported no endometriosis in 56 hysterotomies performed on the 100th day of gestation.

The expected incidence of endometriosis in a breeding colony of primates is low. LAPIN and YAKOVLEVA [3] diagnosed five cases (four macaques, one baboon) in 463 necropsies of female monkeys of various ages.

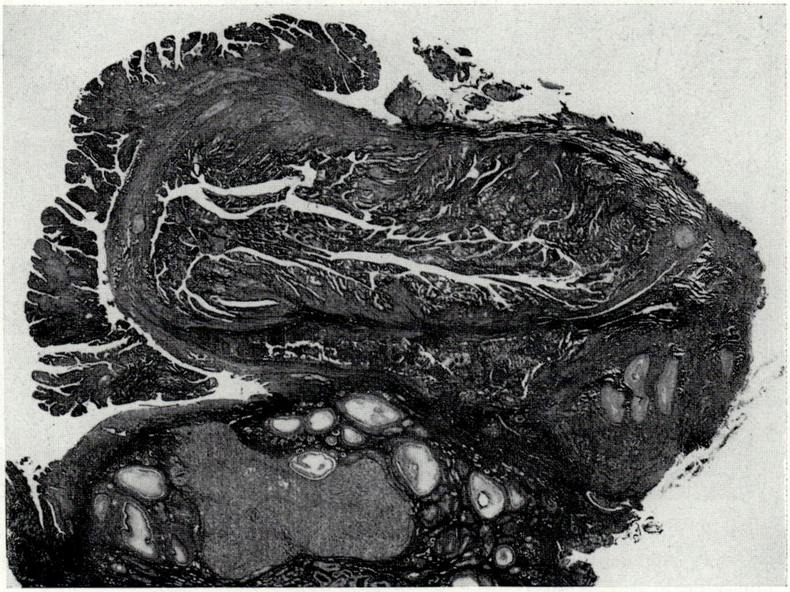

Fig. 8. Endometriosis of the ovary showing typical complex papillary architecture. HE. × 6.7.

9

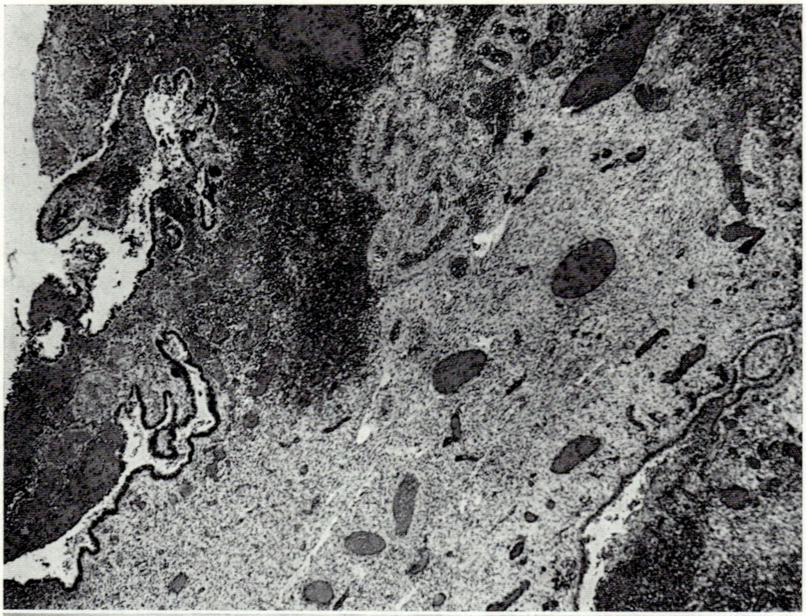

10

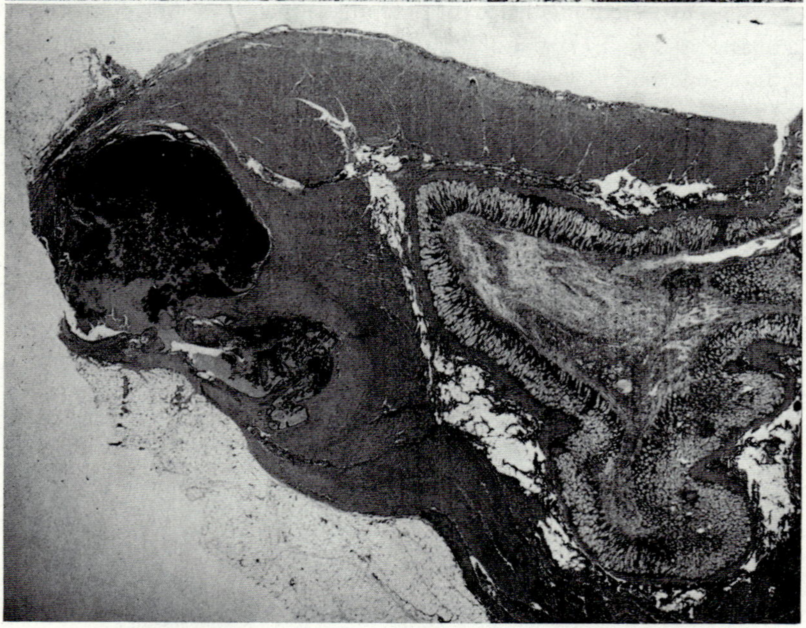

Fig. 9. Photomicrograph of older area of endometriosis showing tortuous arteries, extensive stroma, degenerating mucosa and hemorrhage. This monkey was menstruating at time of death. HE. × 40.

Fig. 10. Endometrial tissue invading muscularis of colon and forming blood filled cysts. HE. × 9.

McCann and Myers [4] reported 21 cases in 63 necropsies of female *M. mulatta* culled from a breeding colony of 250 females. Nineteen of these had had hysterotomies. One of the authors (Valerio) has diagnosed 17 cases in four years from a relatively stable breeding colony of approximately 700 *Macaca* sp. females (table II). None of these had a known history of radiation exposure or hysterotomy.

Prior to this time breeders were imported in large numbers with other monkeys and were therefore not a stable population; nine cases of adenomyosis were diagnosed in monkeys culled for reproductive reasons during that period.

A four-year history of this colony is given as typical of what can be expected. The first year four cases were diagnosed from 44 necropsies; all four had been culled for reproductive failure, abnormal uteri on palpation, etc. Adenomyosis was present in three cases and both adenomyosis and endometriosis in one. The next year, one case of extensive endometriosis with massive adhesions was diagnosed among 29 necropsies. An additional case of endometriosis was found on laparotomy. In the third year, four cases were diagnosed in 33 necropsies; two were reproductive culls with adenomyosis, one animal had endometriosis, and the fourth, culled because of other tumors, had adenomyosis. In the fourth year, three cases were diagnosed in 30 necropsies; one was a reproductive cull, one was a tuberculosis suspect, and the third died of hemorrhage from a ruptured endometrial cyst with widespread abdominal endometriosis. Additional four cases were diagnosed by laparotomy. Thus, in a laboratory that specifically looks for reproductive pathology, the incidence of diagnosed endometriosis and adenomyosis was less than 1% of breeding females per year.

The use of the rhesus monkey as an experimental model for a human disease depends on the similarity of this disease between the two and upon the ease it can be produced at will in the nonhuman species. McCann and Myers [4] have pointed out that there might be basic differences in the pathogenesis of the disease between monkeys and man since no consistent pattern of pathology has been noted in spontaneous cases and that endocrinopathies had been incriminated as the etiology in most cases. Also, the fact that endometriosis is relatively common following cesarean section in the rhesus and decidedly uncommon in women is pointed out by them as evidence that the disease is different. Not all hysterotomies in the rhesus monkey are followed by endometriosis; Wilson [17] observed no endometriosis following 56 hysterotomies. In those reported cases where the authors attempted to incriminate endocrinopathies as the etiology of the disease, it was an assumption based largely upon the symptoms that the animals with the disease had shown. Considering the extensive lesions found at necropsy, it becomes problematic whether the dysmenorrhea, menorrhagia, and infertility described were the result or the cause of the endometriosis. Women with widespread endometriosis show similar symptoms [18].

It should be noted that monkeys have already been used extensively as a model of the human disease [1, 11–14] and methods are available for laboratory production of endometriosis. Scott *et al.* [10] and TeLinde and Scott [16] have produced the disease by diverting by fundectomy the menstrual flow into the abdominal cavity and subcutaneous musculature, and by transplanting endometrial biopsies to the peritoneum and the connective tissue of the abdominal walls. Another suggested site for implantation of endometrial tissue could be the subcutaneous fascia where it could be more easily studied during experimentation by palpation and biopsy.

References

1 Eisenfeld, A. J.; Gardner, W. U., and Wagenen, G. van: Radioactive estradiol accumulation in endometriosis of the rhesus monkey. Amer. J. Obstet. Gynec. *109:* 124–130 (1971).

2 Kluver, H. and Bartelmez, G. W.: Endometriosis in a rhesus monkey. Surg. Gynec. Obstet. *92:* 650–660 (1951).

3 Lapin, B. A. and Yakovleva, L. A.: Comparative pathology in monkeys, pp. 202–206 (Thomas, Springfield 1963).

4 McCann, T. O. and Myers, R. E.: Endometriosis in rhesus monkeys. Amer. J. Obstet. Gynec. *106:* 516–523 (1970).

5 McClure, H. M.; Ridley, J. H., and Graham, C. E.: Disseminated endometriosis in a rhesus monkey *(Macaca mulatta)*. J. med. Ass. Ga. *60:* 11–13 (1971).

6 Merrill, J. A.: Spontaneous endometriosis in the Kenya baboon *(Papis doguera)*. Amer. J. Obstet. Gynec. *101:* 569–570 (1968).

7 Nieberlem, K. and Cohrs, P.: Textbook of the special pathological anatomy of domestic animals, p. 749 (Pergamon Press, New York 1966).

8 Pickering, D. E.: Reproduction characteristics in a colony of laboratory confined mulatta macaque monkeys. Folia primat. *8:* 169–179 (1968).

9 Ruch, T. C.: Diseases of laboratory primates, pp. 454–455 (Saunders, Philadelphia 1959).

10 Scott, R. B.; TeLinde, R. W., and Wharton, L. R., jr.: Further studies on experimental endometriosis. Amer. J. Obstet. Gynec. *66:* 1082–1103 (1953).

11 Scott, R. B. and Wharton, L. R., jr.: The effects of excessive amounts of diethylstibestrol on experimental endometriosis in monkeys. Amer. J. Obstet. Gynec. *69:* 573–591 (1955).

12 SCOTT, R. B. and WHARTON, L. R. jr.: The effect of estrone and progesterone on the growth of experimental endometriosis in rhesus monkeys. Amer. J. Obstet. Gynec. *74:* 852–865 (1957).
13 SCOTT, R. B. and WHARTON, L. R. jr.: The effect of testosterone on experimental endometriosis in rhesus monkeys. Amer. J. Obstet. Gynec. *78:* 1020–1027 (1959).
14 SCOTT, R. B. and WHARTON, L. R. jr.: Effects of progesterone and norethindrone on experimental endometriosis in monkeys. Amer. J. Obstet. Gynec. *84:* 867–875 (1962).
15 SPLITTER, G. A.; KIRK, J. H.; MACKENZIE, W. F., and RAWLINGS, C. A.: Endometriosis in four irradiated rhesus monkeys (submitted for publ.).
16 TELINDE, R. W. and SCOTT, R. B.: Experimental endometriosis. Amer. J. Obstet. Gynec. *60:* 1147–1173 (1950).
17 WILSON, J. G.: Teratological and reproductive studies in nonhuman primates. Proc. Int. Workshop Teratol., vol. 2, pp. 16-33 (1962).
18 WRIGHT, G. P. and SYMMERS, W. S. C.: Systemic pathology, vol. 1, pp. 926–930 (Longmans Green, London 1966).

Authors' addresses: Lt Col WILLIAM F. MACKENZIE and Capt GARY A. SPLITTER, Veterinary Pathology Branch, USAFSAM, *Brooks AFB, TX 78235;* Dr. MARION G. VALERIO, Department of Pathology, Bionetics Research Laboratories, 5510 Nicholson Lane, *Kensington, MD 20795* (USA)

Medical Primatology 1972. Proc. 3rd Conf. exp. Med. Surg. Primates, Lyon 1972, part I, pp. 298–307 (Karger, Basel 1972)

Uterine Blood Flow and Uterine Contractions in Monkeys[1,2]

C. B. MARTIN, jr.

Department of Obstetrics and Gynecology, Los Angeles County-University of Southern California Medical Center, Los Angeles, Cal.

Despite one early report to the contrary [9], there can be no doubt that uteroplacental blood flow is diminished or even halted temporarily during uterine contractions. This relationship, which is apparent clinically from the occurrence of fetal distress during uterine hypertonus and tachysystole, has been demonstrated experimentally by several methods including radioangiography [1, 7, 8], flowmeters [4], and radioisotope-labelled microspheres [5].

The radioangiographic studies showed that the uteroplacental arteries and veins are narrowed and even emptied of contrast material during contractions, suggesting that the vessels are compressed by the contracting myometrium. But other mechanisms may also operate. The vessels may constrict actively during contractions as a result of either sympathetic nerve activity or direct spread of the wave of myometrial contraction to vascular smooth muscle [7].

The studies described here were undertaken to quantitate the changes in uterine blood flow and vascular resistance during uterine contractions, and to investigate the mechanism(s) by which they are produced.

Methods

The pregnancy studies were carried out on 15 rhesus *(Macaca mulatta)*, cynomolgus *(Macaca fascicularis)* and pigtail *(Macaca nemestrina)* macaques.

Blood flow to the nonpregnant uterus was studied in three regularly menstruating

1 Supported in part by USPHS, NIH grants HD-02036 and HD- 04823.

2 A portion of the work described herein was carried out while the author was a member of the Department of Obstetrics and Gynecology, Medical College of Georgia, Augusta, Ga.

rhesus monkeys. The stage of ovarian cycle was determined from daily observation for bleeding and from rectal palpation.

Nitrous oxide-oxygen anesthesia with muscle relaxants was used for implantation of the flow probe and catheters in some of the animals. Others were premedicated with diazepam and anesthetized with pentobarbital or thiopental. After surgery, the depth of anesthesia was permitted to lighten to the minimum required to prevent struggling. The *post partum* observations were carried out under light barbiturate sedation, the sensors having been implanted before delivery. The three monkeys studied during the menstrual cycle had chronically implanted flow probes. In the two of these, the flow observations were obtained with the animal awake in a restraining chair. The remaining animal was tranquillized with phencyclidine.

Uterine artery blood flow was measured by means of a square-wave electromagnetic flowmeter (Carolina Medical Electronics, Inc. King, N.C.). An implantable flow probe (3–5 mm circumference) was placed about the uterine artery midway between pelvic wall and lateral uterine border. Zero flow was determined in some monkeys by occlusion of the artery distal to the probe before and after the observation period, and in others by means of a periaortic occlusion loop.

Pressures were measured by means of open-end polyvinyl catheters and strain gauge transducers. The vascular catheters were introduced via the femoral vessels. Intrauterine pressure was measured by means of a catheter in the amniotic cavity. In the nonpregnant animals, pressure within the endometrial cavity was determined from a catheter inserted via the cervical canal [6].

To study the effect of sympathetic nerve blockade on blood flow during contractions, two animals were given the ganglionic blocker trimethaphan camphorsulfonate i. v. The rate of infusion was increased until maximum hypotensive effect had been obtained. In three other monkeys, spinal (subarachnoid) block was instituted with procaine or tetracaine.

Flow and pressure data were taken from the polygraph records of the pregnancy experiments in order to study statistically the relationships between intrauterine pressure, intravascular pressure and uterine artery blood flow. Mean arterial blood pressure was calculated as the diastolic pressure plus one-third of the pulse pressure. The arteriovenous pressure gradient across the uterus was obtained by subtracting vena cava pressure from mean arterial pressure. In order to take into account any effects on flow resulting from changes in arterial or venous pressures during contractions, uterine artery blood flow was converted to $ml \cdot min^{-1} \cdot mm$ Hg arteriovenous pressure gradient (conductance) for correlation with intrauterine pressure. vascular resistance, the reciprocal of conductance, was also calculated for the entire uteroplacental bed and for two component segments. Since the pressure in the intervillous space is very nearly equal to intraamniotic pressure (except perhaps near the centers of the cotyledons [11]), the resistance of the 'arteries plus cotyledons' portion (arterial resistance) could be approximated from ([mean arterial pressure-intraamniotic pressure]÷flow), and uteroplacental venous resistance could be estimated from ([intraamniotic pressure-venous pressure]÷flow). These values were calculated during uterine relaxation immediately before and after contractions and at the peak intrauterine pressure during contractions. Portions of each record in the pregnancy studies were obtained at paper speeds of 50 mm/min. Data were taken from these records at $^1/_{50}$-min intervals for closer study of changes in uterine artery blood flow and uterine vascular resistance during contractions.

Results

Inspection of the polygraph records of the pregnancy experiments revealed that uterine artery blood flow decreased during uterine contractions, the mean flow tracing forming nearly a mirror image of the intrauterine pressure curve (fig. 1). Contractions of different intensities produced decreases in flow which matched the intensity of the individual contraction. During the prolonged, complex contractions characteristic of late-pregnancy uterine activity in macaques [2], uterine artery blood flow remained depressed as long as intrauterine pressure remained elevated.

The visual impression that the magnitude of blood flow decrease correlated well with contraction intensity was substantiated by graphic and statistical analysis of the relationship between precontraction and peak contraction conductances and corresponding intrauterine pressures. An example of this relationship is shown by figure 2. The relationship was highly significant in all but one of the experiments (No. 126, $0.05 > p > 0.01$).

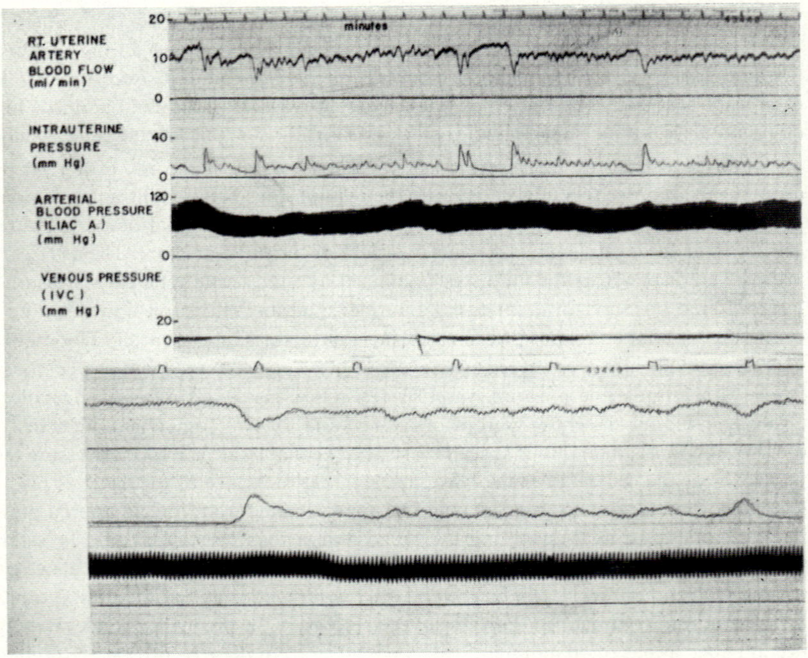

Fig. 1. Uterine artery blood flow during prolonged contractions. Upper record: 10 mm/min; lower record: 50 mm/min. *M. fascicularis,* 104-gram fetus.

For closer analysis of blood flow changes during contraction, uterine vascular conductance was graphed against intrauterine pressure at $^1/_{50}$-min intervals in the form of X-Y plots. In most instances, the results were relatively narrow loops with negative mean slopes (fig. 3). The descending limb of the loop (contracting phase) lay below the ascending limb (relaxing phase) in some contraction, while in others, this relationship was reversed.

X-Y plots were also used to study changes in resistance of the arterial and venous segments of the uterine vascular bed during contractions. When arterial and venous resistance were graphed against intrauterine pressure, the results again were narrow loops, this time with slopes which rose more steeply as intrauterine pressure increased (fig. 4). The resistances rose during all contractions studied, but the magnitude of the increase resulting from

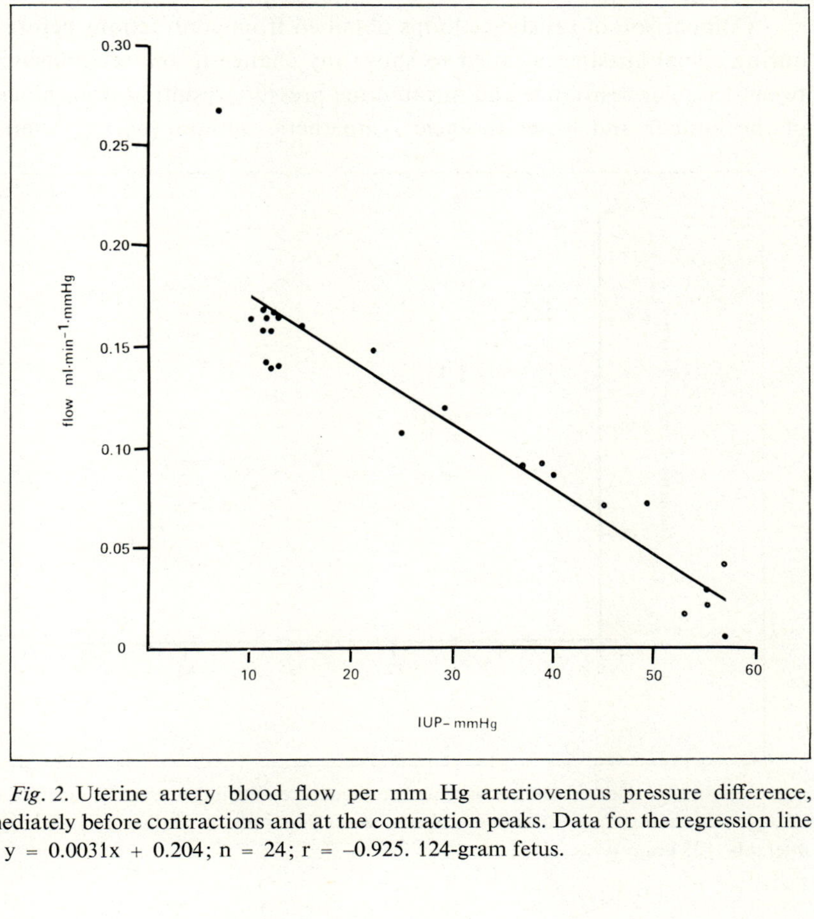

Fig. 2. Uterine artery blood flow per mm Hg arteriovenous pressure difference, immediately before contractions and at the contraction peaks. Data for the regression line are: $y = 0.0031x + 0.204$; $n = 24$; $r = -0.925$. 124-gram fetus.

contractions of equivalent intensity differed markedly from animal to animal (cf. fig. 4 and 5).

Between contractions, resistance to flow in the venous segment of the uteroplacental bed accounted for only a small proportion of the total uterine vascular resistance – less than 10% in all but two of the 15 monkeys. Although both arterial and venous resistance increased with contractions, resistance in the uteroplacental veins became a greater fraction of the total during these times. The stronger the contraction, moreover, the greater was the proportion of total uterine vascular resistance contributed by the uteroplacental veins (fig. 4).

The X-Y plots of uterine vascular resistance and intrauterine pressure for the initial and subsidiary peaks of complex contractions showed no consistent change in resistance at equivalent uterine pressures over the course of the contraction.

Comparison of resistance loops obtained from contractions before and during spinal anesthesia failed to show any change in the relationship between vascular resistance and intrauterine pressure resulting from blockade of the lumbar and lower thoracic sympathetic outflow (fig. 5). Similarly,

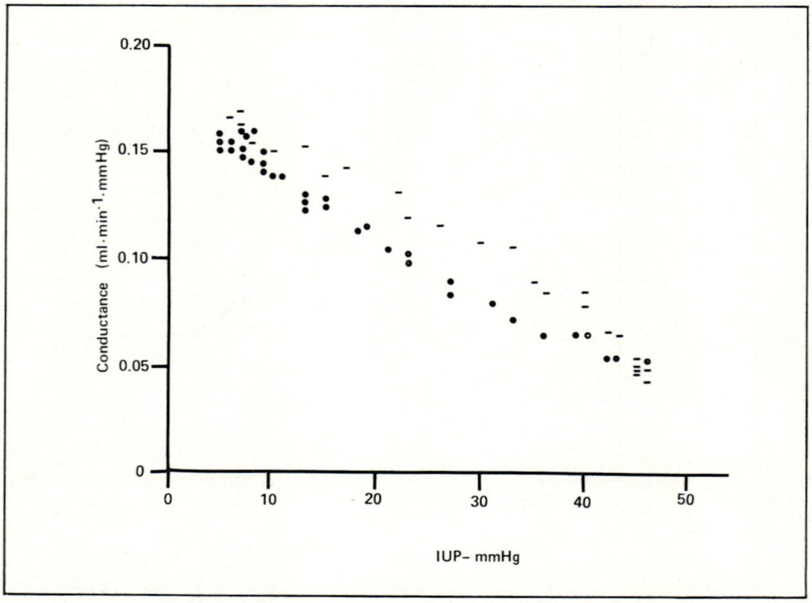

Fig. 3. X–Y plot of uterine vascular conductance and intrauterine pressure during a single contraction. – = Contracting phase; ● = relaxing phase. Data points at $^1/_{50}$-min intervals. 125-gram fetus.

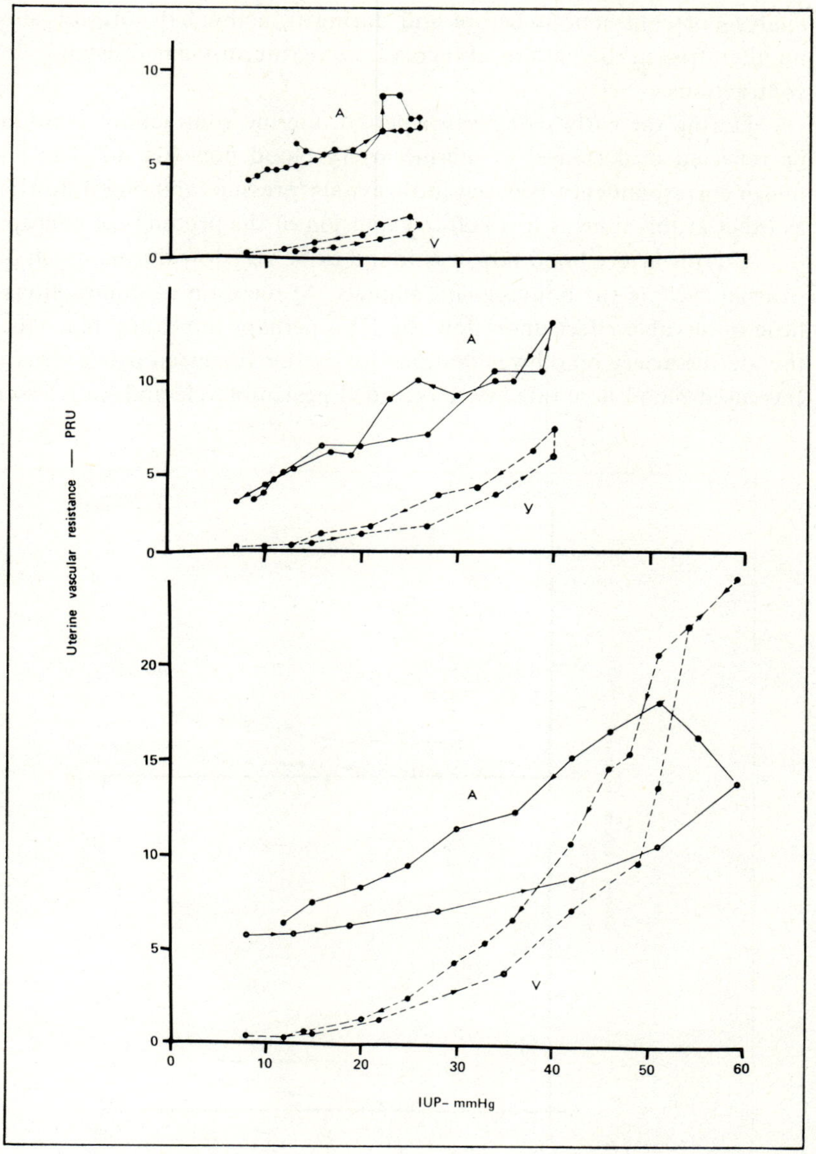

Fig. 4. X–Y plots of uterine arterial and venous reistance and intrauterine pressure during three contractions of differing intensity in the same animal. A = arterial resistance; V = venous resistance. Data points at $^1/_{25}$-min intervals. PRU = peripheral resistance units (mm Hg · ml^{-1} · min).

analysis of contractions before and during trimethaphan infusion showed no alteration in the pattern of increase in uterine vascular resistance during contractions.

During the early *post partum* period, uterine contractions continue to be reflected in decreases in uterine artery blood flow (fig. 6). The mirror image correspondence between intracavitary pressure and blood flow is not as exact at this time as it is before expulsion of the products of conception.

Uterine artery bood flow was found to be very low during much of the ovarian cycle in the nonpregnant animals. At these times, contractions had little observable effect upon flow (fig. 7b), perhaps indicating that much of the uterine artery blood was destined for extramyometrial pelvic structures. Increased blood flow rates were recorded near midcycle and, to a lesser ex-

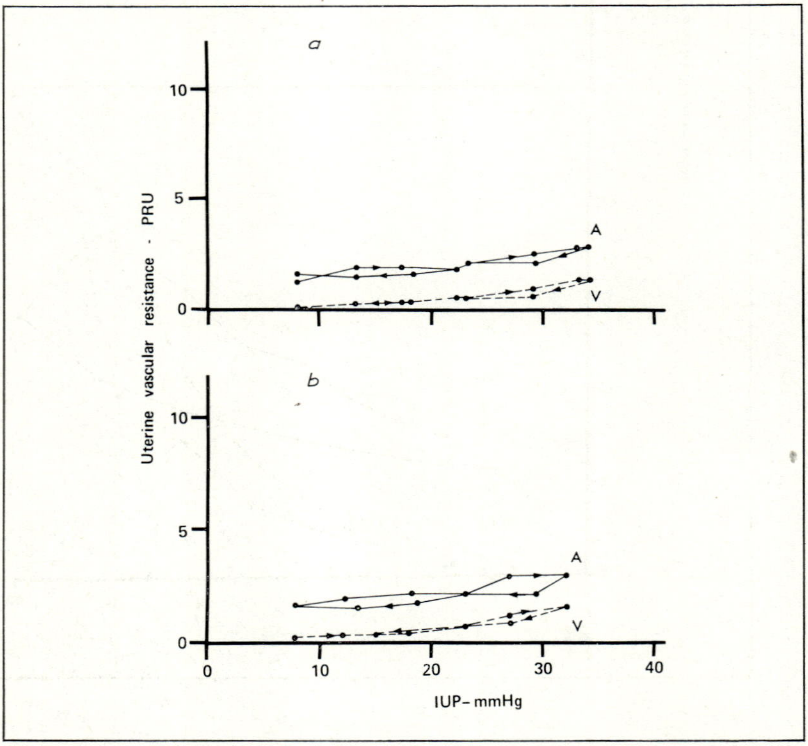

Fig. 5. X–Y plots of uterine vascular resistance and intrauterine pressure during single contractions *a* before and *b* during spinal anesthesia. A = arterial resistance; V = venous resistance. 388-gram fetus. PRU = peripheral resistance units (mm Hg · ml^{-1} · min).

tent, during the late premenstrual phase. During these periods, transient decreases in flow were observed coinciding with the stronger uterine con-

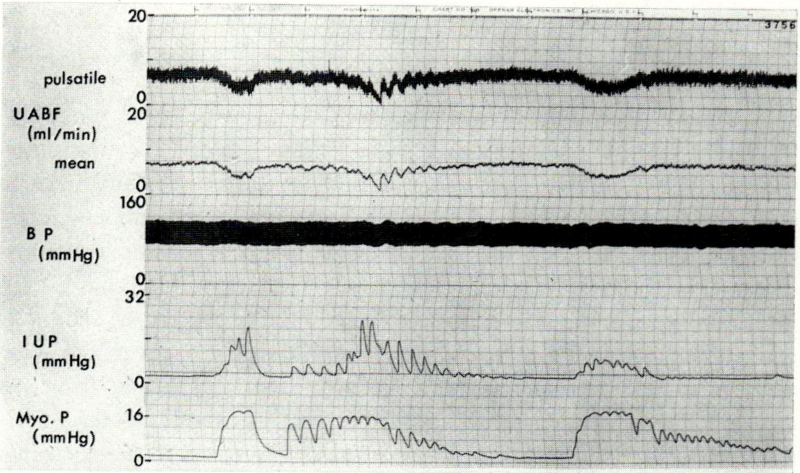

Fig. 6. Uterine artery blood flow record obtained six hours *post partum.* UABF = uterine artery blood flow; IUP = intrauterine pressure; Myo. P = intramyometrial pressure (anterior corpus).

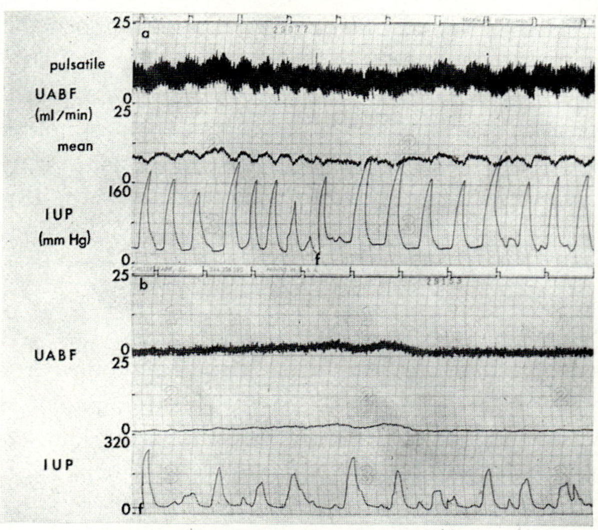

Fig. 7. Uterine artery blood flow records obtained on (a) the third and (b) the eighth days after ovulation. Menstruation began on the 16th postovulatory day. f = Flushing catheter, 0.1 ml.

tractions (fig. 7a), while lesser contractions were frequently not reflected in the flow tracing.

Discussion

These observations demonstrate that the principal (and probably sole) mechanism accounting for the decrease in uterine blood flow during uterine contractions is compression of the uteroplacental arteries and veins by the contracting myometrium. The close temporal correspondence between changes in uterine blood flow and the development of tension by the myometrium strongly indicate a direct mechanical relationship. Failure of spinal sympathetic or ganglionic blockade to modify the pattern of change would seem to exclude vasomotor activity. In high doses, moreover, trimethaphan exerts a direct vascular-relaxing effect [3]. Since the rate of trimethaphan infusion was increased until no further fall in blood pressure occurred, it is probable that vascular paralysis was induced in these experiments – further evidence that active vasoconstriction is not involved in the uterine blood flow decrease during contractions.

The less precise correspondence between intrauterine pressure increase and flow decrease observed in the *post partum* and nonpregnant animals is entirely compatible with the concept that the vessels are compressed by the contracting myometrium. During pregnancy, the amniotic fluid tends to distribute the pressure developed by myometrial contraction hydraulically throughout the uterus. The occurrence of uneven vascular compression – in location and time – resulting from asynchronous myometrial activity is thus reduced. As soon as the products of conception have been expelled, this buffer is no longer present. Without it, asynchronous contractions may compress vessels and affect flow to an extent which is not reflected by the pressure within the uterine cavity.

Underlying the calculation of separate uterine arterial and venous resistances in this report is the assumption that venous outflow from the uteroplacental vascular bed equals arterial inflow. The radiographic studies of RAMSEY and co-workers [8] have shown that, within the limits of discrimination of that technique, the quantity of blood in the intervillous space does not change during contractions. More recently, SCHEFFS *et al.* [10] have presented data from radioisotopic tracer studies which show an increase in placental blood volume during contractions in normal human gravidas. The arterial and venous resistances calculated from the present experiments also suggest that an increase in uteroplacental blood volume might occur during contractions exceeding 50 mm Hg intensity. At these pressures, uterine venous resistance exceeded that of the arteries. Should outflow indeed become less than inflow during contractions, true uterine venous resistance would be greater than that calculated here.

References

1 Borell, U.; Fernstrom, I.; Ohlson, L., and Wiqvist, N.: Effect of uterine contractions on the human uteroplacental blood circulation. Amer. J. Obstet. Gynec. *89:* 881 (1964).

2 Corner, G. W. jr.; Ramsey, E. M., and Stran, H.: Patterns of myometrial activity in the rhesus monkey in pregnancy. Amer. J. Obstet. Gynec. *85:* 2 (1963).

3 Goodman, L. S. and Gilman, A.: The pharmacological basis of therapeutics, 4th ed., p. 594 (Macmillan, New York 1970).

4 Greiss, F. C. jr.: Effect of labor on uterine blood flow. Amer. J. Obstet. Gynec. *93:* 917 (1965).

5 Lees, M. H.; Hill. J. D.; Ochsner, A. J. III; Thomas, C. J., and Novy, M. J.: Maternal placental and myometrial blood flow of the rhesus monkey during uterine contractions. Amer. J. Obstet. Gynec. *110:* 68 (1971).

6 Martin, C. B. jr. and Eckstein, P.: Transcervical uterine catheterization in rhesus monkeys. Amer. J. Obstet. Gynec. *94:* 415 (1966).

7 Ramsey, E. M.; Corner, G. W., jr., and Donner, M. W.: Serial and cineradiographic visualization of the maternal circulation in the primate (hemochorial) placenta. Amer. J. Obstet. Gynec. *86:* 213–225 (1963).

8 Ramsey, E. M.; Martin, C. B. jr.; McGaughey, H. S. jr.: Kaiser, I. H., and Donner, M. W.: Venous drainage of the placenta in rhesus monkeys. Radiographic studies. Amer. J. Obstet. Gynec. *95:* 948 (1966).

9 Robson, J. M. and Schild, H. O.: Effect of drugs on blood flow and activity of the uterus. J. Physiol., Lond. *92:* 9 (1938).

10 Scheffs, J.; Vasicka, A.; Li, C.; Solomon, N., and Siler, W.: Uterine blood flow during labor. Obstet. Gynec., N.Y. *38:* 15 (1971).

11 Reynolds, S.R.M.; Freese, U.E.; Bienarz, J.; Caldeyro-Barcia, R.; Mendez-Bauer, C., and Escarcena, L.: Multiple simultaneous intervillous space pressures recorded in several regions of the hemochorial placenta in relation to functional anatomy of the fetal cotyledon. Amer. J. Obstet. Gynec. *102:* 1128 (1968).

Author's address: Dr. C. B. Martin, jr., Department of Obstetrics and Gynecology, Los Angeles County-University of Southern California Medical Center, *Los Angeles, CA 90033* (USA)

Medical Primatology 1972. Proc. 3rd Conf. exp. Med. Surg. Primates, Lyon 1972, part I, pp. 308–316 (Karger, Basel 1972)

Evaluation of *Macaca mulatta* as an Experimental Model for Studies of Primate Reproduction

ELIZABETH M. RAMSEY

Department of Obstetrics and Gynecology, University of Virginia School of Medicine, Charlottesville, Va.

Well into the early years of the present century one basic problem confronted students of reproductive anatomy and physiology, namely, the scarcity of material for study. Pregnant human uteri came to hand but rarely and quite fortuitously and nonpregnant uteri free of pathology even more infrequently. As a result systematic study of sequential stages was precluded.

In this dilemma recourse to study of animals, as in other such cases, was not the answer, for both uterus and placenta in lower animals differ greatly from their human counterparts. Furthermore, it was originally thought that man was the only mammal having a menstrual cycle and associated structural and physiological changes.

However, in the late 1890's, HEAPE of England [9, 10] and VAN HERWERDEN of Germany [20] recorded pioneer observations of cyclic bleeding in monkeys which they had studied in the wild and in zoos. And then, in 1923, CORNER [4] established a small breeding colony of rhesus monkeys at the Johns Hopkins Medical School in Baltimore, USA, specifically for the study of sex cycles, thus opening a new era in the study of primate reproduction. Since then, as we know, the use of rhesus monkeys has yearly become more widespread both in terms of the number of workers employing them and in the types of study to which they are being devoted.

Gratifying as this development is we must not allow ourselves to be carried away by the useful 'similarity'. Resemblance must not be equated with identity nor the assumption made that observations in rhesus necessarily hold true in the human. Today, indeed, there is increasing evidence that this is not so. Species differences do exist and our use of rhesus will only be helpful and our deductions will only be valid in so far as we recognize the differences and evaluate them.

It is the purpose of this article to describe and weigh some of these differences. Since the total field is vast [6], consideration will be restricted to the circulatory system, nonpregnant and pregnant, and certain related aspects of the behavior of the trophoblast.

The basic vascular pattern of the uterus of rhesus monkey and man is indeed very much alike, especially as regards the endometrial spiral arteries which, in their response to hormonal stimuli, determine a large proportion of the changes occurring during the menstrual cycle. BARTELMEZ [1, 2] made a particularly intensive study of these vessels, the results of which are summarized in figure 1a. Figure 1b shows the results of Schmidt-Matthiesen's similar studies in the human [3, 19]. That there is little difference between the two is readily apparent.

The changes which the endometrial arteries undergo during pregnancy [14] are summarized for the monkey in the upper half of figure 2. Several points may be noted for comparison with parallel phenomena in the human. In addition to continued growth and coiling of the arteries in the first half of gestation, it will be seen that a terminal dilatation or sac develops in the artery at its junction with the base of the placenta. The marked straightening of the arteries in the final half of gestation is also noteworthy. This occurs as the coils are 'paid out' when the uterus stretches [18]. The lower half of figure 2 summarizes the changes which human arteries undergo during pregnancy [8]. Qualitatively they are the same as those which occur in rhesus, but it will be observed that the terminal sacs, which gradually become less prominent in the monkey, persist full blown to term in man. Furthermore, there is less obliteration of arterial coils in man than in monkey. Explanations of these two divergencies are to be found, for the first in the different action of trophoblast on arterial walls, as will be discussed below, and for the second in the difference in thickness of the myometrium in the two species, that of the monkey being much thinner than that of the human, so that stretching is greater.

Here then are certain definite species differences. The important question in the present context is: How significant are they? Do they materially influence the mechanism of circulation?

To find an answer, attention must be turned to physiological studies and, since the uteroplacental circulation is influenced by myometrial activity, it is necessary in the first place to compare the patterns of uterine contractility in monkey and man. It will markedly influence interpretation of circulatory patterns if myometrial activity is basically different in the two species.

If representative recordings of monkey and human uterine activity at

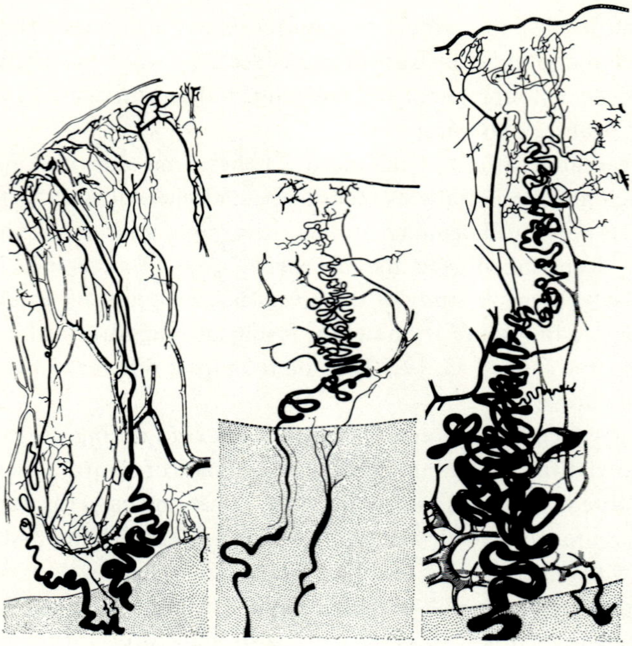

1a

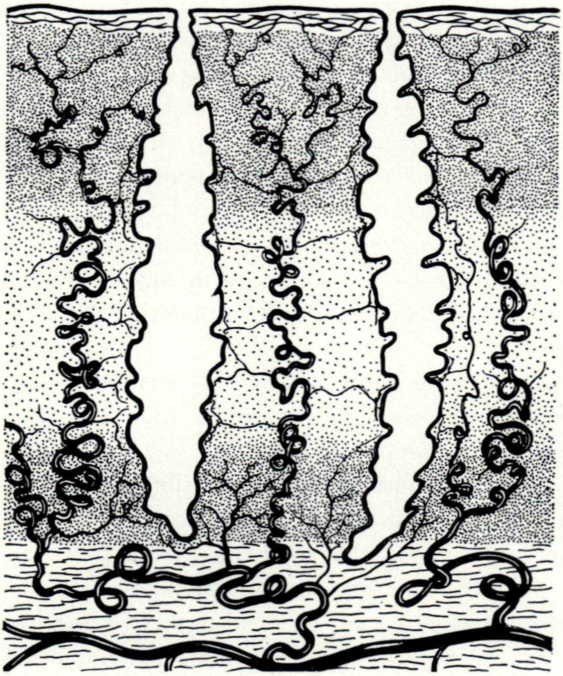

1b

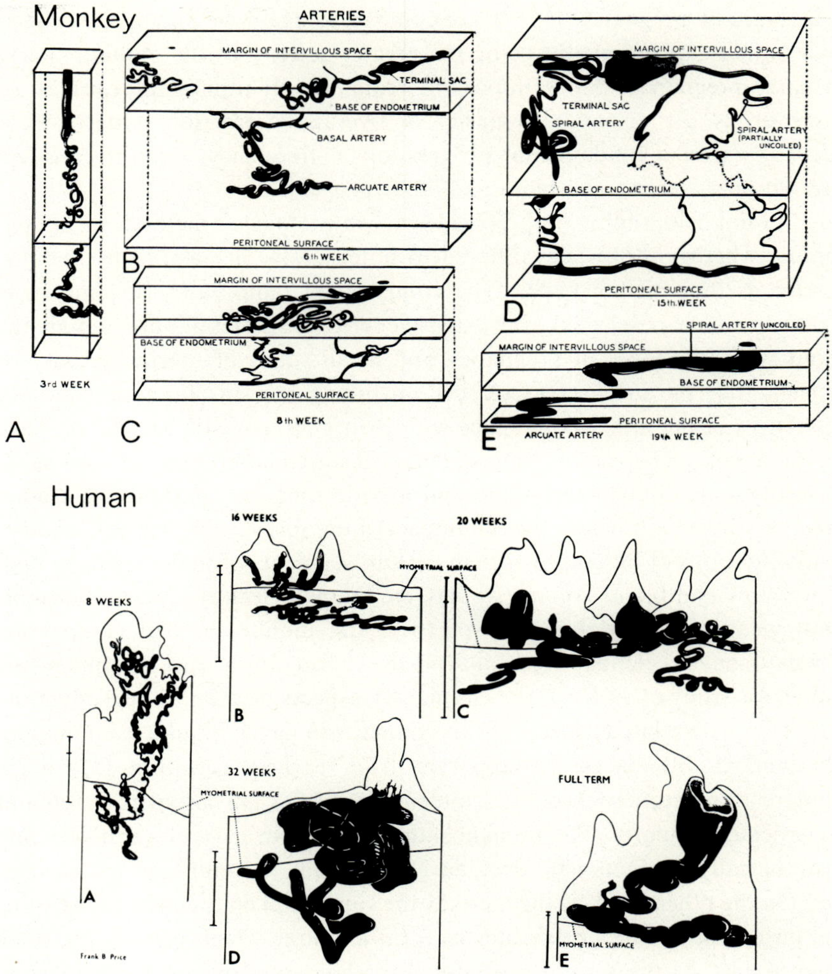

Fig. 2. Diagrammatic representations of the course and configuration of the utero-placental arteries in the rhesus monkey and man at comparable stages of gestation. Reprinted with permission from RAMSEY and HARRIS [17].

Fig. 1. a Camera lucida drawings of the vascular bed at three stages of the menstrual cycle of the rhesus monkey. Left: postmenstrual; center: postovulatory; right: late secretory. Myometrium stippled. *b* A similar drawing of the vascular bed of the human endometrium in the luteal phase of the menstrual cycle. Figure 1a reprinted with permission from BARTELMEZ [2] and 1b from BOYD and HAMILTON [3] after SCHMIDT-MATTHIESEN [19].

key stages of pregnancy [5, 17] are compared, again we find species differences, the most prominent being the greater activity of the monkey uterus in early pregnancy. More important from the standpoint of circulation, however, is the essential similarity of overall contraction pattern. Both species exhibit a fundamental progression in frequency, intensity and coordination as term approaches and labor progresses.

Radioangiography [16] provides a physiological method for determining whether the species differences noted above are as inconsequential *in vivo* as they are on paper. The comparative radiograms [17] shown in figure 3a and b are selected from extensive series of studies of both human and monkey at all stages of pregnancy and in all states of uterine activity. It is clear that the uterine vasculature, including the spiral arteries and the interstices of the intervillous space with their characteristic 'spurts' or 'jets', fill in a completely analogous manner. Radioangiography, of course, is basically a qualitative technique and a wide range of quantitative studies must be carried out before the full impact of the species differences is known: determinations of blood pressure in various segments of the circulatory bed; blood flow and blood volume studies, etc. Nonetheless, the analysis of morphological and physiological aspects of the monkey-human comparison which we have sketched so far confirms the validity of the rhesus monkey as a model for study of, at least, the circulatory aspects of primate reproduction.

There are related aspects of this comparison which require examination. Anatomical differences are apparent from earliest gestation [15] when human implantation becomes interstitial while the monkey's remains superficial. Similarly, the human endometrium forms a pronounced and histologically characteristic decidua [15, 17], but the monkey stroma does not. On the other hand, in the monkey the superficial epithelium surrounding the implantation site proliferates to form a plaque which persists for some three weeks. Nothing similar occurs in the human. It is noteworthy that in

---

Fig. 3. a and *b* Radiographs made following injection of contrast material into the femoral artery in monkey (a) and man (b). Carnegie monkey 1B, 117th day of pregnancy. Human specimen, University of Virginia, No. 37 26 56, 32nd week of pregnancy (courtesy of the late Dr. HARRY S. MCGAUGHEY jr.). Reprinted with permission from RAMSEY and HARRIS [17]. *c* and *d* Uteroplacental arteries near the placental base to show trophoblastic penetration and replacement of vessel wall. In the monkey (c), note absence of trophoblast in endometrial stroma and stout arterial wall. Injected India ink is present in the lumen of the artery. In the human (d), trophoblast is present in the arterial lumen and in the periarterial stroma. Note single cell at 6 o'clock which is displacing endothelium toward the center of the vessel lumen. Carnegie monkey C-629, 53rd day of pregnancy.

a *b*

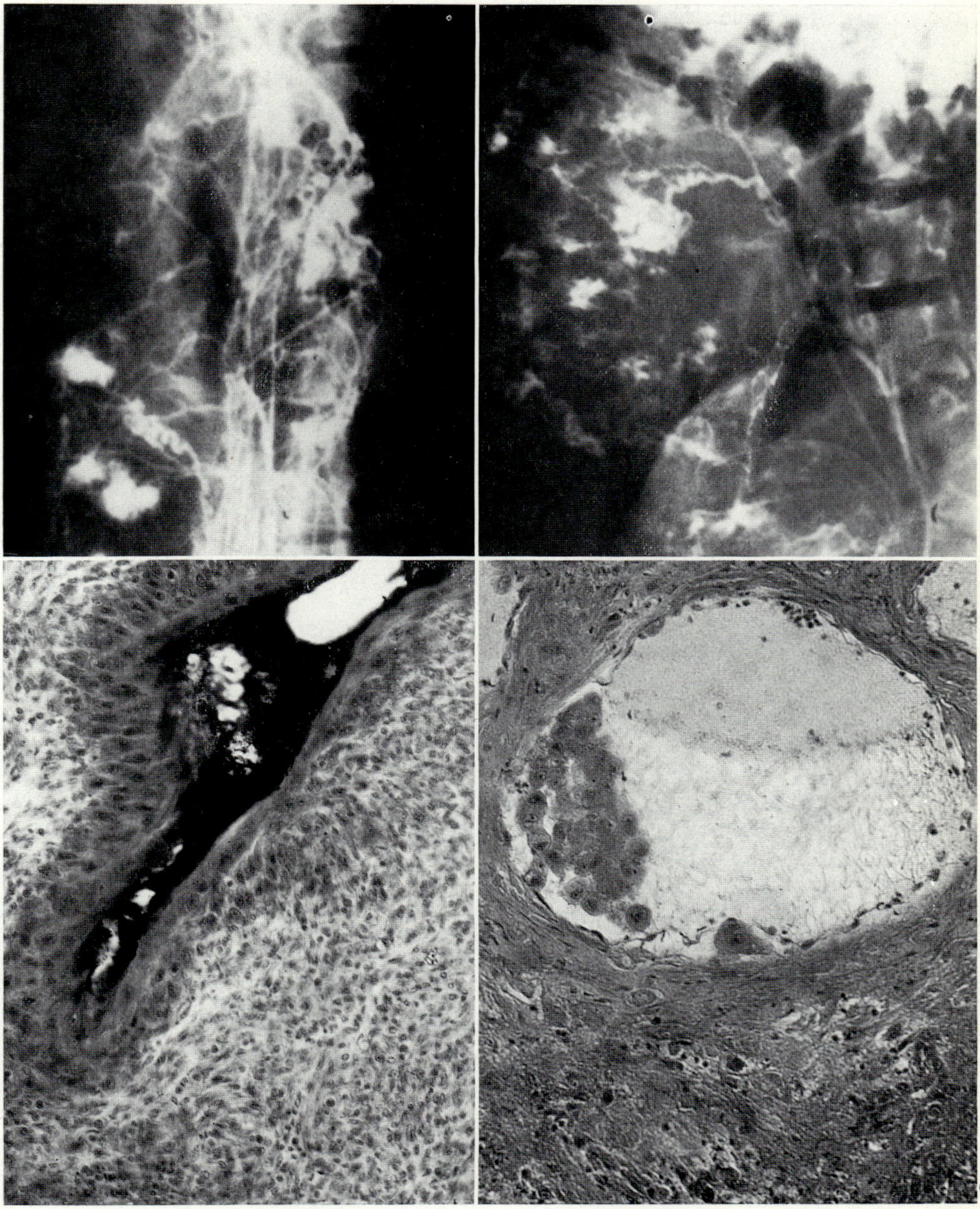

c *d*

Section 100, × 200. Human specimen, Carnegie Collection No. 2328, 69th day of pregnancy. × 200. Figure 3c reprinted with permission from RAMSEY [15] and 3d from HARRIS and RAMSEY [8].

the three respects just noted the baboon resembles the human rather than the monkey [12, 13]. A little later in gestation, when the placenta has taken definite form and a maternal-fetal junctional zone has become established, it can be noted that the junction is a fairly smooth band in the rhesus monkey but a strikingly jagged one in the human [15].

The latter feature leads to consideration of the varying invasive potential of human *versus* monkey trophoblast. That human trophoblast is appreciably the more active is further attested by the widespread infiltration of single trophoblastic giant cells throughout the endometrium and even into the myometrium (fig. 4a), a phenomenon familiar to obstetrical pathologists but not duplicated in rhesus.

Even more significant from the point of view of the circulation is the action of trophoblast upon arterial walls [7, 8, 17]. At an early stage the monkey shows accumulation of cytotrophoblastic cells within the lumina of arteries which connect with the intervillous space. These intraluminal accumulations gradually progress deeper and deeper into the endometrial and even myometrial segments of the arteries, 'like candle wax dripping down the inner wall' in the graphic phrase of HARRIS' own [7]. The same

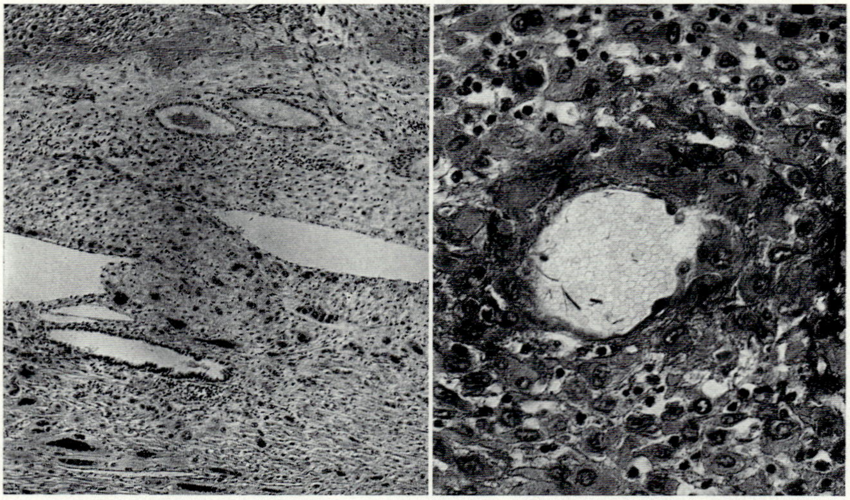

a b

Fig. 4. a Section of human uterine wall beneath the site of placental attachment showing extensive penetration of trophoblastic giant cells. Carnegie Collection No. 2274. × 50. Courtesy of the Carnegie Institution of Washington. *b* A uteroplacental artery at the placental base in a human specimen. A weak ring of trophoblast replaces the normal arterial wall. Heavy trophoblastic infiltration of periarterial stroma. Carnegie Collection No. 10117, 85th day of pregnancy. × 400. Reprinted with permission from RAMSEY [15].

phenomenon occurs in the human but not until some weeks later [8]. Still later, in both species, the trophoblast invades and replaces the arterial wall. The sharp difference lies in the mechanism of replacement. In rhesus the penetration is entirely from within the lumen (fig. 3c). In man, penetration is both from within by the intraluminal cells and from without by the wandering trophoblastic giant cells previously mentioned (fig. 3d). In the monkey the trophoblast forms a fairly competent wall, although muscular and elastic components have been eliminated so that normal vascular reactivity is impossible – hence the early terminal dilatation or sac which is later minimized. In man the trophoblastic action, while far from uniform, is essentially more destructive and no solid trophoblastic wall is formed (fig. 4b). This in turn is responsible for the persistence, and even augmentation, of the terminal sac as pregnancy advances.

Since it was concluded in a previous section that the uteroplacental circulatory systems of rhesus and man are closely similar, the question may be asked why these additional considerations have been raised. To answer prompts reiteration of the thesis of this article: There are species differences; we have examined just a few of them. It appears that within certain contexts (here the mechanism of uteroplacental circulation) they do not impair the reliability of rhesus as a model. Nevertheless, in a different context, they might entirely preclude transfer of findings from monkey to man. Therefore, it is imperative to be fully familiar with the species differences and to evaluate them carefully if data are to be useful and deductions valid.

References

1 Bartelmez, G. W.: The human uterine mucous membrane during menstruation. Amer. J. Obstet. Gynec. *21:* 623–643 (1931).

2 Bartelmez, G. W.: The form and the functions of the uterine blood vessels in the rhesus monkey. Contrib. Embryol. Carneg. Inst. *36:* 153–181 (1957).

3 Boyd, J. D. and Hamilton, W. J.: The human placenta (Heffer & Sons, Cambridge 1970).

4 Corner, G. W. jr.: Ovulation and menstruation in *Macacus rhesus*. Contrib. Embryol. Carneg. Inst. *15:* 73–101 (1923).

5 Corner, G. W. jr.; Ramsey, E. M., and Stran, H. M.: Patterns of myometrial activity in the rhesus monkey in pregnancy. Amer. J. Obstet. Gynec. *85:* 179–185 (1963).

6 Gruenwald, P.: Lobular structure of hemochorial primate placentas, and its relation to maternal vessels (in press).

7 Harris, J. W. S.; Houston, M. L., and Ramsey, E. M.: Intravascular trophoblast in human, baboon and monkey uteri. Anat. Rec. *169:* 334 (1971).

8 HARRIS, J. W. S. and RAMSEY, E. M.: The morphology of human uteroplacental vasculature. Contrib. Embryol. Carneg. Inst. *38:* 43–58 (1966).

9 HEAPE, W.: The menstruation of *Semnopithecus entellus*. Philos. Trans. B *185:* 411–471 (1894).

10 HEAPE, W.: The menstruation and ovulation of *Macacus rhesus*, with observations on the changes undergone by the discharged follicle. Philos. Trans. B *188:* 134–166 (1897).

11 HOUSTON, M. L.: The villous period of placentogenesis in the baboon *(Papio* sp.*)*. Amer. J. Anat. *126:* 1–15 (1969).

12 HOUSTON, M. L.: The development of the baboon *(Papio* sp.*)* placenta during the fetal period of gestation. Amer. J. Anat. *126:* 17–29 (1969).

13 PANIGEL, M.: Structure et ultrastructure comparées de la membrane placentaire chez certains primates non humains *(Galago demidovii, Erythrocebus patas, Macaca irus* [*fascicularis*], *Macaca mulatta* et *Papio cynocephalus)*. Bull. Ass. Anat. *145:* 319–337 (1969).

14 RAMSEY, E. M.: The vascular pattern of the endometrium of the pregnant rhesus monkey *(Macaca mulatta)*. Contrib. Embryol. Carneg. Inst. *33:* 113–147 (1949).

15 RAMSEY, E. M.: Placental circulation in rhesus and man; in MACK Prenatal life, pp. 37–53 (Wayne State University Press, Detroit 1970).

16 RAMSEY, E. M.; CORNER, G. W. jr., and DONNER, M. W.: Serial and cineradioangiographic visualization of maternal circulation in the primate (hemochorial) placenta. Amer. J. Obstet. Gynec. *86:* 213–225 (1963).

17 RAMSEY, E. M. and HARRIS, J. W. S.: Comparison of uteroplacental vasculature and circulation in the rhesus monkey and man. Contrib. Embryol. Carneg. Inst. *38:* 59–70 (1966).

18 REYNOLDS, S. R. M.: Uterine accomodation of the products of conception. Physiologic considerations. Amer. J. Obstet. Gynec. *53:* 901–913 (1947).

19 SCHMIDT-MATTHIESEN, H.: The normal human endometrium (McGraw-Hill, New York 1963).

20 VAN HERWERDEN, M.: Beitrag zur Kenntnis des menstruellen Cyclus. Mschr. Geburtsh. Gynäk. *24:* 730–748 (1906).

Author's address: Dr. ELIZABETH M. RAMSEY, Department of Obstetrics and Gynecology, University of Virginia School of Medicine, *Charlottesville, VA 22901* (USA)

Medical Primatology 1972. Proc. 3rd Conf. exp. Med. Surg. Primates, Lyon 1972, part I, pp. 317–324 (Karger, Basel 1972)

Studies of Menstrual Physiology Using the Uterocutaneous Fistula Model of *Macaca mulatta*[1]

Applications to Contraceptive Research

S. T. SHAW, jr., D. L. MOYER, C. H. POON and T. P. W. NOGUEIRA

Section of Experimental Pathology, Departments of Obstetrics and Gynecology and Pathology, University of Southern California School of Medicine, Los Angeles, Cal.

Excessive uterine bleeding, as menorrhagia, metrorrhagia, or both, is the most important single complication of contraception with the intrauterine device (IUD) in humans [TIETZE, in press]. This complication is under current study in our laboratory from the standpoints of mechanism, quantitative definition of the problem, and prevention. The major emphasis of this work is on uterine tissue responses. In regard to investigations concerning mechanism and prevention, experimental observations on human subjects are limited by legal, social, and ethical restraints. For this reason, we are employing the female rhesus monkey *(Macaca mulatta)* as a model for the study of genital bleeding associated with IUD contraception in women. This paper will be devoted to a justification of this approach in two types of rhesus models.

There are many similarities between rhesus monkeys and human females in their reproductive functions. Figure 1 is a histogram illustrating the distribution of menstrual cycle lengths among the two species. Our data (the bottom histogram) and the data of HARTMAN [1932] and AREY [1939, graphed by HARTMAN, 1962] are taken from adult subjects in apparent good health, and indicate similar means and distributions for humans and *Macaca mulatta*. The 1923 data of CORNER [from HARTMAN, 1932] is from adolescent monkeys and shows a wider distribution of cycle lengths, which is also seen in adolescent human females. In addition to these similarities, we have recently reported a comparable range of blood loss occurs during menses in the rhesus monkey and human when the menstrual blood volume is related

1 Supported by a grant from the Ford Foundation.

to body weight (ml/kg) [SHAW *et al.*, 1972]. Patterns of blood concentrations for luteinizing hormone (LH), estradiol, and progesterone during the menstrual cycles of both species indicate that pituitary-ovarian function is also similar. LH and progesterone apparently follow nearly identical patterns in rhesus monkeys and humans [NEILL *et al.*, 1967; MONROE *et al.*, 1970], although minor differences in estradiol pattern may exist [MISHELL *et al.*, 1971; HOTCHKISS *et al.*, 1971; ABRAHAM *et al.*, 1972].

Studies on uterine reaction to an IUD in rhesus monkeys have also been carried out. Morphologic response of endometrium of both intact [KAR and CHANDRA, 1965; KELLY *et al.*, 1969a] and castrated-hormonally stimulated [TAMADA and STURGIS, 1966] monkeys appears to be similar to that in humans [ISRAEL and DAVIS, 1966; MOYER and MISHELL, 1971]. In addition, the IUD has been shown to have a definite contraceptive effect in

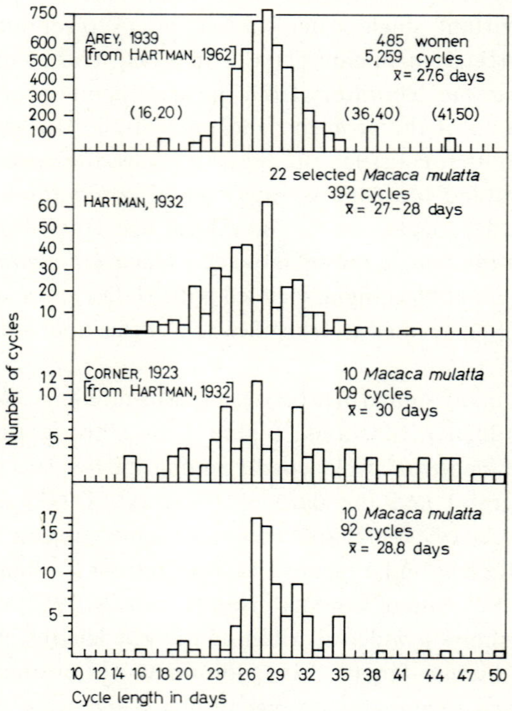

Fig. 1. Histograms of menstrual cycle length distribution in women and rhesus monkeys. (The histograms by HARTMAN 1932 and CORNER 1923 have been submitted by courtesy of Carnegie Institution of Washington.)

rhesus monkeys [KELLY *et al.*, 1969b], as is true in women. Finally, the low cost of procuring and maintaining a colony of rhesus monkeys relative to that of other subhuman primates (e.g. baboons and chimpanzees) makes this species very useful as a model for studying many human responses.

Certain aspects of the *Macaca mulatta* species, however, give rise to significant disadvantages for such studies. For instance, ovulation is subject to considerable seasonal variation and other fluctuations, even in individually caged animals under controlled lighting and temperature [KEVERNE and MICHAEL, 1970; RIESEN *et al.*, 1971]. Periods of amenorrhea without ovulation are frequent in the summer months [RIESEN *et al.*, 1971]. Our experience with ten animals during 13 months demonstrates, in addition, periods of amenorrhea in some animals at times of the year other than summer, and considerable variation of cycle length within individual animals (table I). Whereas a few monkeys were quite regular in the frequency of their menstruation (e.g. No. 182 and 185), others showed considerable variation in cycle length (e.g. No. 180, varying from 25 to 45 days, and 183, varying from 20 to 35 days), while others menstruated only sporadically during the year (e.g. No. 181, 186 and 187). These animals were maintained in individual cages in a room containing caged male monkeys; and temperature and light were controlled. Another disadvantage of the rhesus monkey as a research model for the human is that access to the uterine cavity through the cervical canal is difficult in nearly all animals, and essentially impossible in many. Figure 2 illustrates the anterior colliculus which distorts the cervical canal

Table I. Menstrual cycle length (days)

Period	Monkey No.									
	180	181	182	183	185	186	187	189	190	191
April–May 1971	28	28	31	20	–	–	28	29	20	25
May–June	27	27	27	26	28	32	–	–	35	21
June–July	30	–	27	26	31	31	–	–	–	43
July–August	26	–	27	35	30	31	–	–	–	–
August–September	28	–	32	32	27	–	–	–	–	–
September–October	26	–	23	29	30	–	–	31	–	–
October–November	29	–	26	29	27	–	27	27	30	–
November–December	25	27	29	29	31	–	–	31	34	24
December–January 1972	45	25	29	–	28	–	–	27	23	26
January–February	28	–	28	–	28	26	28	31	32	28
February–March	27	–	27	29	28	25	27	19	31	–
March–April	35	–	33	30	29	–	–	–	32	–
April–May	35	–	28	28	34	–	–	–	27	29

in its upper half. In addition to the canal being of minute caliber as it approaches the internal os, it seems to possess blind tracts into which a probe is more likely to pass than into the uterine cavity. Attempts at dilatation of the endocervical canal have often resulted, in our hands, in perforation of the upper cervical or lower uterine wall.

In an attempt to gain easier access to the uterine cavity for endometrial biopsies, GOOD and MOYER [1966] developed a uterocutaneous fistula model of the rhesus monkey. This procedure involves exposing the uterus through a midline abdominal incision; cutting a hole through the wall of the uterine fundus into the upper uterine cavity; elevating the upper uterus into the abdominal incision, sewing first the peritoneum to the upper uterine corpus, and then the anterior rectus sheath to the uterine fundus; and, finally, folding the skin into the hole in the uterus and sewing the skin edges to the endometrium in order to form a skin-lined fistula into the cavity (fig. 2). The fistula permits introduction of biopsy curets, intrauterine devices, and other instruments into the uterine cavity with ease and accuracy. Reliable samples may be removed without either anesthesia or undue trauma to the animal. The original fistula colony developed by GOOD and MOYER [1966] was ovariectomized in order to study the relationships between estradiol, progesterone, and endometrial response. An injection schedule of estradiol and progesterone which produced normal histologic maturation in the endometrium of these animals was thereby established [GOOD and MOYER, 1968].

The ovariectomized-fistula model has been used for our research primarily because hormone injections can provide reliable and synchronous

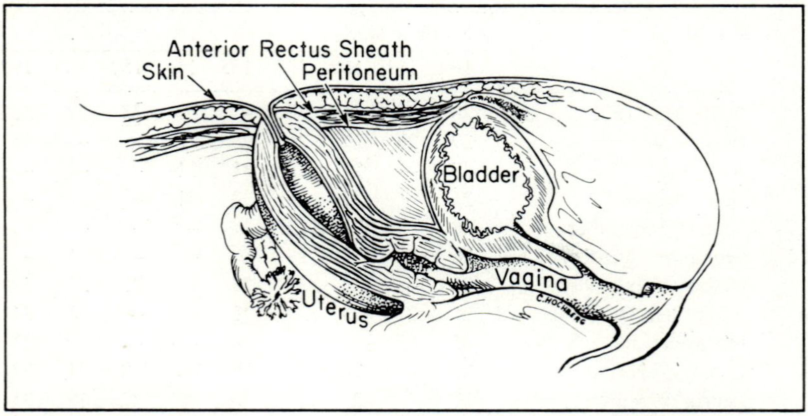

Fig. 2. Diagram of midsagittal section through reproductive organs of *Macaca mulatta* to show narrow, bending cervical canal and uterocutaneous fistula.

menstrual cycles in a large group of animals. Based on the work of GOOD and MOYER [1968], we currently employ the following injection schedule which is simultaneously administered to approximately 25 ovariectomized-fistula monkeys: 25 μg estradiol-17β in propylene glycol i.m. daily for 22 consecutive days, and 1.5 mg progesterone in sesame oil i.m. daily for the last 12 consecutive days of estradiol administration. At the end of the 22-day schedule, hormones are withdrawn for one week to allow for menstruation, which regularly occurs in all animals. Thus, a large number of monkeys may be artificially synchronized by this schedule in order that samples representing the same cycle day for all animals may be accurately collected at one time. The primary advantage of this hormone-stimulated model over the intact animal is, in our opinion, that it provides a better basis for making inter- and intraindividual comparisons of data. A secondary advantage is enhanced laboratory convenience and efficiency.

Preliminary study of peripheral blood levels of estradiol and progesterone in our ovariectomized-fistula model [unpubl. study with Dr. I. THORNEYCROFT] indicates that the progesterone pattern of the rhesus menstrual cycle may be reasonably simulated by the outlined injection schedule, but the estradiol pattern is not. In an effort to produce a more physiologic model, therefore, we have formed uterocutaneous fistulae in ten adult animals leaving the ovaries intact. Menstrual cycle duration and blood loss have been studied prior to and following fistula surgery in order to evaluate the effect, if any, of this anatomical alteration on ovarian function. Because of permanent uterine displacement caused by the procedure, impairment of ovarian blood supply seemed possible. Regarding cycle duration, the grand mean and standard deviation of the ten monkeys followed for 47 cycles prior to operation were 29.00 and 2.41 days, respectively; whereas the grand mean and standard deviation for 45 cycles after the operation were 28.47 and 1.59 days, respectively. These means and standard deviations were not significantly different by the t- and F-tests, respectively ($t = 1.24$, $F = 1.51$). In regard to menstrual blood loss changes after fistula surgery, table II contains values for nine of the ten animals measured before and after fistulation, respectively. Note that the ranges of blood loss are similar pre- and postoperatively. The mean values appear to differ, however, exhibiting a decrease in menstruation after surgery. Marked variations of menstrual loss in ovary-intact animals indicates that the 23 cycles measured prior to surgery primarily during fall and winter months were an insufficient number of seasonally biased measurements to serve as a basis for comparison with postsurgery cycles. Additional animals should probably be measured during one year before and one year following surgery in order to test the effect of fistulation on menstrual blood loss. The ovary intact-fistula animals show a very similar mean and range (on a volume per body weight basis) compared to that determined for healthy, nonanemic humans (table II).

The ranges of menstrual blood loss (in ml/kg body weight) in healthy, nonanemic human subjects and all three monkey groups are similar (table II). On the other hand, the mean loss (ml/kg) for these human subjects appears to be considerably higher than for the ovariectomized rhesus model, and lower than the mean loss for intact rhesus monkeys (before fistula). These apparent differences merit further study. With baseline values for menstrual blood loss established in the fistula models, nonetheless, and with some assurance that monkey endometrium reacts in a fashion similar to that of the woman, quantitative studies of uterine bleeding in response to various IUD materials and designs may be pursued.

Finally, in an attempt to define mechanisms which may promote hemorrhage from endometrium exposed to the intrauterine device, studies of proteolytic factors, particularly those in the plasminogen-plasmin system, have been undertaken in the fistula rhesus models. Radioactive fibrin contained in sealed Millipore chambers has been incubated inside the uterine cavities of fistula animals in late luteal phase (day 21). Control chambers incubated *in vitro* lost only 6.1 ± 1.6% of radioactivity. Chambers incubated for the same time period in animals not wearing an IUD lost 32.9 ± 10.2%, while those from IUD-wearing animals lost 56.7 ± 8.0%. Thus, there appears

Table II. Menstrual blood loss

	Rhesus monkey			Human	
	10 ovari-ectomized fistula (n = 71)[1]	9 ovary-intact before fistula (n = 23)	9 ovary-intact after fistula (n = 44)	healthy nonanemic[2] 283 subjects (n = 523)	Göteborg, random sample[3] 458 subjects
X, ml	2.11	4.69	2.80	31.12	43.40
Range, ml	0.28–10.51	0.82–12.86	0.52–15.79	3–130	550
X, ml/kg	0.39	0.84	0.50	0.54	0.75
Range, ml/kg	0.05–1.78	0.14–1.89	0.09–2.32	0.05–2.24	9.48

1 Calculated from 71 menstrual periods measured.

2 Combined data from BALDWIN *et al.* [1961], HALLBERG and NILLSON [1964]. TAUXE [1962], JACOBS and BUTLER [1965], HYTTEN [1964], and HALLBERG *et al.* [1966].

3 From HALLBERG *et al.* [1966].

to be a stimulation of fibrin proteolysis caused by the IUD in monkeys [SHAW *et al.*, 1970; CIHAK and SHAW, in press]. Current work in the monkey is directed to establishing a relationship, if any, between proteolytic activity of endometrial fluid and tissue and the magnitude of blood loss associated with intrauterine devices.

References

ABRAHAM, G. E.; ODELL, W. D.; SWERDLOFF, R. S., and HOPPER, K.: Simultaneous radioimmunoassay of plasma FSH, LH, progesterone, 17-hydroxyprogesterone, and estradiol-17β during the menstrual cycle. J. clin. Endocrin. *34:* 312 (1972).

AREY, L. B.: The degree of normal menstrual irregularity. An analysis of 20,000 calendar records from 1,500 Individuals. Amer. J. Obstet. Gynec. *37:* 12 (1939)

BALDWIN, R. M.; WHALLEY, P. J., and PRITCHARD, J. A.: Measurements of menstrual blood loss. Amer. J. Obstet. Gynec. *81:* 739 (1961).

CIHAK, R. W. and SHAW, S. T. jr.: Measurement of fibrin proteolysis *in vivo* by a radioactive fibrin-millipore tube technique. Thromb. Diath. haemorrh. (in press).

GOOD, R. G. and MOYER, D. L.: Technique of serial endometrial biopsy in the monkey through a utero-cutaneous fistula. J. Reprod. Fertil. *12:* 573 (1966).

GOOD, R. G. and MOYER, D. L.: Estrogen-progesterone relationships in the development of secretory endometrium. Fertil. Steril. *19:* 37 (1968).

HALLBERG, L. and NILSSON, L.: Constancy of individual menstrual blood loss. Acta Obstet gynec. Scand. *43:* 352 (1964).

HALLBERG, L.; HOGDAHL, A.; NILSSON, L., and RYBO, G.: Menstrual blood loss. A population study. Acta Obstet. gynec. Scand. *45:* 320 (1966).

HARTMAN, C. G.: Studies in the reproduction of the monkey macacus (pithecus) rhesus, with special reference to menstruation and pregnancy. Contrib. Embroyol. Carneg. Inst. *23:* 1 (1932).

HARTMAN, C. G.: Science and the safe period, p. 130 (Williams & Wilkins, Baltimore 1962).

HOTCHKISS, J.; ATKINSON, L.E., and KNOBIL, E.: Time course of serum estrogen and luteinizing hormone (LH) concentrations during the menstrual cycle of the rhesus monkey. Endocrinology *89:* 177 (1971).

HYTTEN, F. E.; CHEYNE, G. A., and KLOPPER, A. I.: Iron loss at menstruation. J. Obstet. Gynaec. brit. Clth. *71:* 255 (1964).

ISRAEL, R. and DAVIS, H. J.: Effect of intrauterine contraceptive devices on the endometrium. J. amer. med. Ass. *195:* 764 (1966).

JACOBS, A. and BUTLER, E. B.: Menstrual blood loss in iron-deficiency anemia. Lancet *ii:* 407 (1965).

KAR, A. B. and CHANDRA, H.: Uterine bleeding in rhesus monkeys after insertion of an intrauterine contraceptive device. Indian J. exp. Biol. *3:* 269 (1965).

KELLY, W. A.; MARSTON, J. H., and ECKSTEIN, P.: Effect of an intra-uterine device on endometrial morphology and the deciduomal reaction in the rhesus monkey. J. Reprod. Fertil. *19:* 331 (1969a).

KELLY, W. A.; MARSTON, J. H., and ECKSTEIN, P.: Effect of an intra-uterine device on conception and ovulation in the rhesus monkey. J. Reprod. Fertil. *19:* 143 (1969b).

KEVERNE, E. B. and MICHAEL, R. P.: Annual changes in the menstruation of rhesus monkeys. J. Endocrin. *48:* 669 (1970).

MISHELL, D. R. jr.; NAKAMURA, R. M.; CROSIGNANI, P. G.; STONE, S.; KHARMA, K., and NAGATA, Y.: Serum gonadotropin and steroid patterns during the normal menstrual cycle. Amer. J. Obstet. Gynec. *3:* 60 (1971).

MONROE, S. E.; ATKINSON, L. E., and KNOBIL, E.: Patterns of circulating luteinizing hormone and their relation to plasma progesterone levels during the menstrual cycle of the rhesus monkey. Endocrinology *87:* 453 (1970).

MOYER, D. L. and MISHELL, D. R. jr.: Reactions of the human endometrium to the intrauterine foreign body. II. Long-term effects on the endometrial Histology and cytology. Amer. J. Obstet. Gynec. *3:* 66 (1971).

NEILL, J. D.; JOHANSSON, E. D. B., and KNOBIL, E.: Levels of progesterone in peripheral plasma during the menstrual cycle of the rhesus monkey. Endocrinology *81:* 1161 (1967).

RIESEN, J. W.; MEYER, R. K., and WOLF, R. C.: The effect of season on occurrence of ovulation in the rhesus monkey. Biol. Reprod. *5:* 111 (1971).

SHAW, S. T. jr.; CIHAK, R. W., and MOYER, D. L.: Fibrin Proteolysis in the monkey uterine cavity. Variations with and without IUD. Nature *228:* 1097 (1970).

SHAW, S. T. jr.; EL SAHWI, S. Y., and MOYER, D. L.: Menstrual blood quantitation in the rhesus monkey. An experimental tool for improving intrauterine contraceptive devices (IUDS). Fertil. Steril. *23:* 257 (1972).

TAMADA, T. and STURGIS, S. H.: The endometrium of hormone-stimulated castrated monkeys bearing an intrauterine device. Obstet. Gynec. *28:* 793 (1966).

TAUXE, W. N.: Quantitation of menstrual blood loss. A radioactive method utilizing a counting dome. J. nucl. Med. *3:* 282 (1962).

TIETZE, C.: Intrauterine Devices. Clinical aspects; in HAFEZ and EVANS Human fertility. Conception and contraception (Harper & Row, London, in press)

Author's address: Dr. S. T. SHAW, Section of Experimental Pathology, Department of Obstetrics and Gynecology, University of Southern California School of Medicine, *Los Angeles, CA 90033* (USA)

Genetics, Phylogenetics and Evolution

Medical Primatology 1972. Proc. 3rd Conf. exp. Med. Surg. Primates, Lyon 1972, part I, pp. 327–335 (Karger, Basel 1972)

Serum Protein Polymorphisms in Rhesus Monkeys

K. BERG[1] and H. BALNER

Institute of Medical Genetics, University of Oslo, Oslo and Radiobiological Institute, TNO, Rijswijk

The purpose of the present report is to summarize results of a search we have made for polymorphism in serums from the rhesus monkey colony at Rijswijk. We have looked for antigenic polymorphisms in β-lipoprotein by means of precipitating antisera against inherited human antigens, and for electrophoretic protein variations by a high voltage agarose gel electrophoresis technique [BERG *et al.*, 1970]. With the latter technique, we studied albumin, transferrin and the complement component C'3.

The Lp Polymorphism

The technique used for Lp type determination has been described [BERG, 1964].

When sera from rhesus monkeys *(Macaca mulatta)* were tested against a rabbit antiserum against human Lp(a) antigen, positive and negative reactions were observed. The precipitin bands obtained with those rhesus monkey sera which gave a positive reaction, were in general quite distinct and strong and the presence or absence of this antigen clearly represents a dimorphism also in this species.

When serum from a positive rhesus monkey was tested next to a human Lp(a+) serum, the precipitin band obtained with the latter serum spurred over that obtained with rhesus monkey serum, producing a reaction of partial identity. This finding, which is in agreement with a previous observation [BERG, 1969] indicates that the Lp antigen of the rhesus monkey is

1 Supported by The Norwegian Research Council for Science and the Humanities and by a gift from the life insurance company Norske Liv.

antigenically deficient, compared to that of man. Because of this difference, we shall refer to Lp positive rhesus monkeys as 'reactors' rather than as being Lp(a+), and to Lp negative rhesus monkeys as 'non-reactors'.

The results of testing sera from 87 not closely related rhesus monkeys are summarized in table I. The estimated frequency (0.5712) of the dominant Lp gene is higher than observed in any human population sample. It is, however, of the same order of magnitude as that observed previously in a smaller series of rhesus monkeys [BERG, 1969].

Serum samples from members of 16 rhesus monkey family groups were also tested. Four different males had each been mated to from three to five different females (totalling 16 females) and produced a total of 54 offspring. The results of serum screening of these families are shown in table II. The distribution of reactors and non-reactors in the family groups provides strong evidence that the Lp antigen is governed by an autosomal dominant mode of inheritance also in the rhesus monkey.

Table I. Occurrence of Lp antigen in a colony of rhesus monkeys

		Number	Frequency
Reactors		71	0.82
Non-reactors		16	0.18
Total		87	1.00
Gene frequencies			
Dominant gene	0.5712		
Recessive gene	0.4288		

Table II. Reactions of sera from members of 16 rhesus monkey family groups with antiserum to human Lp(a) antigen

Father	Mother	Offspring
–	+	+
		+
		–
	+	–
		–
		–
		+
	+	+
		–
		+
		–
		+

Table II, (continued)

Father	Mother	Offspring
	−	−
		−
		−
+	+	+
		+
		+
	+	+
		+
		+
		+
	+	+
		+
		+
		+
	+	+
		+
		+
		+
	+	+
		+
−	+	+
		−
		−
		+
		−
	+	+
		+
		−
	+	+
		+
		−
	+	+
		−
+	+	+
		+
		+
		−
	+	+
		−
		+
	+	−
		+

The demonstration of the Lp serum lipoprotein polymorphism in the rhesus monkey may suggest that the event which gave rise to the Lp polymorphism occurred before the separation of the human evolutionary line from those of other primates.

Search for the Ag(x) and Lt(a) Antigens of Serum β-Lipoprotein

The technique and the reagents used to test for these antigens have been described [BERG and BEARN, 1970].

We tested the same series of rhesus monkey sera with human antiserum to the human Ag(x) lipoprotein antigen. No positive reactions were obtained. Thus, in this colony of rhesus monkeys no antigen sufficiently similar to human Ag(x) antigen to react with a precipitating antiserum appears to be present.

We also tested the rhesus monkey sera with antiserum of human extraction to the inherited human β-lipoprotein antigen which we refer to as Lt(a) [see BERG and BEARN, 1970]. Positive and negative reactions occurred. However, many of the reactions were extremely weak and hard to score. Of the scoreable sera from not closely related rhesus monkeys, about 80% possessed an antigen reacting with anti-Lt(a) serum. The limited family data available indicate that the antigen is inherited in an autosomal dominant fashion also in this species.

However weak many of the reactions were, the results clearly indicate that an Lt polymorphism is present also in the rhesus monkey.

The C'3 Polymorphism

The details of the technique of high voltage electrophoresis in agarose gel have been reported [BERG *et al.*, 1970].

In a minority of the rhesus monkey sera, the C'3 zones were sufficiently distinct to permit scoring of possible electrophoretic variants in this complement protein. As illustrated in figure 1, differences in electrophoretic pattern were observed in the C'3 area. Since these protein zones could be recognized with certainty in only a few samples, we can present no data at this stage as to the distribution of different C'3 types among unrelated individuals or in families. However, the data appear to confirm the observation by ALPER *et al.* [1971] that the C'3 component of complement is polymorphic also in the rhesus monkey.

Search for Genetic Variants of Albumin

The electrophoretic technique used also permits scoring of possible variants of albumin. We have looked for such variation in all 141 samples tested but found no example of double albumin or of a single albumin zone having an atypical electrophoretic mobility. Thus, electrophoretic albumin variants were not observed in this population of rhesus monkeys.

The Transferrin Polymorphism

The transferrin region is easily recognized with the electrophoretic technique employed. A high degree of polymorphism in this protein soon became evident. We identified the transferrin zones by specific staining, and the transferrin nature of several of the variants was confirmed by immunologic tests. In the first experiments, it was obvious that transferrin zones with several different mobilities could occur and we found it interesting to try to find out whether protein zones with only slightly different electrophoretic mobility in fact reflected the presence of different genes. Because

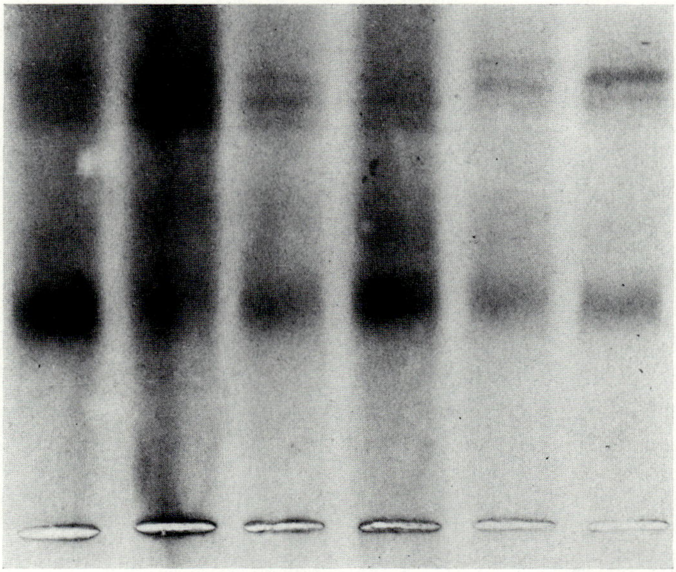

Fig. 1. Electrophoretic patterns (top set of zones) of the complement component C'3 obtained with sera from six different rhesus monkeys. Three different phenotypes are discernible.

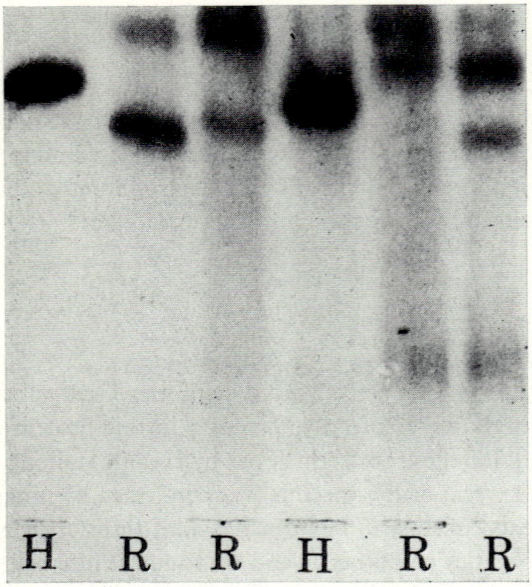

Fig. 2. Electrophoretic patterns of transferrin obtained with sera from four different rhesus monkeys (R) compared to that of human transferrin C (H).

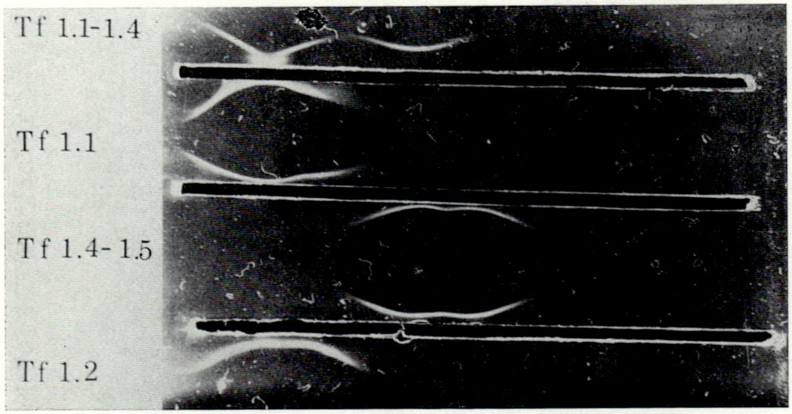

Fig. 3. Immunological identification of protein zones as transferrin variants. Following electrophoresis in agarose, antiserum to human transferrin were introduced into troughs parallel to the direction of protein migration. The figures indicate length of migration relative to human transferrin C. The anode was to the right.

Table III. Transferrin types[1] in 16 ıhesus monkey family groups

Father	Mother	Offspring
1.24–1.4	1.2	1.2–1.24
		1.2–1.24
		1.2–1.24
	1.1–1.5	1.4–1.5
		1.4–1.5
		1.4–1.5
	1.2	1.2–1.24
		1.2–1.24
		1.2–1.24
		1.2–1.4
		1.2–1.4
	1.1–1.4	1.4
		1.1–1.4
		1.24–1.4
1.2	1.0–1.2	1.2
		1.0–1.2
		1.2
	1.2–1.3	1.2
		1.2–1.3
		1.2
		1.2
	1.2–1.3	1.2
		1.2–1.3
		1.2
		1.2
	1.2	1.2
		1.2
		1.2
		1.2
	1.2	1.2
		1.2
1.1	1.1–1.5	1.1
		1.1
		1.1–1.5
		1.1
		1.1
	1.1	1.1
		1.1
		1.1
	1.1–1.5	1.1
		1.1
		1.1
	0.9–1.2	1.1–1.2
		1.1–1.2

Table III. (continued)

Father	Mother	Offspring
1.2–1.24	1.2	1.2
		1.2–1.24
		1.2
		1.2–1.24
	1.24–1.4	1.2–1.4
		1.24–1.4
		1.2–1.4
	1.24–1.4	1.24–1.4
		1.2–1.4

1 Each figure indicates presence of a transferrin component and its electrophoretic mobility, upon agarose electrophoresis, relative to human transferrin C.

we could analyse the occurrence of these protein zones in sera from members of family groups, we had the opportunity to study this.

We tested the samples from all family members next to samples of human serum containing the human transferrin C and measured the electrophoretic mobility of each individual transferrin zone relative to transferrin C. This relative mobility of transferrin in individual samples exhibited a high degree of reproducibility when the same samples were tested repeatedly. Examples of different transferrin zones are shown in figure 2. Immunological confirmation of protein zones as transferrin is illustrated in figure 3, where figures indicating relative mobility of the components are included.

These figures have been arrived at by dividing the length of migration of the rhesus transferrin zone in question by the length of migration of the human transferrin C zone. Thus, the phenotypes exhibiting only one protein zone are symbolized by one single figure, those with two transferrin zones by two figures. We employ this system of notation merely for easy identification of electrophoretic zones.

It turned out in the family analysis that electrophoretic zones defined as described behaved as if determined by codominant alleles at a single autosomal locus. The family data are presented in detail in table III. Transmission from parent to offspring was observed for the following variants: 1.0, 1.1, 1.2, 1.24, 1.3, 1.4, 1.5. Transferrin variants slower than human transferrin C were also observed, but no family data are as yet available for the former. The possible heterogeneity of the precipitation arc representing the sample designated Tf 1.2 in figure 3 might reflect still further genetic heterogeneity.

Genetic variation in rhesus monkey transferrin has been reported by several workers previously, but the amount of family data available is still limited. In this study, evidence for genetic transmission of seven different variants was obtained. Additional variants were observed. Our results agree with those of GOODMAN and POULIK [1961] who more than ten years ago called attention to the high degree of polymorphism in rhesus monkey transferrin. Recent data [FRELINGER, 1972] indicating that powerful selective pressure may be operating to maintain the transferrin polymorphism in other species add interest to the remarkable polymorphism of transferrin in the rhesus monkey.

References

ALPER, C. A., ROBIN, N. I., and REFETOFF, S.: Genetic polymorphism in rhesus C3 and Gc globulin. J. Immunol. *107:* 96–98 (1971).

BERG, K.: Studies on the reaction between Lp(a+) human sera and anti-Lp(a) sera from rabbits. Acta path. microbiol. scand. *62:* 613–622 (1964).

BERG, K.: The use of nonhuman primates in the study of genetic antigenic markers of human serum proteins. Ann. N.Y. Acad. Sci. *162:* 189–194 (1969).

BERG, K. and BEARN, A. G.: Antibodies to inherited β-lipoprotein antigens in the serum of multiply transfused patients. Clin. Genet. *1:* 104–120 (1970).

BERG, K., NOER, G., and HEIBERG, A.: Studies on the genetic polymorphism of the C'3 of human serum. Clin. Genet. *1:* 177–180 (1970).

FRELINGER, J. A.: The maintenance of transferrin polymorphism in pigeons. Proc. nat. Acad. Sci., Wash. *69:* 326–329 (1972).

GOODMAN, M. and POULIK, E.: Serum transferrins in the genus *Macaca*. Species distribution of nineteen phenotypes. Nature, Lond. *191:* 1407–1408 (1961).

Authors' addresses: Prof. KÅRE BERG, MD, Director, Institute of Medical Genetics, University of Oslo, Blindern, *Oslo 3* (Norway); Dr. HANS BALNER, Radiobiological Institute of the Organization for Health Research TNO, 151 Lange Kleiweg, *Rijswijk Z.H.* (The Netherlands)

Medical Primatology 1972. Proc. 3rd Conf. exp. Med. Surg. Primates, Lyon 1972, part I, pp. 336–341 (Karger, Basel 1972)

Chromosomal Abnormalities in Primates

J. Egozcue

Instituto de Biología Fundamental, Universidad Autónoma de Barcelona, Barcelona

Chromosomal abnormalities are rarely observed in wild animals, either because the affected individuals become the easy victims of predators, or succumb to disease or to the hardships of capture and transportation. However, some of these animals do reach the laboratories where they can be studied. This is especially true in the case of carriers of balanced translocations.

During the past few years, we have observed six cases of chromosomal abnormalities in primates. Three other reports have appeared in the literature. In this paper, we comment on our cases, and review the published reports.

Case Descriptions

So far, the chromosomal abnormalities seen in primates include three cases of numerical aberrations, three cases of balanced translocation, one intrachromosomal rearrangement and two cases of sex chromosome abnormalities.

Numerical Aberrations

As far as numerical abnormalities are concerned, two cases of hyperploidy and one case of hypoploidy are known.

We studied one male ouakari *(Cacajao rubicundus)* which was consistently found to have only 45 chromosomes, instead of the normal 46 [Chu and Bender, 1961]. Since Chu and Bender's report did not include a description of the karyotype, identification of the missing chromosome was difficult. However, all submetacentric chromosomes in the complement could be classified; on the other hand, the animal studied, an apparently normal male, could hardly be considered as an XO. Thus, the case has tentatively been classified as an autosomal monosomy affecting one of the acrocentric chromosomal pairs.

In 1965, HSU described one female *Alouatta villosa* with 53 chromosomes. Since no description of the karyotype is available, and *Alouatta caraya* has 52 chromosomes, the case can be interpreted as a trisomy.

A better documented and most interesting case of chromosomal hyperploidy was published by MCCLURE *et al.* [1969, 1971]. The affected animal was a female chimpanzee *(Pan troglodytes)* with clinical and behavioral features similar to that seen in Down's syndrome. Cytogenetic studies revealed a trisomy for one of the smaller acrocentric pairs.

Balanced Translocations

Three cases of balanced translocation are known. One male *Ateles geoffroyi* studied by us [EGOZCUE *et al.*, 1969] had a reciprocal translocation affecting the long arms of a member of pair No. 3 and the short arms of a chromosome No. 9. The animal was apparently normal.

One female *Saimiri* sp. had an insertion of the heterochromatic region of one member of pair No. 4 into its homologue. The translocation produced one chromosome with a heterochromatic region twice the normal size, and a metacentric chromosome without the heterochromatic zone (fig. 1).

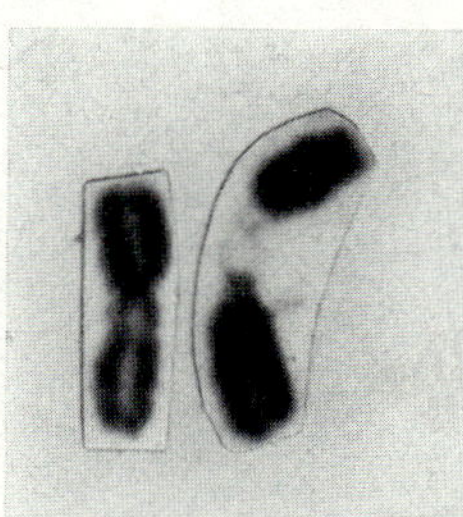
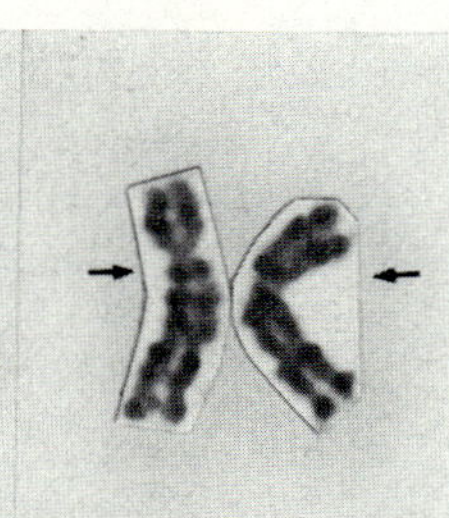

2 1

Fig. 1. Insertion of the heterochromatic region of chromosome No. 4 of *Saimiri* into its homologue. The inserted chromosome shows an achromatic zone twice the normal size.

Fig. 2. Pericentric inversion in *Ateles geoffroyi.* The first band of the long arms of the normal chromosome (left) is in the short arms of the inverted chromosome (right).

Recently, DE BOER [1971] has described one female *Aotus trivirgatus* with 53 chromosomes instead of the normal 54, and a centric fusion affecting two nonhomologous acrocentrics.

Intrachromosomal Rearrangements

The only case of chromosomal rearrangement observed so far is that of a female *Ateles geoffroyi* with a pericentric inversion in a chromosome No. 1. Giemsa-banding studies have confirmed the presence of the inversion (fig. 2).

Sex Chromosome Abnormalities

Two cases of sex chromosome abnormalities are known. In 1965, blood samples were obtained from one apparently normal baby chimpanzee *(Pan troglodytes)*. However, since the chromosomal complement of the chimpanzee was well known [YEAGER *et al.*, 1940; YOUNG *et al.*, 1960; BENDER and

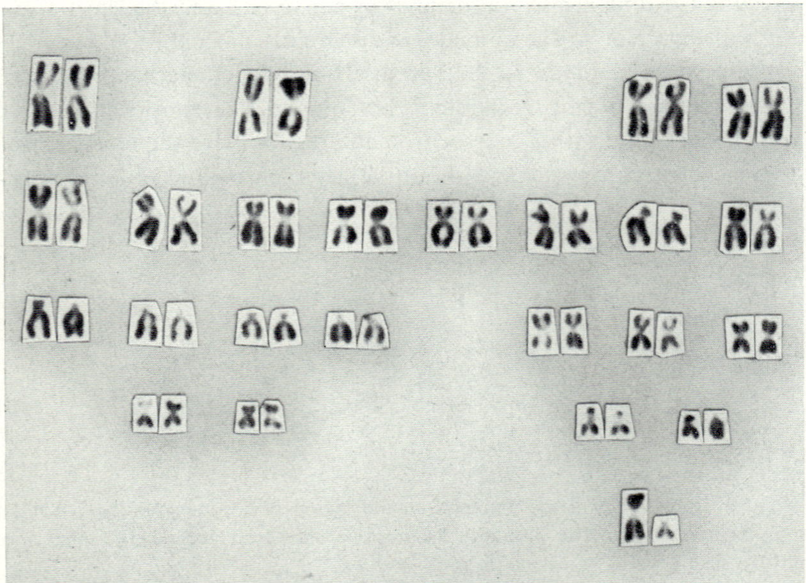

Fig. 3. XY karyotype of mosaic XX/XY chimpanzee.

METTLER, 1961; CHIARELLI, 1962] the preparations were not analyzed until recently. The animal, a male, turned out to be an XX/XY mosaic (fig. 3, 4) with a predominance of XX cells (60 %). All efforts to locate this chimpanzee have been unsuccessful so far.

We have also studied one 'male' *Presbytis entellus* whose karyotype had been first reported by USHIJIMA *et al.* [1964]. Since the karyotype reported by these authors did not coincide with the descriptions of CHIARELLI [1966], SHARMA and KAKATI [1966], GOLDSTEIN and BIRDWELL [1967] and HSU and BENIRSCHKE [1970], we restudied the case, and came to the conclusion that the animal must have been an XX male with a deletion of the short arms of one autosome [EGOZCUE, 1971].

Discussion

The small number of chromosomal abnormalities found so far in primates is probably the reflection of the scant possibility of survival of the affected animals.

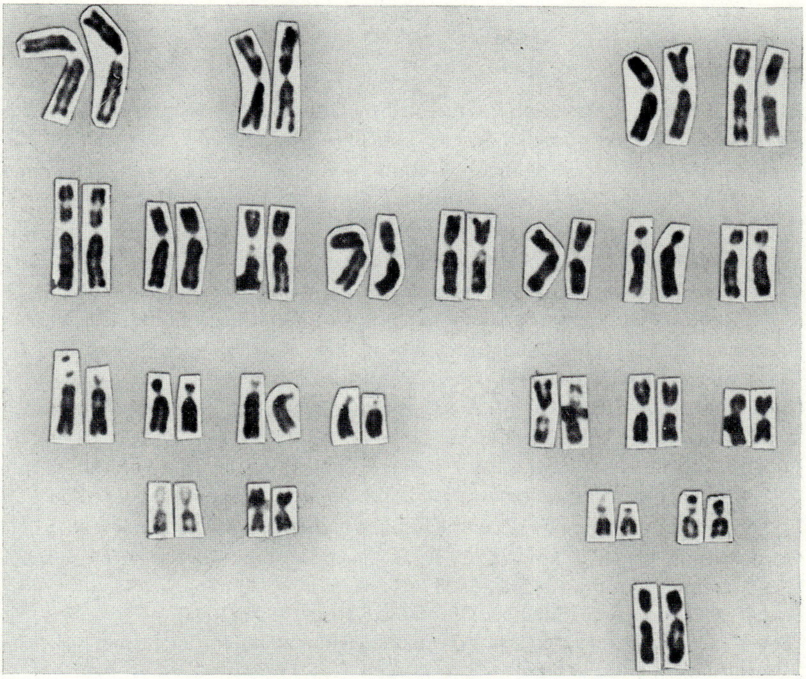

Fig. 4. XX karyotype of mosaic XX/XY chimpanzee.

Some of the abnormalities described could be considered as a reflection of the well-documented evolutionary drive to increase chromosomal size and/or metacentricity. However, when such changes have not yet been established in a population, their classification as 'abnormalities' seems warranted, at least for the time being. Such is the case of the centric fusion described by DE BOER [1971] in *Aotus trivirgatus*. When the report was published, it was the only known case of this type in *Aotus;* meanwhile, BRUMBACK *et al.* [1971] demonstrated that such a karyotype was an intermediate step between the diploid numbers of 2n = 54 and 2n = 52 found in this species.

The mongoloid-like condition reported by MCCLURE *et al.* [1969] in a female chimpanzee is of special interest, since the cytogenetic findings coincide with the chromosomal anomaly found in Down's syndrome in man. Recently [EGOZCUE, unpublished] we have studied the Q- and G-banding patterns of the chromosomes of the chimpanzee *(Pan troglodytes);* the patterns obtained are identical to those found in man. The results suggest that a high degree of genetic homology must exist between the two species.

References

BENDER, M. A. and METTLER, L. E.: Cited by CHU and BENDER Chromosome cytology and evolution in primates. Science *133:* 1399–1405 (1961).

BOER, L. E. M. DE: A case of abnormal chromosome number in a female *Aotus trivirgatus* (Humboldt, 1811) *(Cebidae,* Primates). Genen Phaenen *14:* 51–55 (1971).

BRUMBACK, R. A.; STATON, R.D.; BENJAMIN, S. A., and LANG, C. M.: The chromosomes of *Aotus trivirgatus* Humboldt 1812. Folia primat. *15:* 264–273 (1971).

CHIARELLI, B.: Comparative morphometric analysis of primate chromosomes. I. The chromosomes of anthropoid apes and of man. Caryologia *15:* 99–121 (1962).

CHIARELLI, B.: Caryology and taxonomy of the Catarrhine monkeys. Amer. J. phys. Anthrop. *24:* 155–169 (1966).

CHU, E. H. Y. and BENDER, M. A.: Chromosome cytology and evolution in primates. Science *133:* 1399–1405 (1961).

EGOZCUE, J.; PERKINS, E. M.; HAGEMENAS, F., and FORD, D. M.: The chromosomes of some Platyrrhini *(Callicebus, Ateles* and *Saimiri).* Folia primat. *11:* 17–27 (1969).

EGOZCUE, J.: A possible case of centric fision in a primate. Experientia *27:* 969–970 (1971).

GOLDSTEIN, P. R. and BIRDWELL, T. R.: Cytogenetic studies in the purple-faced langur. Mammal. Chromos. Newsl. *8:* 197–201 (1967).

HSU, T. C.: Foreword. Mammal. Chromos. Newsl. *15:* 98 (1965).

HSU, T. C. and BENIRSCHKE, K. An atlas of mammalian chromosomes, vol. 4, p. 199 (Springer, Berlin 1970).

MCCLURE, H. M.; BELDEN, K. H.; PIEPER, W. A., and JACOBSON, C. B.: Autosomal trisomy in a chimpanzee. Resemblance to Down's syndrome. Science *165:* 1010–1012 (1969).

McClure, H. M.; Pieper, W. A.; Keeling, M. E., and Jacobson, C. B.: Mongoloid-like condition in a chimpanzee; in Biegert and Kummer Proc. 3rd Int. Congr. Primatology, Zurich 1970, vol. 2, pp. 110–115 (Karger, Basel 1971).

Sharma, T. and Kakati, S. Chromosomes of indian common langur. Mammal. Chromos. Newsl. *20:* 70 (1966).

Ushijima, R. N.; Shininger, F. S., and Grand, T. I.: Chromosome complements of two species of primates: *Cynopithecus niger* and *Presbytis entellus*. Science *146:* 78–79 (1964).

Yeager, C. H.; Painter, T. S., and Yerkes, R. M.: The chromosomes of the chimpanzee. Science *91:* 74–75 (1940).

Young, W. J.; Mergt, T.; Ferguson-Smith, M. A., and Johnston, A. W.: Chromosome number of the chimpanzee *(Pan troglodytes)*. Science *131:* 1672–1673 (1960).

Author's address: Dr. J. Egozcue, Instituto de Biología Fundamental, Universidad Autónoma de Barcelona, *Barcelona* (Spain)

Medical Primatology 1972. Proc. 3rd Conf. exp. Med. Surg. Primates, Lyon 1972, part I, pp. 342–348 (Karger, Basel 1972)

Sexual Differences in Adult Skulls of *Pan paniscus*

R. FENART and R. DEBLOCK

Laboratoire de Craniologie comparée, Faculté libre de Médecine, CNRS, Lille

The choice of this primate for testing the sexual differences between skulls is motivated by the fact that the dwarf chimpanzee is the closest relative to mankind among contemporary anthropomorphs. This fact is proven by many anatomical, physiological and biological characteristics. To know the 'sexual parameters' of its skull is to approach in an indirect way, the phylogenetic way, the problem of the sexual differences in the human skull. It is indeed justified to wonder whether the corresponding measurements performed on *Homo sapiens,* which too rarely are clearly significant, compare with the data on *Pan paniscus.*

Material and Methods

Fifty-nine skulls of adult *Pan paniscus* (28 males and 31 females) have been examined, as much for their classical measurements as for those drawn from their orientation according to the vestibular axes. For each specimen, 326 parameters have been noted: metric, angular relations, index, volumes and weights. Normal statistical methods were used for establishing the average and the standard deviation, for the males and females, and this led to the evaluation of Students' 't' for each dimension. (Further calculations with the help of a discriminant function are planned.)

Results

99% significance has been obtained with 37 sexual parameters of skulls (of 326), therefore with a little more than $^1/_{10}$. If one goes down to 95%, 28 other parameters may be added.

However, we limit ourselves to the 37 highly significant parameters which are listed below. (The abbreviations in the table refer to the averages, standard deviations and 't'.)

Aa	upper oblique width of the ascending branch
Ba	basion
Ba.C.a.	distance: anterior sifted-basion
b.al.	minimal width or alar pinching
b.As.	biasteric width
b.cr.	expanded 'bicrest' width
b.m.m.	maximal bimastoïdian width
b.z.m.	maximal bizygomatic width
C.a.-C.p.-Ba	angle: anterior sifted-posterior sifted-basion
C.a.-S-Ba	angle: anterior sifted-centre of the turcic saddle-basion
CC'	bicondylian width
CD'	coronoïdian height on the mandibular plane
EF'	condylian height on the mandibular plane
Eg	distance: gonion-condylian point
EL'	distance: condyle-symphysal lower point
E.n.h.	height of the nasal opening
gg'	bigoniac width
Go	gonion (ord.)
Hg	oblique width of the ascending branch
$\frac{100\text{ Hg}}{\text{R'R''}}$	index of relative, oblique width of the ascending branch
(Kl	point of Klaatsch, at the junction of Gl-L and Ba-Br)
$\frac{100\text{ Kl-L}}{\text{Gl-L}}$	index: Kl-lambda/Glabel-lambda
l.c.s.	mesio-distal width of the crown of the upper canine tooth
L-I-Op	angle: lambda-inion-opisthion
l.t.o.	maximal width of the occipital hole
O.e.	external orbital point (absc.)
P.mt.	metafacial point (ord.)
Post	hindmost point of the skull (ord.)
P.p.	posterior palatine point (ord.)
P.V.I.	maximal internal prevestibular length
Py	pterygoalveolar point (ord.)
QR"	maximal width of the mandibular condyle
(R'R"	maximal bicondylian width of the mandible)
S	centre of the turcic saddle (ord.)
S.i.	symphysal inferior point (ord.)
v	distance between the centre of the axes and the point of intersection of the vestibular horizontal and the orbitomandibular line
Vol	cranial capacity
ρ	occipital basilar angle
Σ	angle of orbital frontalisation

Table I

	Adult females		Adult males		't'
Parameters	M	σ	M	σ	
Mandible					
Aa	35.48	2.05	37.43	2.74	2.88
CD'	52.50	2.82	56.47	3.76	4.28
EF'	48.59	3.86	52.58	5.06	3.17
Eg	49.61	3.23	52.37	4.30	2.60
EL'	90.29	4.59	94.33	4.57	3.12
Hg.	32.42	3.56	35.69	3.41	3.35
$\frac{100\ Hg}{R'R''}$	34.36	3.82	36.99	3.14	2.65
QR"	18.82	1.38	20.23	1.81	3.14
CC'	78.09	3.86	82.01	4.84	3.20
gg'	69.68	5.35	74.98	4.86	3.70
Coordinates					
abs. O.e.	–65.65	3.20	–68.71	3.14	3.64
ord. Ba.	–11.42	1.91	–13.29	2.01	3.60
ord. Go.	–55.04	5.31	–60.81	7.35	3.26
ord. Py.	–44.66	4.46	–47.57	3.50	2.72
ord. S.i.	–97.38	6.18	–102.40	4.29	3.38
ord. S.	0.23	2.23	–1.93	2.56	3.37
ord. P.mt.	–20.55	3.21	–21.77	2.98	5.14
ord. P.p.	–37.68	3.47	–40.34	3.21	2.99
ord. Post.	30.35	5.86	26.54	4.63	2.70
Cranium and face					
b.cr.	77.80	22.3	61	16.20	3.15
E.n.h.	22.39	1.42	24.24	1.34	4.92
l.t.o.	22.12	1.30	23.21	1.34	3.02
$\frac{100\ Kl\text{-}L}{Gl\text{-}L}$	45.10	2.50	43.46	2.19	2.60
Vol.	340.06	34.15	363.82	31.39	2.73
C.a.-Ba	70.19	3.34	72.86	3.82	2.82
v	75.11	4.04	78.20	3.03	3.13
P.V.I.	67.70	3.51	71	2.80	3.87
C.a.-C.p.-Ba	193.23	6.74	186.98	8.87	2.98
C.a.-S-Ba	144.03	6.40	148.82	6.05	2.87
p	32.15	5.10	28.57	3.19	3.14
L-I-Op	123.87	6.02	127.54	5.35	2.80
l.c.s.	8.74	0.45	10.99	0.51	15.91
b.As.	76.85	3.04	73.77	3.62	3.49
b.m.m.	104.97	3.78	108.03	4.01	2.96
b.z.m.	110.12	5.05	116.65	5.60	4.53
b.al.	67.01	3.26	63.79	2.04	4.42
Σ	155.17	4.68	158.43	3.96	2.77

Interpretation

(1) It is advisable to consider separately (fig. 1) the mesiodistal width of the crown of the upper canine whose 't' is 15.9, a very large value that is easily understood when we consider the histogram where the male values are placed against the female values; the classes do not cross. The value of 10 mm is a very sure limit when it is taken on a teleskull in profile, drawn at 4.50 m.

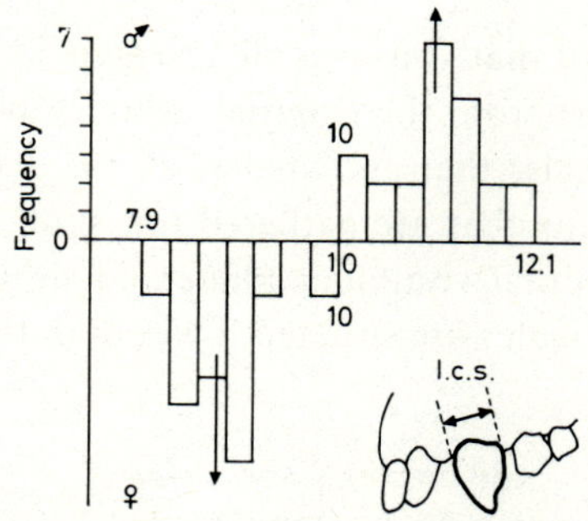

Fig. 1. Histograms of distribution of frequencies concerning the width (teleradiographic) of the upper canine tooth. Comparison between males and females.

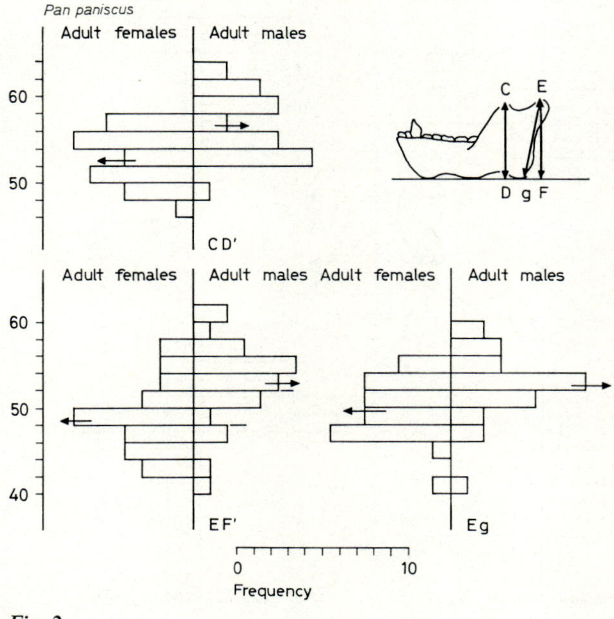

Fig. 2

The other sexual parameters have a far lower 't' (4.92 for the height of the nasal opening and 5.14 for the ordinate of the metafacial point, are the highest parameters registered). These parameters are found again with almost equal frequency in the craniofacial and mandibular classical measurements, and in the vestibular coordinates of craniometrical points, in sagittal projection. The endocranial capacity with a 't' of 2.73 is worth noting.

As a guide, you will find histograms in figures 2–4, showing some examples of frequency distributions for measurements taken on skull and mandible.

(2) The analysis of the table shows that, on a skull oriented in the vestibular axes, and seen in profile, apart from the external orbitary point which is noticeably more advanced in males than in females, all the sexual differences bear on the ordinates of points that are gathered (on skull and mandible) in the area near the verticofrontal vestibular plane and situated mostly in front of it. All those points of males are situated lower than those

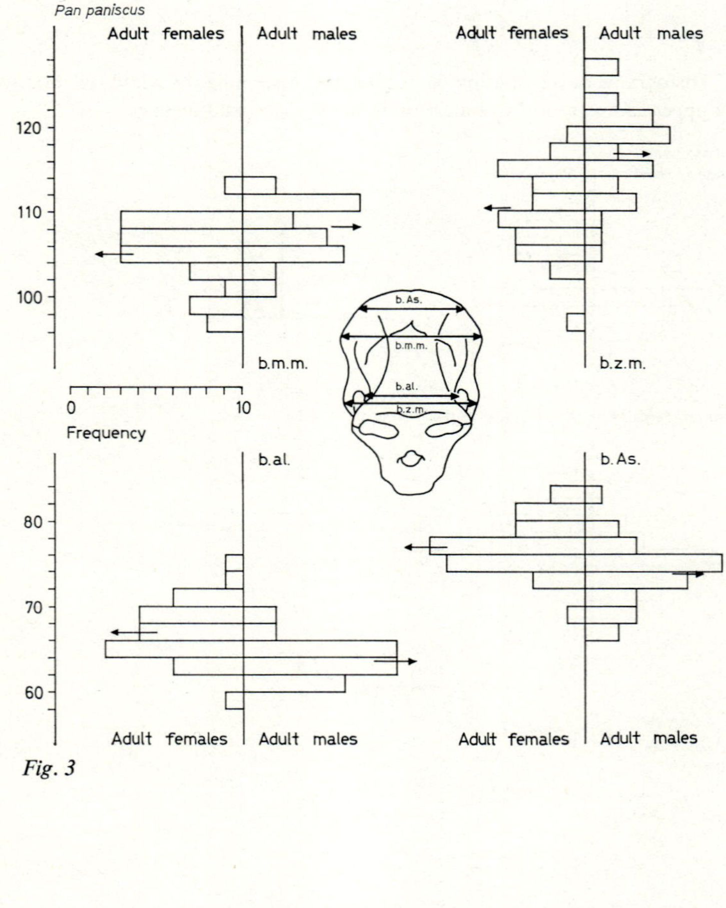

Fig. 3

of females (splanchnocranial higher). It is the same for the hindmost point of the skull, which is higher in males than in females.

(3) Several length parameters of the prevestibular segments (in antero-posterior or oblique direction) are significantly longer in males than in females. We find this again in the neurocranium, the base of skull and mandible (especially the width of the ascending branch); but it should be noted that the most advanced parts of profile (in front of a line joining the orbit and the inferior symphysary point) are not taken into account.

(4) Along the vertical dimension, the tendency to length is very clear for the ascending branch of the mandible.

(5) Transversally, it is in the parietooccipital region that one finds the sexual parameters that are bigger in males than in females in some respects (the bizygomatic maximum) but smaller in others (especially the biasteric, the bialar, and the distance between the temporal crests). That characteristic is curious and worth noting. The main posterior mandibular widths are higher in males; thus, the mandible ascending branches demonstrate clearly a sexual difference on *Pan paniscus* because of the superior dimensions of males.

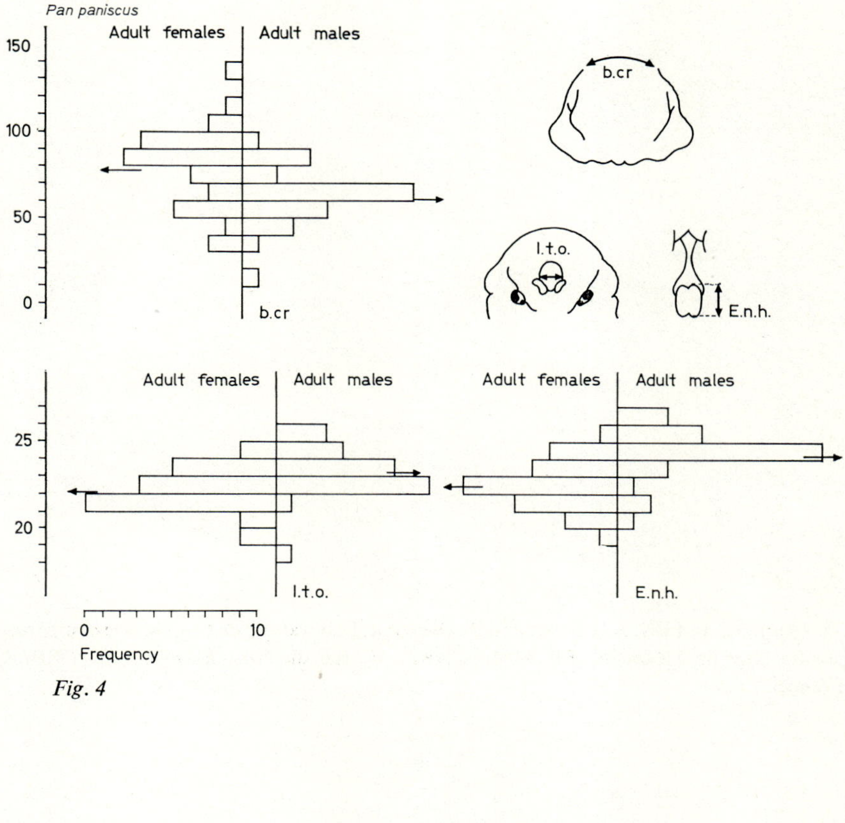

Fig. 4

(6) Among the angles with a 't' beyond the 99-percent limit, three concern the base of skull, one the occipital (flatter in males) and another the frontal (showing a projection of external orbital rims more marked in males than in females).

(7) It should be noted, finally, that neither the foraminian angle nor the hiatic angle are included in our list of significant parameters (99 or 95%).

Conclusion

In conclusion, we may say that in *Pan paniscus,* the sexual differences in the skull are apparently mainly in the posterosuperior part in the width, in the two-third posterior prevestibular part in the length and in the posterior part of the splanchnocranium and the base of the skull in the height. But it is the width of the canine that is the major distinguishing feature. In the near future, these results will be compared with those observed on the common chimpanzee *(Pan troglodytes),* in order to compare the two species in relation to each other.

Authors' address: R. FENART and R. DEBLOCK, Laboratoire de Craniologie comparée, Faculté libre de Médecine, ER 93 du CNRS, 56, rue du Port, *F-59046 Lille CEDEX* (France)

Medical Primatology 1972. Proc. 3rd Conf. exp. Med. Surg. Primates, Lyon 1972, part I, pp. 349–360 (Karger, Basel 1972)

Identification of Three Blood Factors in the Marmoset, *Saguinus fuscicollis* ssp.[1,2]

N. GENGOZIAN and MARY L. PATTON

Our ability to differentiate marmosets by their red blood cell antigens is becoming increasingly important in view of the now-established experimental role of this primate. Studies in viral oncology, dental disease, transplantation, reproduction, hematology, and infectious diseases have recently demonstrated successful exploitation of this primate family [9]. The present report represents an extension of our earlier studies [5, 6] concerning the development of isoimmune reagents in the species *Saguinus fuscicollis* and documents the existence of at least three blood factors. Breeding studies suggest the identified blood factors are inherited as simple Mendelian dominants and although each appears to be limited to one subspecies of *S. fuscicollis,* matings between two subspecies can result in the expression of each factor in the hybrid offspring.

Materials and Methods

Animals. Marmosets of the species *S. fuscicollis* were used as donors and recipients in our isoimmunization studies. HERSHKOVITZ [10] identifies 13 subspecies in this group, four of which are maintained in our colony: *S.f. illigeri, S.f. leucogenys, S.f. lagonotus, S.f. nigrifrons.* Intra- and intersubspecies matings produced the offspring used to test the inheritance of the blood factors. Other species used in this study were *S. oedipus, S. nigricollis,* and *S. mystax.* Maintenance and handling procedures have been described in detail elsewhere [4].

1 Marmoset Research Center, Medical Division, Oak Ridge Associated Universities, Oak Ridge, Tenn., under contract with United States Atomic Energy Commission.
2 Supported by USPHS, NIH grant RO1 AM 09289-08 AIB.

Immunizations. Intra- and intersubspecies red cell immunizations were performed with *S.f. illigeri* and *S.f. lagonotus* marmosets. The primary immunization series consisted of three i.m. injections given on alternate days, each inoculum containing 0.1 ml washed packed red cells suspended in 0.1 ml complete Freund's adjuvant. Subsequent booster stimuli consisted of 0.25 ml of a 50-percent saline suspension of washed cells, also given i.m. on three alternate days. Animals were bled at seven- to ten-day intervals after the last injection of a series.

Antibody detection and titration. Routine serological techniques were used for detection and titration of antibodies. Our previous studies established that three kinds of antibodies may be formed: cold agglutinins reactive at 4°C (cold Ab), agglutinins reactive at room temperature (RT Ab), and incomplete antibodies detected with a Coombs reagent (incomplete Ab) [5]. Titration of antibodies was made by twofold serial dilutions of 0.1 ml antiserum in 1% saline; to each diluted sample was added 0.1 ml of a 1-percent washed red cell suspension. Conditions appropriate for the detection of the three antibodies were then pursued with the diluted antiserum [5]. The Coombs reagent was made by immunization of a rabbit (or goat) with marmoset globulins and was absorbed free of any hemagglutinating antibodies prior to use.

Quantitation of red cell chimerism. Two methods were used to quantitate red cells bearing a specific antigen in a cell suspension. The first was to count the free, unagglutinated cells with a Coulter Counter (Model B) after reaction of the cell suspension with the specific immune reagent. Control studies with cell mixtures containing known percentages of red cells reactive with the antisera revealed the technique to be sensitive at the 5-percent level; i.e., mixtures containing the specific antigen in concentrations from 5 to 95% could readily be detected and counted. The second procedure used fluorescein-conjugated immune reagents. Since the isoimmune serum used in these tests contained incomplete antibodies, globulin fractions of this reagent and a goat antimarmoset γ-globulin serum (Coombs reagent) were conjugated with fluorescein isothyocyanate (FITC) [16]. After appropriate treatment with these two antisera, the washed red cells were resuspended completely in buffered glycerin (pH 9.3) and examined as a wet preparation on a microscope slide. A Zeiss fluorescent microscope was used, fitted with a dark-field condenser and two light sources allowing illumination alternately from a tungsten bulb and an Osram HBO 200 high-pressure mercury lamp. For fluorescent illumination a BG12 exciter filter was used in conjunction with 50/44 barrier filters. With a 40× (oil) objective and 10× ocular containing a grid insert, the preparations were viewed alternately with dark-field and fluorescent illumination. Several areas of the slide were examined until a minimum of 400 cells was counted and the percentage of labelled cells determined. Negative controls in these experiments included reaction of the conjugated immune reagents with red cells known to be lacking the specific antigen, and conversely, reacting antigen-bearing red cells with normal immunoglobulins that had been conjugated with FITC. In neither situation was there any nonspecific staining.

Blood factor notation. In a previous communication we had identified a blood factor in *S.f. illigeri* [6]. Because of its apparent limitation to this subspecies, it was noted tentatively as ${}_{s.f.}A^{i}$ were the subscript *s.f.* identified the genus and species and the superscript *i* indicated subspecies specificity. Subsequent blood factors in this and other races of *S. fuscicollis* can be noted in the same manner, using a capital letter symbol (B, C, etc.) for the new factor and the appropriate superscript to identify the subspecies. This has been done for two factors now identified in *S.f. lagonotus* and are noted as $_{s.f.}\mathbf{B}^l$ and $_{s.f.}\mathbf{C}^l$. When data on *S. fuscicollis* are presented, the notation may be abbreviated by omitting the subscript, e.g. $\mathbf{A}^i$. Those animals lacking the specific factors are identified with corresponding lower case symbol, i.e., ä, b, or c. Animals having a chimeric population of red cells, i.e., A and ä, are identified with the blood factor symbol in parenthesis, e.g. (A^i)[3].

Results

Blood Factor $\mathbf{A}^i$

Identification of factor $\mathbf{A}^i$ was made with an isoimmune reagent (IR-K) derived from an intrasubspecies immunization of *S.f.illigeri* [6]. The excellent discrimination offered by this serum and its use in demonstrating red cell chimerism prompted us to reimmunize the donor ('K') to increase the reagent pool. However, response to booster injections (two series) of the immunizing red cell antigen was poor; therefore, additional *S.f. illigeri* marmosets lacking blood factor $\mathbf{A}^i$, i.e. ä, were injected with this antigen to establish another reagent donor. Included among this group of six animals was the chimeric cotwin of 'K' and two full sibs. Surprisingly, none of the, sibs formed any isohemagglutinins after each of three series of injections. One animal in this group, however, did respond to the $\mathbf{A}^i$ factor, producing both an incomplete Ab and a complete agglutinating RT Ab. Parallel tests with this new reagent, IR-42A (animal's number and letter designation), and IR-K with a panel of cells from A^i and ä-type animals showed complete agreement on specificity, i.e. no reactivity with ä-type cells and strong and equivalent reactivity with A^i cells. Further, reagent IR-42A did not react with other subspecies of *S. fuscicollis* and three other species of *Saguinus* behaving similarly to IR-K (table I). The two antibodies in IR-42A showed the same specificity, but routinely we have used the incomplete Ab in subsequent tests because of the greater titer.

Blood Factor $\mathbf{B}^l$

Immunization of an *S.f. illigeri* marmoset with red cells from an *S.f. lagonotus* donor yielded an antiserum containing incomplete Ab which

3 Symbols for blood factors and their corresponding antibodies are printed in bold-face type, symbols for genes and genotypes in italics, symbols for agglutinogens, phenotypes, and blood group systems in regular type. The lower case letter ä is so designated to avoid confusion in the text.

Table I. Tests for blood factors $\mathbf{A}^i$, $\mathbf{B}^l$ and $\mathbf{C}^l$ in an imported population of *S. fuscicollis* ssp., *S. mystax, S. oedipus* and *S. nigricollis*[1]

Species	Number of animals with phenotypes					
	A^{i}[2]	ä	B^l	b	C^l	c
S. fuscicollis ssp.						
illigeri	26	85	0	35	0	31
lagonotus	0	35	7	18	4	21
nigrifrons	0	24	0	20	0	22
leucogenys	0	10	0	8	0	8
illigeri / *leucogenys* [3]	0	10	0	5	0	5
S. mystax	0	18	0	10	0	7
S. oedipus	0	20	0	20	0	17
S. nigricollis	0	9	0	9	0	7

1 Does not include animals chimeric for blood factors since genotypes could not be established for all.
2 Identical results obtained with immune reagents IR-K and IR-42A (see text).
3 Taxonomic classification of these animals uncertain.

Table II. Reaction of isoimmune sera with red cells of *S.f. lagonotus* marmosets identifying blood factors $\mathbf{B}^l$ and $\mathbf{C}^l$

Immune reagent[1]	Animal No.	Serum dilution $(\log_2)^{-1}$									Phenotype designation
		2	3	4	5	6	7	8	9	10	
IR-23	1[2]	4	4	4	4	4±	3	2	1	–	B^l
	2	4±	3	1	–						b
	3	3	3	1	–						b
	4	–	–	–							b
	5	–	–	–							b
	6	1±	–								b
	7	4	4	4	4	4	3	1±	–		B^l
	8	4	4	4	4	4	3	2	1	–	B^l
	9	2	1+	–	–						b
IR-52	1[2]	4	4	4	4	4	4	3	2	1	C^l
	2	–	–								c
	3	–	–								c
	4	–	–								c
	5	4	4	4	4	4	4±	3	2	1	C^l
	6	4	4	4	4	4	4±	3	2	1	C^l
	7	–	–	–							c
	8	–	–	–							c
	9	–	–	–							c

1 IR-23 and IR-52 made by immunization of *S.f. illigeri* marmosets with *S.f. lagonotus* red cells.
2 Note this animal has factors $\mathbf{B}^l$ and $\mathbf{C}^l$ on its red cells.

showed specificity only for *S.f. lagonotus* cells. However, in contrast to the monospecificity demonstrated with blood factor $\mathbf{A}^i$, this reagent contained antibodies having varying activities for *S.f. lagonotus* cells (table II). Thus, while several cells in the test panel showed strong and equal reactivity with this reagent (identified as IR-23), the majority reacted only at the lower dilutions or were completely negative. Absorption of the serum with a pool of packed red cells having weak reactivity removed these minor variants without reducing the activity for the major antigenic specificity more than one dilution ($\log_2$). This factor has tentatively been designated $\mathbf{B}^l$ and its absence by b. As with blood factor $\mathbf{A}^i$, it could not be detected in other subspecies of *S. fuscicollis* and other species of *Saguinus* (table I)

Since the above immune reagent, IR-23, was made by an intersubspecies immunization, an attempt was made to obtain a similar antiserum by isoimmunizing within the *S.f. lagonotus* race; i.e. injecting red cells with factor $\mathbf{B}^l$ into *S.f. lagonotus* marmosets typed as b. Five animals receiving a series of three immunizations failed to produce any antibody.

Blood Factor $\mathbf{C}^l$

The death of both the original red cell donor (*S.f. lagonotus)* and the producer of IR-23 *(S.f. illigeri)* prompted us to immunize an additional group of *S.f. illigeri* marmosets with red cells bearing factor $\mathbf{B}^l$ to obtain additional reagent specific for this antigen. Of five animals immunized, two produced isohemagglutinins (incomplete Ab) of high titer. However, in testing these sera against a panel of red cells from *S. f. lagonotus* marmosets a specificity different from that of factor $\mathbf{B}^l$ was found. Also, in contrast to the detection of minor variants associated with factor $\mathbf{B}^l$, the new reagent (IR-52) demonstrated excellent monospecificity (table II). Note the complete absence of reactivity with the majority of the red cells in this panel, and conversely the strong and equal reactivity toward those bearing the specific antigen. As with the other factors identified, this new factor, designated $\mathbf{C}^l$, appeared only in one subspecies of *S. fuscicollis* and also could not be detected in other species of *Saguinus* (table I).

Inheritance of Blood Factors $\mathbf{A}^i$, $\mathbf{B}^l$, and $\mathbf{C}^l$

Our breeding program in the marmoset has provided us with families with which to test the inheritance pattern of the blood factors identified above. Since our colony consists predominantly of *S.f. illigeri* marmosets, data concerning inheritance of factor $\mathbf{A}^i$ are readily available; conversely, matings of *S.f. lagonotus* marmosets with blood factors $\mathbf{B}^l$ and $\mathbf{C}^l$ are limited. The following points may be noted in table III. Of 15 mating pairs of ä × ä animals, all 93 offspring were phenotypically ä. In ten pairings of A^i × ä, a total of 15 offspring were ä, 16 were A^i, and 10 were (A^i), i.e. chimeras. Only three pairs of $A^i \times A^i$ matings were available, and among these, two offspring were ä, eight were A^i, and two (A^i). Note that in the single mating combination of an ä × (A^i) mating there were four ä offspring and four A^i, indicating that the chimeric father (A^i) was genotypically A^i. Although it is not apparent from the tabulations in table III, all (A^i) young, i.e. chimeras, were members of a cotwin pair or of triplets. This is best illustrated in figure 1 showing diagramatically the offspring resulting from the mating of an ä × A^i and an A^i × (A^i) pair. Finally, matings of *S.f. illigeri* and *S.f. lagonotus* marmosets have produced several hybrid offspring in our colony; however,

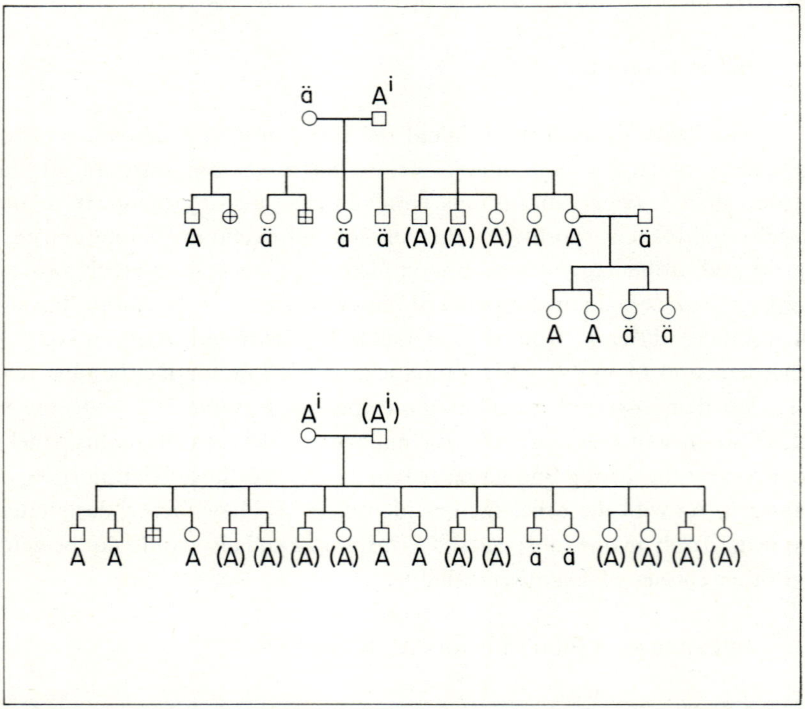

Fig. 1. Phenotypes of offspring resulting from the mating of A^i, ä, or (A^i) animals. A^i, positive for blood factor $\mathbf{A}^i$; ä, negative for blood factor $\mathbf{A}^i$; (A^i), mixture of A^i and a cells, i.e. chimera. + Indicates animal died before tests could be made.

only two pairs have involved animals bearing their respective blood factors (last two mating pairs of table III). Mating of animals with factors $\mathbf{B}^l$ and $\mathbf{A}^i$ resulted in a cotwin pair with each having both antigens on their red cells. Note the phenotype of the single-born animal produced by mating of an *S.f. lagonotus* marmoset having factors $\mathbf{B}^l$, $\mathbf{C}^l$ with an A^i-type S.*f. illigeri*. The chimerism of this young for blood factor $\mathbf{C}^l$ suggests that this animal had a cotwin which died *in utero*.

Red Cell Chimerism

Hemopoietic chimerism in the marmoset has been demonstrated exclusively through sex differentiation of cells: chromosome identification in metaphase plates, or the 'drumstick' of the mature neutrophil [1,3,7, 8] The availability of monospecific reagents for marmoset blood antigens permitted tests for chimerism independent of sex, yet also revealed the relationship between red and white cell chimerism in heterosexual chimeras. The results with blood factor $\mathbf{A}^i$ in *S. f. illigeri* marmosets are shown in table IV.

Table III. Phenotypes of offspring produced by mating of marmosets with blood factors $\mathbf{A}^i$, $\mathbf{B}^l$ and $\mathbf{C}^l$

S. fuscicollis ssp.	Mating combination	Number mating pairs	Offspring number and phenotype
S.f. illigeri	ä × ä	15	93 ä
	A^i × ä[1]	10	15 ä, 16 A^i, 10 (A^i)
	A^i × A^i	3	2 ä, 8 A^i, 2 (A^i)
	A^i × (A^i)[2]	6	3 ä, 15 A^i, 17 (A^i)
	(A^i) × (A^i)	1	1 A^i, 2 (A^i)
	ä × (A^i)	1	4 ä, 4 A^i
S.f. lagonotus	bc × bc	4	8 bc
	bc × B^lc[3]	3	3 bc, 7 B^lc,
	bc × bC^l	1	2 bc, 7 bC^l
	B^lc × bC^l	1	2 B^lc, 2 B^lC^l, 2 bC^l
	bc × B^lC^l	1	2 bc, 1 b(C^l)
S.f. illigeri	bc × A^i	1	2 A^ibc
and	A^i × B^lc	1	2 A^iB^lc
S.f. lagonotus	B^lC^l × A^i	1	1 äB^l(C^l)[4]

1 Includes reciprocal matings ä × A^i.
2 Includes reciprocal matings (A^i) × A^i.
3 Includes reciprocal matings B^lc × bc.
4 Single-born animal with chimerism for blood factor $\mathbf{C}^l$.

It may first be noted that the two methods of quantitating the percentage of A^i-type cells in a mixture, counting of free unagglutinated ä-type cells in a Coulter Counter, and counting of fluorescein-tagged A^i red cells, agreed remarkably well. Although a maximum difference of 6% was noted between the two procedures in two animals, these preliminary tests suggest they complement each other. A photomicrograph shows the typical 'ring' fluorescence of red cells reacting with the conjugated immune reagents (fig. 2). In four of six heterosexual blood chimeras there appeared to be a direct correlation between the degree of white and red cell chimeiism. For example, animal No. 1 was an 18-percent white blood cell chimera and approximately 17% of its red cell population was phenotypically ä. Conversely, animal No. 3, which had 24% white cell chimerism, showed 79% of its cells to be phenotypically ä; therefore, the chimeric white cell population appears to be associated with the 21-percent red cell population typed as A^i. Data for

Table IV. Quantitation of red cell chimerism in heterosexual and isosexual blood chimeras

Animal No. and sex	Percentage white cell chimerism[1]	Percentage red cells type ä[2]	Percentage red cells type A^i [3]
1 Male	18	17	83 (84)
2 Female } [4]	68	60	40 (41)
3 Male }	24	79	21 (20)
4 Male	4	86	14 (20)
5 Male	45	20	80 (82)
6 Female	12	64	36 (39)
7 Female }	–	58	42 (43)
8 Female }	–	59	41 (45)
9 Male }	–	ND[5]	– (95)
10 Male }	–	ND	– (97)
11 Male [6]	–	18	82 (88)
12 Male	–	ND	– (96)

1 Percentage of cells of opposite sex found in blood lymphocytes by sex-chromosome analysis. Minimum of 100 metaphase plates examined for each animal.

2 Percent cells unreactive with IR-K (or IR-42A) as determined by counting free, unagglutinated cells with the Coulter Counter.

3 Percent cells reactive with IR-K (or IR-42A). Determined by subtracting percent ä-type cells from 100%. Values in parentheses obtained by counting A^i cells tagged with fluorescein-conjugated reagents.

4 Double brackets indicate cotwin pairs.

5 Not done.

6 Single bracket indicates cotwin died before testing.

two other heterosexual chimeras (No. 2 and 4) may be interpreted to suggest that the percentage white and red cell chimersism may be the same; on this premise, therefore, the probable phenotypes of the chimeric red cell population and the animal's own red cells can be estimated for these four. However, in animals No. 5 and 6 there is no apparent correlation between the white cell chimerism and the percentage of either A^i or ä-type cells. In these, and also in the first four, positive identification of animals bearing the genotype for blood factor $\mathbf{A}^i$, or its absence ä, can be obtained only through breeding studies (see for example, mating combination ä × [A^i], table III). The last six animals, representing four sets of isosexual cotwins (two young died before testing), demonstrate red cell chimerism via the blood factor $\mathbf{A}^i$.

Discussion

This report documents the existence of at least three blood factors in the marmoset, *S. fuscicollis*, each appearing to be restricted to one subspecies in this group. The numbers of imported animals available for testing were limited, however, placing some reservation on the latter observation. We have not yet found blood factors common to the four subspecies of

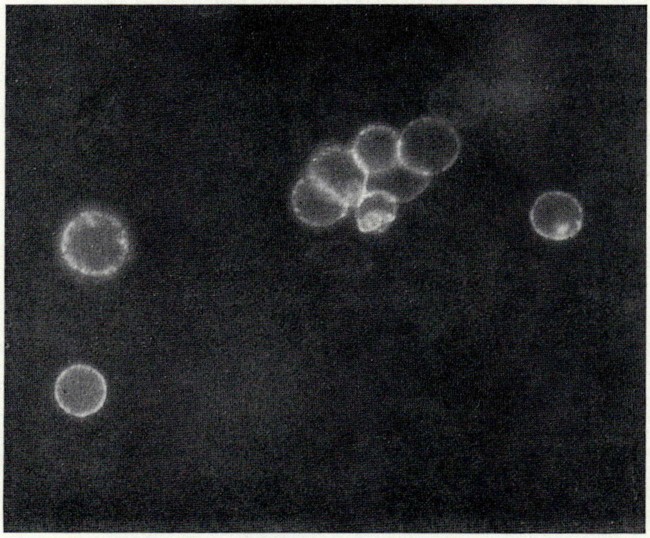

Fig. 2. 'Ring' fluorescence of A^i red blood cells following their reaction with FITC-conjugated immune reagents.

S. fuscicollis in our colony, although our previous studies suggest they exist [5, 6]. Their identification would facilitate blood chimerism studies in the marmoset, which until now have been dependent upon sex differentiation of cells. Much of our effort in detecting blood factors has involved producing reagents by immunization within the species, and it may be necessary to utilize more genetically disparate combinations to reveal a common antigen. Heteroimmunization of cattle with marmoset blood followed by extensive absorptions has suggested this to be feasible [2]. Similarly, studies involving interspecies immunizations, e.g. *S. fuscicollis. S. mystax*, have already produced reagents specific for *S. fuscicollis* red cells but individual discrimination has been lacking [17].

Quantitation of red cell chimerism via the blood factor $\mathbf{A}^i$ suggests that red and white cell chimerism in individual animals do not necessarily correlate (table IV). Similar observations have been made in chimeric cattle twins [15]. This may reflect (a) differential seeding of hemic precursors in a cotwin during embryonic development, or (b) an extrinsic influence on the differentiation and proliferation of the various hemopoietic cell lines. Striking alterations in white blood cell chimerism (sex-chromosome analyses of lymphocytes) have been observed in our laboratory [3]; similarly, STONE *et al.* [14] have reported changes in red cell chimerism in cattle twins. Simultaneous studies on both red and white cell chimerism as a function of age may reveal the validity of the second alternative, but not to the exclusion of the first. The predominance of one cell population, host or cotwin, in chimeric marmoset and cattle twins has been established, i.e., a 50 :50 mixture of cells occurs infrequently [3, 11–14]. Further, in marmosets, male cells seem to be favored in heterosexual chimeras [3, 8]; however, the limited data we have accumulated with isosexual cotwin chimeras (table IV and unpubl.) would indicate that factors other than or in addition to sex, are involved in creating the disproportionate mixtures commonly observed in these natural blood chimeras.

The blood factors identified appear to be inherited as simple Mendelian dominants; the family data compiled with factor $\mathbf{A}^i$ certainly supports this interpretation and the limited information for factors $\mathbf{B}^l$ and $\mathbf{C}^l$ suggest a similar mode of inheritance. Animals typed as A^i, B^l, or C^l may, therefore, be homozygous or heterozygous for the factor, e.g. A^i/A^i or A^i/a, while animals lacking the respective antigens may be designated as homozygous recessives, e.g. a/a, b/b, and c/c. Mating of such recessive homozygous animals has always yielded young lacking the respective blood factors; subsequent mating of these animals with those bearing the antigen has frequently resulted in young with the corresponding blood factor. As anticipated, the availability of more than one blood factor among a mating pair increases the probability of detecting chimerism. This was best demonstrated in the chimerism of the single-born animal resulting from the mating of an *S. f. illigeri* (type A^i) with an *S. f. lagonotus* (type B^l C^l); genotypically, the former animal must be heterozygous for factor $\mathbf{A}^i$, i.e., A^i/a, and the latter heterozygous for factor $\mathbf{C}^l$, i.e., C^l/c, to have produced the young with the phenotype ä B^l (C^l). Chimerism in single-born animals has been detected before by sex-chromosome analysis and has been attributed to the death and resorption of a cotwin *in utero* [8]. The present observation again emphasizes the universality of twinning and chimerism in the marmoset [3].

References

1 Benirschke, K.; Anderson, J.M., and Brownhill, L.E.: Marrow chimerism in the marmosets. Science *138:* 513–515 (1962).

2 Datta, S.P. and Gengozian, N.: Unpubl.

3 Gengozian, N.: Male and female cell populations in the chimeric marmoset; in Goldsmith and Moor-Jankowski Medical primatology 1970. Selected papers from the 2nd Conference on Experimental Medicine and Surgery in Primates, New York 1969, pp. 926–938 (Karger, Basel 1971).

4 Gengozian, N.: Marmosets. Their potential in experimental medicine. Ann. N.Y. Acad. Sci. *162:* 336–367 (1969).

5 Gengozian, N.: Inter- and intrasubspecies red cell immunizations in the marmoset, *Saguinus fuscicollis* ssp. J. med. Prim. *1:* 172–192 (1972).

6 Gengozian, N.: A blood factor in the marmoset, *Saguinus fuscicollis*. Its detection, mode of inheritance, and species specificity. J. med. Prim. *1:* 272–286 (1972).

7 Gengozian, N.; Batson, J.S., and Eide, P.: Hematologic and cytogenetic evidence for hematopoietic chimerism in the marmoset *Tamarinus nigricollis*. Cytogenetics *3:* 384–393 (1964).

8 Gengozian, N.; Batson, J.S.; Greene, C.T., and Gosslee, D.G.: Hemopoietic chimerism in imported and laboratory-bred marmosets. Transplantation *8:* 633–652 (1969).

9 Goldsmith, E.I. and Moor-Jankowski, J. (ed.): Medical primatology 1970. Selected papers from the 2nd Conference on Experimental Medicine and Surgery in Primates, New York 1969 (Karger, Basel 1971).

10 Hershkovitz, P.: Taxonomic notes on tamarins, genus *Saguinus (Callithricidae,* primates), with description of four new forms. Folia primat. *4:* 381–396 (1966).

11 Kanagawa, H.; Kawata, K., and Ishikawa, T.: Chromosome studies on heterosexual twins in cattle. II. Significance of sex chromosome chimerisme (XX/XY) in early diagnosis of freemartin. Jap. J. vet. Res. *13:* 43–49 (1965).

12 Owen, R.D.: Immunogenetic consequences of vascular anastomoses between bovine twins. Science *102:* 400–401 (1945).

13 RENDEL, J.: Studies of cattle blood groups. III. Blood grouping as a method of diagnosing the zygosity of twins. Acta Agric. scand. *8:* 162–190 (1958).
14 STONE, W.H.; FRIEDMAN, J.M., and FREGIN, A.: Possible somatic cell mating in twin cattle with erythrocyte mosaicism. Proc. nat. Acad. Sci., Wash. *51:* 1036–1044 (1964).
15 STONE, W.H.: Personal commun.
16 THE, T.H. and FELTKAMP, T.E.: Conjugation of fluorescein isothiocyanate to antibodies. II. A reproducible method. Immunology, Lond. *18:* 875–881 (1970).
17 YURCONIC, M. and GENGOZIAN, N.: Unpubl.

Authors' address: Dr. N. GENGOZIAN and Dr. MARY L. PATTON, Medical Division, Oak Ridge Associated Universities, *Oak Ridge, TN 37830* (USA)

Medical Primatology 1972. Proc. 3rd Conf. exp. Med. Surg. Primates, Lyon 1972, part I, pp. 361–362 (Karger, Basel 1972)

The Leukaemogenic Role of LSD in Nonhuman Primates

J. Grozdea, P. Colombies and C. Bouloux

Unité de Recherches (U-100 INSERM) and UER médicale A, Toulouse

There is general disagreement between various authors as to the teratogenic and leukaemogenic effects of LSD in man.

We have had the opportunity to study the action of high doses of LSD administered to baboons *(Papio papio)*. Normal controls were available. Our biological investigations comprised a chromosome study, lymphocyte reactivity to phytohaemagglutinin (PHA) and a quantitative analysis of leukocyte enzymes (myeloperoxydase, alkaline phosphatases and PAS).

Materials and Methods

Seventeen clinically and haematologically normal baboons made up the control group. A total of 110–240 μg of LSD/kg were administered to a series of five animals and 300 μg/kg to two other animals. Karyotyping and lymphocyte reactivity were carried out using the techniques currently used in our laboratory [Ducos and Colombies, 1967; Ruffie *et al.*, 1970]. The quantitative assay of leukocyte myeloperoxydase, alkaline phosphatases and PAS was determined on blood films using a score technique based on 100 cells examined.

The techniques of Graham and Knoll [1952], for the quantitative assay of peroxydases, and of Hayoe *et al.* [1964] for alkaline phosphatases and PAS, were used.

Results

(1) Lymphocyte reactivity to PAS. In the control group the rate of transformation varied between 42 and 53 % for a hundred lymphocytes, 42.7

and 53.5% in the animals having received 110–240 μg/kg of LSD, 1.2–14.9% in those receiving large doses (300 μg/kg).

(2) A slight increase in the percentage of hypodiploid cells was observed in the intoxicated animals.

(3) Cytochemical studies revealed the following:

MPO score	*16–40*	(110–240 μg/kg)	(normal rate 19–29).
	3–106	(300 μg/kg)	
PAL score	*159–205*	(110–240 μg/kg)	(normal rate 91–190).
	109–221	(300 μg/kg)	
PAS score	*11–18*	(110–210 μg/kg)	(normal score 10–27).
	5–16	(300 μg/kg)	

Discussion

LSD causes only a slight increase of the number of hypodiploid cells. As far as its leukaemic effects are concerned, the transformation tests showed the presence of abnormal and malign-looking cells with a corresponding increase of PAL as encountered in human leukaemia. High dose of LSD caused a fall in lymphocyte reactivity and concomitant enzyme anomalies (moderate increase of PAL and MPO) witnessing immune depression as in leukaemia. In conclusion, our experimental work following LSD administration to primates has enabled us to demonstrate great analogy with the abnormal biological pattern observed in human pathology.

References

DUCOS, J. et COLOMBIES, P.: Influence du volume total du milieu de culture et de sa concentration en protéines sur la transformation blastique des lymphocytes. C. R. Soc. Biol. *161:* 1452 (1967).

GRAHAM, J. et KNOLL, A.: Réactions de la peroxydase. Planches Hématol. Sandoz, vol. 1, p. 28 (1952).

HAYOE, F.; QUAGLINO, D., and DOLLER, R.: Cytology and cytochemistry of acute leucemias. *1:* 16–23 (1964).

RUFFIE, J.; COLOMBIES, P.; GINOUX-MOUNIE, CH. et CARLES, E.: Etude cytogénétique de quatre espèces de primates. Comparaisons avec le caryotype humain. Ann. Génét. *13:* 4–10 (1970).

Authors' address: Dr. J. GROZDEA, Dr. P. COLOMBIES and Dr. C. BOULOUX, Unité de Recherches (U-100 INSERM) et UER médicale A, *F-31000 Toulouse* (France)

Medical Primatology 1972. Proc. 3rd Conf. exp. Med. Surg. Primates, Lyon 1972, part I, pp. 363–371 (Karger, Basel 1972)

Polymorphic Enzymes in Rhesus Monkeys and Chimpanzees

P. Meera Khan and H. Balner

Department of Human Genetics, University Medical Center, Leiden

Introduction

The electrophoretically detectable intraspecific macromolecular variations appear to be of universal occurrence amongst the natural populations of diploid organisms [Harris, 1966; Lewontin and Hubby, 1966; Weiss and Goodman, 1972]. The occurrence of such polymorphisms would incidentally endow the species concerned with a series of built-in markers. It has been suggested that the genetic markers expressed both in erythrocytes and leucocytes would provide practical means to follow the repopulation events of different forms of blood cells in radiation chimeras that survived bone marrow transplantation [Giblett, 1969; Meera Khan *et al.*, 1972].

The unrestricted usefulness of the isozyme approach depends upon the availability of a number of enzyme markers, and the occurrence of their variations with such a frequency that any chosen donor-recipient pair[1] has at least one usable marker.

The present report deals with an electrophoretic study of 20 different red cell enzyme systems in 57 chimpanzees *(Pan troglodytes)* and 164 rhesus monkeys *(Macaca mulatta)* maintained at the Rijswijk (TNO) Primate Center.

Materials and Methods

Venous bloods were drawn in ACD vacutainers and hemolysates were prepared as described elsewhere [Meera Khan *et. al.*, 1972]. They were analyzed for the enzymes listed

1 This approach is obviously useless in the case of allografts involving identical twin pairs.

in table I by zone electrophoresis on cellulose acetate gel (Cellogel of Chemetron, Milan). The zones of enzyme activities were visualised on Cellogel by suitably modifying the staining recipes [see GIBLETT, 1969; BREWER, 1970; MEERA KHAN, 1971].

Results

The enzymes found to be polymorphic in chimpanzees and rhesus monkeys are indicated by a plus sign in table I; the number following the plus sign represents the presumed number of alleles involved in the corresponding polymorphism.

Table I shows that seven of 20 enzyme systems in chimpanzees and six of 19 in rhesus monkeys are polymorphic.

Table I. Red cell enzyme systems studied zymographically in chimpanzees and rhesus monkeys maintained at Rijswijk (TNO) Primate Center

Enzyme (trivial name)	Locus	Chimpanzee		Rhesus monkey	
		poly.	number of alleles	poly.	number of alleles
1 Adenosine deaminase	*ADA*	–	1	–	1
2 Adenylate kinase	*AK*	–	1	–	1
3 NADH-diaphorase	*Dia. A*	–	1	+	3
4 NADPH-diaphorase	*Dia. B*	+	2	+	2
5 Glucose-6-phosphate dehydrogenase	*G6PD*	+	3	–	1
6 Glutamate oxaloacetate transaminase	*GOT*	–	1	–	1
7 Indophenol oxidase A	*IPO A*	–	1	–	1
8 NADP-isocitrate dehydrogenase	*IDH*	+	3	–	1
9 Lactate dehydrogenase A	*LDH A*	–	1	–	1
10 Lactate dehydrogenase B	*LDH B*	–	1	–	1
11 NAD-malate dehydrogenase	*MDH*	–	1	–	1
12 Peptidase A	*Pep. A*	+	3	–	1
13 Peptidase B	*Pep. B*	–	1	–	1
14 Peptidase C	*Pep. C*	+	3	+	2
15 Phosphohexose isomerase	*PHI*	–	1	+	2
16 Phosphoglucomutase$_1$	PGM_1	+	2	+	2
17 Phosphoglucomutase$_2$	PGM_2	–	1	–	1
18 Phosphogluconate dehydrogenase	*PGD*	+	2	+	2
19 Phosphoglycerate kinase	*PGK*	–	1	–	1
20 Red cell acid phosphatase	*AcP*	–	1	not tested	

The + sign indicates the occurrence of electrophoretically detectable structural polymorphism and the – sign its absence.

The electrophoretic patterns of various phenotypes of diaphorase B (Dia. B), peptidase A (Pep. A), peptidase C (Pep. C), phosphoglucomutase (PGM)[2], and phosphogluconate dehydrogenase (PGD) occurring in the present series of chimpanzees are shown in figures 1–5. The phenotypic frequencies of these polymorphic enzymes are presented in table II.

The data at hand indicate that the chimpanzee Pep. A polymorphism is determined by three alleles at an autosomal locus. The phenotype 2 appears to be less active than the phenotype 1. The phenotype 3 has no

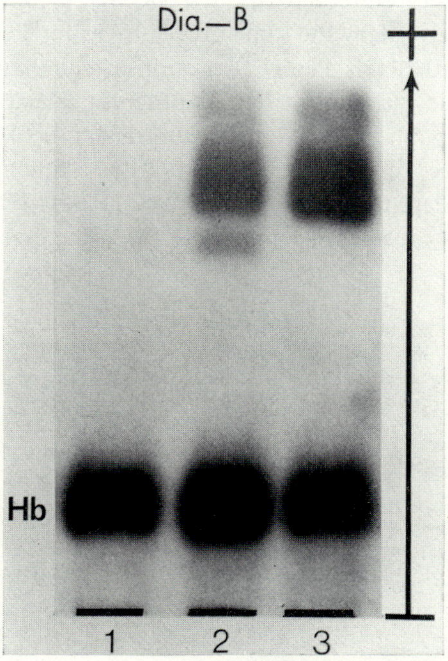

Fig. 1. Electrophoretic patterns of NADPH-diaphorase (Dia. B) phenotypes in chimpanzees. The homozygous variant phenotype Dia. B 2 (ch. 1) is grossly deficient compared to the activity of homozygous normal phenotype Dia. B 1 (ch. 3). The heterozygote Dia. B 2–1 (ch. 2) has an intermediate activity. Note that the mutant enzyme moiety (the sharp rear band) of the heterozygote 2–1 (ch. 2) is apparently activated. In all the chimpanzee zymograms appearing in figures 1–5, the origin of each sample is marked by a line at the bottom and the corresponding channel (ch.) is numbered. Hb = hemoglobin. The migrations of all the isozymes are towards the anode (+).

2 For reasons peculiar to the present material, the notations used to designate the various chimpanzee PGM_1 phenotypes in this publication contrasts with those of GOODMAN and TASHIAN [1969] and somewhat agrees with those of SCHMITT *et al.* [1970].

detectable activity (fig. 2). The presumed phenotypes 3–1 and 3–2 are not readily distinguishable from the phenotypes 1 and 2, respectively, on zymograms. Therefore, all the individuals were scored into the confidently discernible phenotypes 1, 2, 2–1 and 3. Based on the autosomal triallelic hypothesis, the gene frequencies for the alleles *1, 2* and *3* were calculated and from these gene frequencies the expected phenotype frequencies reported in table II were derived [LI, 1955].

The available evidence suggests that the chimpanzee Pep. C variation has been determined by three alleles at an autosomal locus, one of which is a deficiency variant (ch. 3, fig. 3). Therefore, the expected phenotypic frequencies for Pep. C (table II) were derived by the same procedure as for Pep. A above. It is interesting to note that this Pep. C variation in chimpanzees somewhat resembles the triallelic variation of Pep. C observed in Babinga pygmies of Central African Republic [SANTACHIARA BENERECETTI, 1970].

The Pep. C 3 of chimpanzees (ch. 3, fig. 3) appears to be a functional analogue to that of the human Pep. C 4 reported by POVEY *et al.* [1972]. It is likely that this phenotype has considerable activity in the white or other nucleated cells.

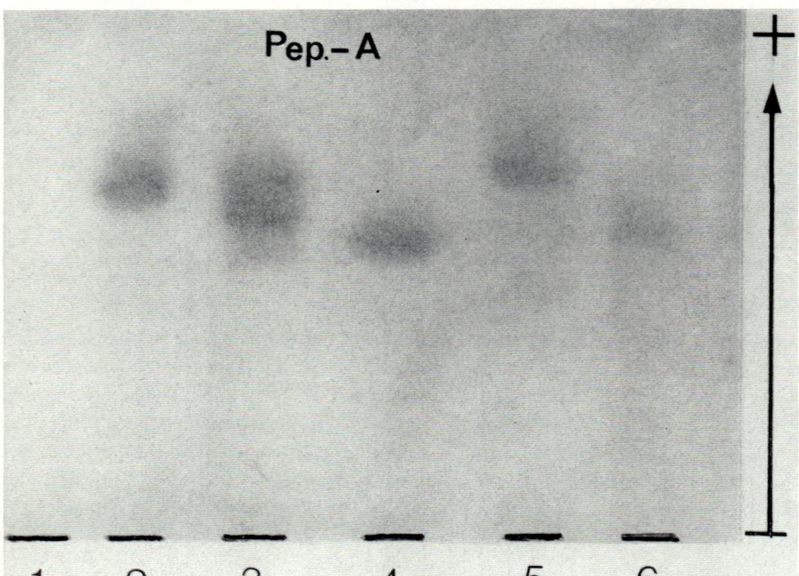

Fig. 2. Electropherogram of Pep. A variants showing phenotypes 3 (ch. 1), 1 (ch. 2 and 5), 2–1 (ch. 3) and 2 (ch. 4 and 6) of chimpanzees.

The occurrence of a single hybrid band with an intermediate mobility, seen in the heterozygote phenotype PGD 2–1 (fig. 5) indicates that the functional molecule of PGD is a dimer and strengthens the suggestion that the gene for PGD is autosomal in chimpanzees. This holds also for Pep. A (ch. 3, fig. 2).

Discussion

Our earlier work has shown that the loci for PGD, phosphohexose isomerase (PHI) and phosphoglucomutase$_1$ (PGM$_1$) are autosomal and polymorphic in *Macaca mulatta* [MEERA KHAN *et al.*, 1972]. An extension of this study has revealed that the loci for NADH-diaphorase (Dia, A), NADPH-diaphorase (Dia. B) and peptidase C (Pep. C) are also autosomal and are highly polymorphic in this species [LOS and MEERA KHAN, unpubl.].

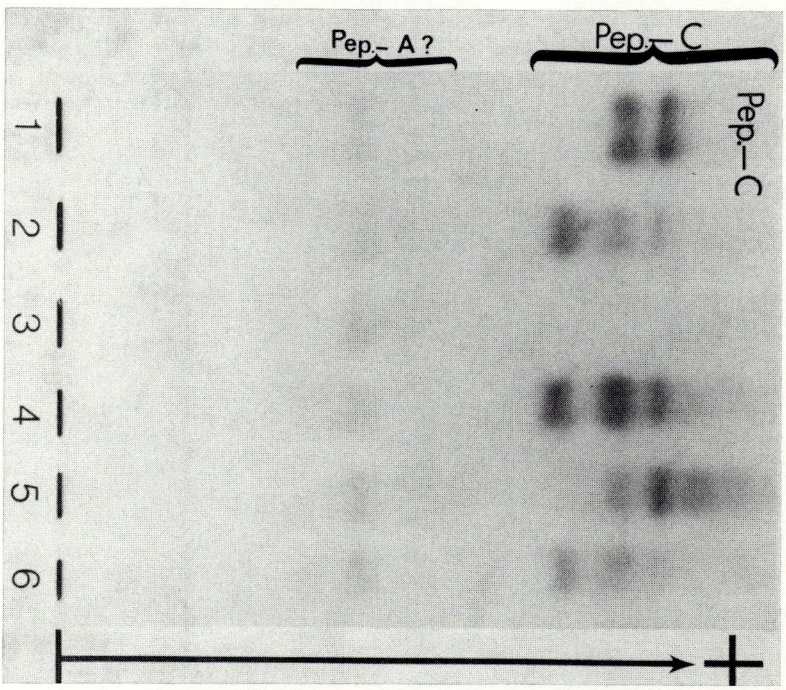

Fig. 3. Zymograms of the red cell Pep. C variants in chimpanzees, showing Pep. C 1 ch. 1), Pep. C 2 (ch. 2), Pep. C 3 (ch. 3), Pep. C 2–1 (ch. 4), presumed Pep. C 3–1 (ch. 5) and presumed Pep. C 3–2 (ch. 6) patterns. The zone in between the origin and Pep. C was thought to be due to Pep. A activity.

The present work indicates that about one-third of all the loci investigated for electrophoretically detectable allelic variations exhibit polymorphisms in *Macaca mulatta* and *Pan troglodytes*. This fraction agrees with those reported for man [HARRIS, 1966], *Macaca fascicularis* [WEISS and GOODMAN, 1972], and *Drosophila pseudoobscura* [LEWONTIN and HUBBY, 1966]. This finding implies that, on an average, one locus for every three investigated will have a high probability of being polymorphic in these or in any other natural population of diploid organisms.

Data on hand suggest that the G6PD variation is determined by three alleles at an x-linked locus in the present series of chimpanzees and that there exist two loci for chimpanzee IDH, one of which is triallelic and highly polymorphic.

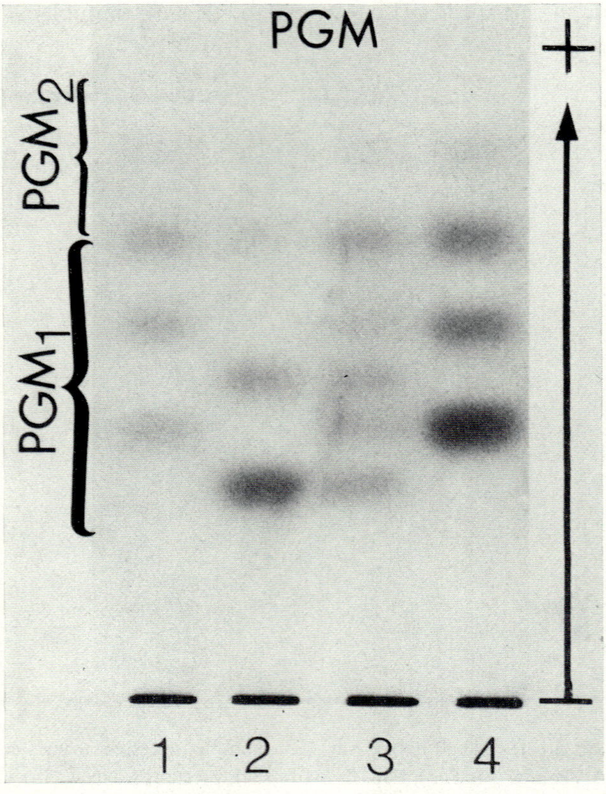

Fig. 4. The electrophoretic patterns of PGM_1 1 (ch. 1 and 4), PGM_1 2 (ch. 2), and PGM_1 2–1 (ch. 3) phenotypes in chimpanzees. The PGM_2 appears to be monomorphic.

The polymorphic loci for *Dia. B, Pep. A, Pep. C,* PGM_1 and *PGD* whose alleles are codominant, appear to be autosomal in chimpanzees. The data summarized in table II clearly suggest that the present population of chimpanzees is not homogeneous at least for PGM_1, *PGD* and *Dia. B* genes. The apparent homogeneity suggested by the numbers for Pep. A in table II is misleading, in view of the fact that six of the seven animals known to be variants for PGM_1 (with the phenotype 2 or 2–1) are the only individuals in the entire series who carried the variants, 2–1, 2 or 3 of *Pep. A*. Even though there is a 50-percent chance for a randomly chosen individual to be a variant for Pep. C, all the seven individuals with the variant PGM_1 were found to be of Pep. C 1 phenotype.

There is no evidence to suggest that there is molecular interdependence between PGM_1, Pep. A and Pep. C. Therefore, it is very likely that the chimpanzees at the Rijswijk colony represent a genetically heterogeneous group. Similar nonhomogeneity was found by Moor-Jankowski *et al.* [1966] in the Holloman colony of 91 chimpanzees, in which subspecies could then be demonstrated using morphological characteristics defined by Osman Hill

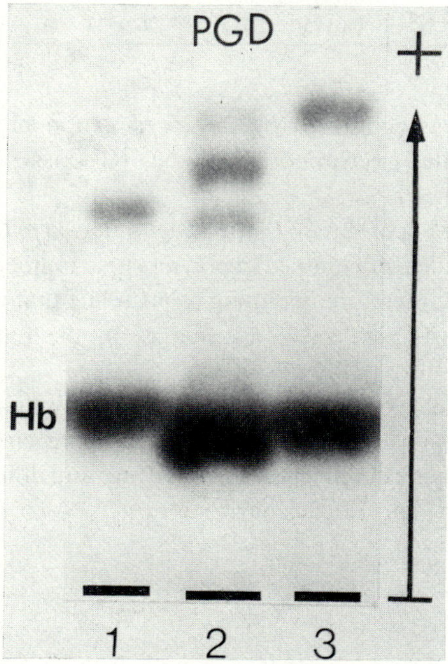

Fig. 5. Zymograms of chimpanzee PGD phenotypes, PGD 1 (ch. 1), PGD 2–1 (ch. 2) and PGD 2 (ch. 3). Hb = hemoglobin.

Table II. Phenotypic distribution of polymorphic enzymes in 57 chimpanzees maintained at the Rijswijk (TNO) Primate Center

Enzyme	Phenotype	Number observed	Observed frequency	Expected frequency	χ^2	*d.f.*	p
NADPH-diaphorase	Dia. B 1	45	0.7895	0.7542			
(diallelic)	Dia. B 2–1	9	0.1579	0.2285			
	Dia. B 2	3	0.0526	0.0173	9.549	2	<0.01 >0.001
Peptidase A	Pep. A 1	51	0.8947	0.8927			
(triallelic)	Pep. A 2–1	4	0.0702	0.0739			
	Pep. A 2	1	0.0175	0.0145			
	Pep. A 3	1	0.0175	0.0190	0.09995	3	>0.99
Peptidase C	Pep. C 1	28	0.4912	0.5056			
(triallelic)	Pep. C 2–1	21	0.3684	0.3430			
	Pep. C 2	7	0.1228	0.1389			
	Pep. C 3	1	0.0175	0.0116	0.6671	3	>0.99
Phosphoglucomutase$_1$	PGM_1 1	50	0.8772	0.7849			
(diallelic)	PGM_1 2–1	1	0.0175	0.2021			
	PGM_1 2	6	0.1053	0.0130	84.25	2	<0.001
Phosphogluconate	PGD 1	54	0.9474	0.9311			
dehydrogenase	PGD 2–1	2	0.0351	0.0677			
(diallelic)	PGD 2	1	0.0175	0.0011	26.0533	2	<0.001

(1966). This nonhomogeneity was confirmed by the investigation of IgG factors [ALEPA, 1968] and by later electrophoretic studies [GOODMAN and TASHIAN, 1969] in the same animals.

It is interesting to note that each of three *(Dia. B, Pep. A* and *Pep. C)* of 20 loci possessed one deficiency allele maintained in polymorphic frequencies in the Rijswijk chimpanzees. It is, therefore, tempting to postulate that these deficiencies probably impart certain selective advantages in the natural habitat.

The deficient chimpanzees may be of value in evaluating the physiological role and the biological importance of the hitherto obscure enzyme NADPH-diaphorase, and in studying the effects of various drugs and different forms of stress on the Pep. A or Pep. C deficiency.

Acknowledgements

We are grateful to Miss M. VERVLOET, Miss B. DOPPERT and Mrs. L. WURZER-FIGURELLI for their excellent technical collaboration.

References

ALEPA, P.F.: Antigenic factors characteristic of human immunoglobulin G, detected in the sera of nonhuman primates; in KRATOCHVIL 1st Holloman Symp. on Prim. Immunol. and Molec. Genet. Primates in Medicine, vol. I, pp. 1–9 (Karger, Basel 1968).

BREWER, G.J.: An introduction of isozyme techniques (Academic Press, New York 1970).

GIBLETT, E.R.: Genetic markers in human blood (Blackwell, Oxford 1969).

GOODMAN, M. and TASHIAN, R.E.: A geographic variation in the serum transferrin and red cell phosphoglucomutase polymorphisms of chimpanzees. Hum. Biol. *41:* 237–249 (1969).

HARRIS, H.: Enzyme polymorphisms in man. Proc. roy. Soc. B *164:* 298–310 (1966).

HILL, OSMAN, W.C.: Chimpanzee. Encyclopaedia Britannica, vol. 5, p. 555 (1966).

LEWONTIN, R.C. and HUBBY, J.L.: A molecular approach to the study of genic heterozygosity in natural populations. II. Amount of variation and degree of heterozygosity in natural populations of *Drosophila pseudoobscura.* Genetics *54:* 595–609 (1966).

LI, C.C.: Population genetics (University of Chicago Press, Chicago 1955).

MEERA KHAN, P.: Enzyme electrophoresis on cellulose acetate gel. Arch. Biochem. *145:* 470–483 (1971).

MEERA KHAN, P.; SOMEREN, H. v.; JONG, W.W. DE, and VERVLOET, M.: Red cell enzyme polymorphisms in rhesus monkeys and chimpanzees. Transplant. Proc. *4:* 137–140 (1972).

MOOR-JANKOWSKI, J.; WIENER, A.S.; KRATOCHVIL, C.H., and FINEG, J.: Seroprimatology of chimpanzees. Blood-group distribution as a 'racial' characteristic. Science *152:* 219–220 (1966).

POVEY, S.; CORNEY, G.; LEWIS, W.H.P.; ROBSON, E.B.; PARRINGTON, J.M., and HARRIS, H.: The genetics of peptidase C in man. Ann. hum. Genet. *35:* 455–465 (1972).

SANTACHIARA BENERECETTI, S.A.: Studies in African pygmies. III. Peptidase C polymorphism in Babinga pygmies. A frequent erythrocyte deficiency. Amer. J. hum. Genet. *22:* 228–231 (1970).

SCHMITT, J.; LICHTE, K.H., and FUHRMANN, W.: Red cell enzymes of the Pongidae. Humangenetik *10:* 138–144 (1970).

WEISS, M.L. and GOODMAN, M.: Frequency and maintenance of genetic variability in natural populations of *Macaca fascicularis.* J. hum. Evolut. *1:* 41–48 (1972).

Authors' addresses: P. MEERA KHAN, Department of Human Genetics, University Medical Center, Wassenaarseweg 62, *Leiden;* H. BALNER, Radiobiological Institute, TNO *Rijswijk, Z.H.* (The Netherlands)

Medical Primatology 1972. Proc. 3rd Conf. exp. Med. Surg. Primates, Lyon 1972, part I, pp. 372–378 (Karger, Basel 1972)

Significance of Basicranial Synchondroses in Nonhuman Primates and Man

Maria Michejda

University of Nebraska, Lincoln, Neb. and Kinderspital, Zürich

Increasing use of nonhuman primates in biomedical research during the last decade resulted in the accumulation of data interesting not only for experimental medicine but also for other disciplines. These data provide new approaches which lead to better understanding of many unclear problems of human biology as well as primate biology at large. For example, our data obtained from longitudinal and cross-sectional studies carried out for the last six years on nonhuman primates allowed us to learn much about the growth patterns of the skull in higher primates, particularly those which concern cranial base morphogenesis. Information about the ontogenic mechanism of the cranial base flattening, known as skeletal platybasia, became available. This growth phenomenon of the skull base typical for nonhuman primates can be observed also in various congenital disorders, such as mongolism, hypothyroidism and microcephaly. Thus, the normal flattening of the skull base in nonhuman primates anatomically resembles the abnormal development of the skull base in humans. In the latter case, however, platybasia usually occurs together with the general depression of endochondral bone formation.

The morphogenesis of the human cranial base and other mammalian forms has been studied for many years. However, the problems associated with the flexion of the skull base, such as the timing, mode, and exact site at which the bending processes in the normal human skull and the flattening processes in nonhuman primates occur remained unsolved. Moreover, the role and significance of the cranial base synchondroses in these processes have not been clear and have been largely neglected in the consideration of the mechanism of flexion.

There are three main synchondroses in the cranial base of higher primates including Homo: spheno-occipital, midsphenoidal, and spheno-ethmoidal. However, these three growth centers are not involved in the same manner in the growth phenomena of the cranial base and in its flexion.

Information about the fusion of the spheno-ethmoidal synchondrosis, which unites the presphenoid and cribriform plate, is most conflicting even in man [Ford, 1958; Baume, 1968; Davies and Davies, 1962].

The time of fusion varies widely from early puberty to adulthood. More detailed investigations, however, indicate that in Homo, growth at this site seems to cease and remains stationary after about six to eight years of age. Whether this synchondrosis then or later becomes a synostosis is not clear. However, it is a well-established fact that this growth center contributes to the elongation of the cranial base, particularly in the anterior portion. In non-human primates, fusion of the spheno-ethmoidal synchondrosis takes place earlier than in man. In our neonatal material of *Macaca mulatta,* and of

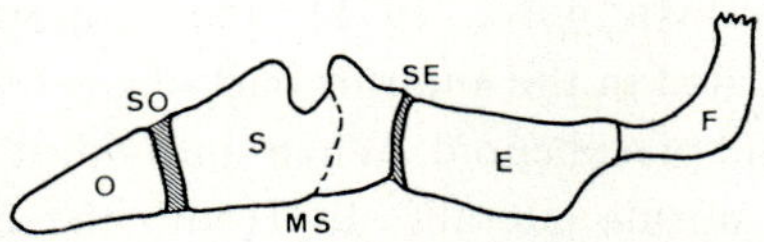

Fig. 1. Diagram of the transverse section of the human cranial base showing the position of three skullbase synchondroses relative to the cranial bones. O = basioccipital; SO = spheno-occipital synchondrosis; S = sphenoid, MS = midsphenoid synchondrosis; SE = spheno-ethmoidal synchondrosis; E = ethmoid; F = frontal bones.

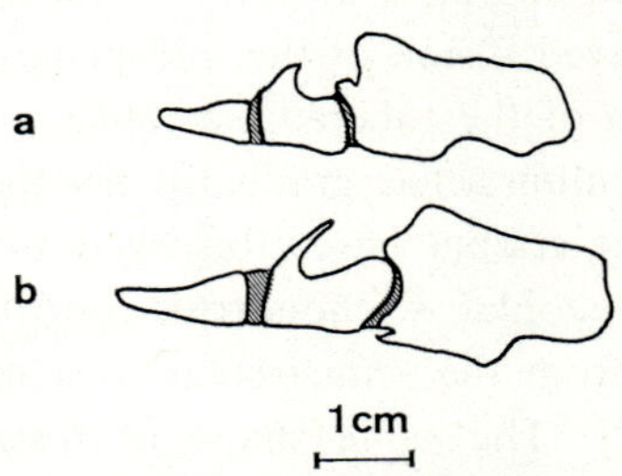

Fig. 2. Generalized diagram of the transverse section of the cranial base in nonhuman primates showing differential growth of the spheno-occipital and midsphenoid synchondroses. *a* Young specimen: endochondral activity present in both growth centers – spheno-occipital and midsphenoid. Well-defined tuberculum sellae present anteriorly to midsphenoid synchondrosis. *b* Adult specimen, wider cartilaginous column of spheno-occipital synchondrosis shows increased endochondral activity of this growth center. The midsphenoid synchondrosis shows first signs of fusion. The tuberculum sellae is resorbed and the fossa has changed from semicircular to more oval, horseshoe shape.

anthropoid apes, no true synchondrosis or defined bony junction was found. Thus, fusion of this growth center occurs in fetal life, and its role in growth phenomena, such as cranial base flexion is completely insignificant.

More significant than spheno-ethmoidal synchondrosis in skull ontogeny is the spheno-occipital synchondrosis between basioccipital and postsphenoid bones. The time of fusion, well established by numerous studies, is similar in most of the primates, being usually related to early adolescence or late puberty [BJORK, 1955; HOYTE, 1966; MOSS and GREENBERG, 1955; IRVIN, 1960; POWELL, 1962; SCOTT, 1958; SICHER, 1965]. The mode of growth is also similar in Homo and nonhuman primates and is limited to elongation of the cranial base, particularly its posterior portion.

Information about the mode and time of fusion of the midsphenoidal synchondrosis is limited and is controversial with regard to these flexional parameters, particularly in nonhuman primates [ASHTON, 1957; FORD, 1956 and 1958; SCOTT, 1958 and 1963]. However, according to our findings, the midsphenoidal synchondrosis contributes most directly to the flexion phenomena and thus skeletal platybasia [MICHEJDA, 1971]. This important growth center of the cranial base is located in the anterior part of the hypophyseal fossa between postsphenoid and presphenoid. When unfused, it can be detected easily on radiograms in norma lateralis. In Homo the midsphenoidal synchondrosis fuses in the last months of prenatal development and under normal conditions is fully obliterated at the end of the first year of postnatal development. As a result of fusion of the pre- and postsphenoid in man, the tuberculum sellae rises [SHAPIRO, 1960]. The tuberculum is well defined and can also be seen on lateral radiograms, anterior to the midsphenoidal synchondrosis. Significant delayed fusion of the midsphenoidal synchondrosis and reverse growth pattern of the tuberculum sellae is observed in nonhuman primates. The tuberculum sellae gradually disappears during the ontogenetic development of the cranial base. Resorbed tuberculum sellae together with unfused midsphenoidal synchondrosis contribute to the change in the hypophyseal fossa from the semicircular to a more oval, horseshoe shape [MICHEJDA, 1972]. The same shape of fossa is frequently observed in mongoloid, hypothyroid and microcephalic skulls.

It seems that both synchondroses, midsphenoidal and spheno-occipital, have rather differing growth patterns and accordingly different functions in the ontogenetic development of the cranial base. In nonhuman primates, the delay of fusion of the midsphenoidal synchondrosis is much more prolonged than the earlier studies of skeletal material indicated. This was observed not only in our histologic material, but also on lateral roentgenograms of living specimens of known age. Moreover, the data obtained from the longitudinal radiographic studies indicated the continuous processes of cranial base flattening even in young adult animals, as expressed by in-

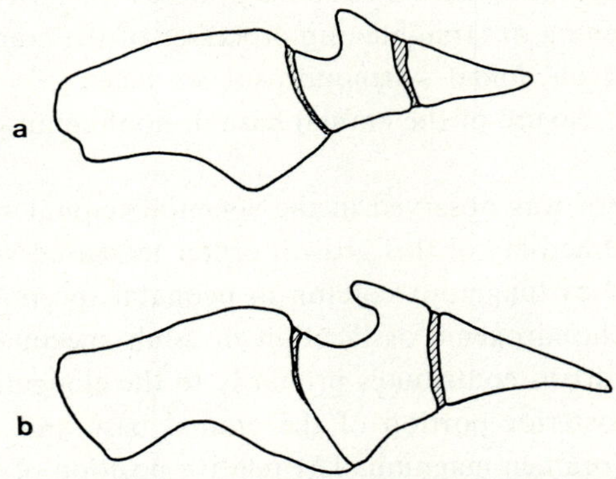

Fig. 3. Diagrams of the transverse section of mongoloid skulls. *a* The cranial base of a 15-month-old girl exhibits unfused midsphenoid synchondrosis, wide cartilagenous column of spheno-occipital synchondrosis, resorbed tuberculum sellae. Pituitary fossa shows early deformation and changes from semicircular to semioval. *b* The cranial base of a 5-year-old girl shows partly fused midsphenoid synchondrosis, wide open spheno-occipital synchondrosis, resorbed tuberculum sellae and deformation of the fossa hypophysea toward an oval, horseshoe-like shape.

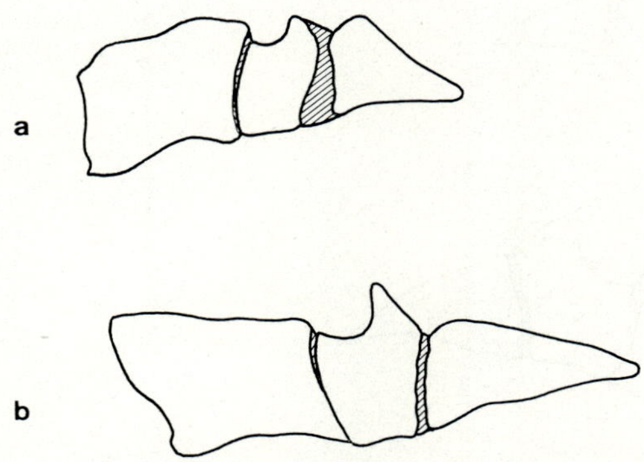

Fig. 4. Diagrams of the transverse section of microcephalic skulls. *a* The cranial base of a 1-month-old girl shows unfused midsphenoid synchondrosis, very wide cartilaginous column of spheno-occipital synchondrosis, not completely developed basi-occipital and basiosphenoid bones. Tuberculum sellae is largely resorbed, some change in fossa from the normal semicircular shape. *b* The cranial base of a 5-year-old girl shows partly fused midsphenoid synchondrosis, wide, cartilaginous column of spheno-occipital synchondrosis, resorbed tuberculum sellae, deformed, shallow semioval shaped fossa.

creasing magnitude of the cranial base angle [MICHEJDA and LAMEY, 1971]. However, the flattening shows the tapering off of the rate in these animals which exhibit the first signs of fusion or partial fusion of the midsphenoidal synchondrosis. Moreover, when the complete fusion is attained, the growth changes, particularly the flattening or straightening processes of the cranial base, are completed. The midsphenoidal synchondrosis, we believe, is the primary factor in the mode of flexure of the cranial base in nonhuman primates.

The reverse growth pattern was observed in the spheno-occipital synchondrosis. The endochondral activity of this growth center increased with age from a nonactive, narrow cartilaginous column in neonatal specimens to a broad band with high chondrogenic ossification in adult specimens. The spheno-occipital synchondrosis contributes primarily to the elongation of the clivus and thus the posterior portion of the cranial base and the ontogenetic migration of the foramen magnum. The relative position of the spheno-occipital synchondrosis with respect to the coronal and lambdoid sutures influences the growth of the cranial vault. This is apparently a secondary factor in the flattening processes of the cranial base.

At this stage of our radiographic investigations of skeletal abnormalities

a

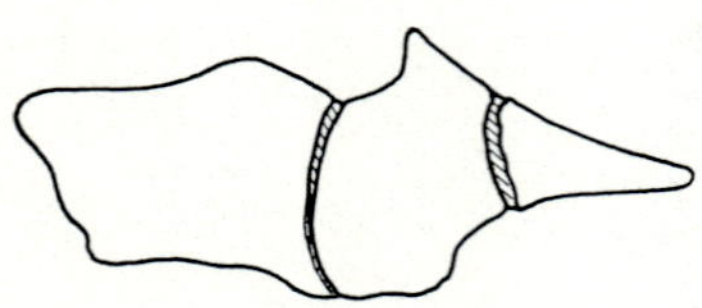

b

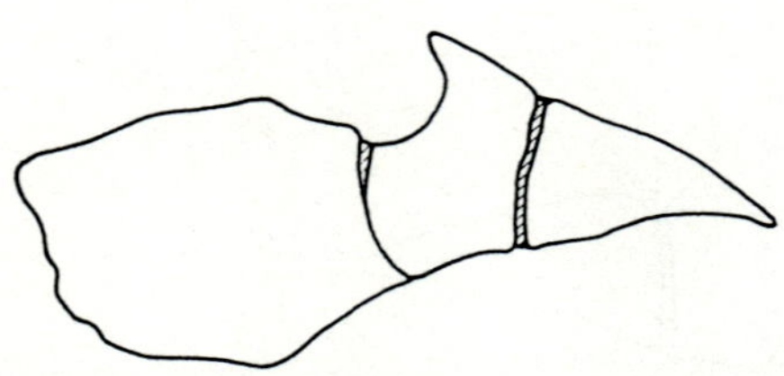

Fig. 5. Diagrams of transverse section of hypothyroid skulls. *a* The skull base of a 10-month-old girl shows unfused midsphenoid synchondrosis, widened spheno-occipital synchondrosis, partly resorbed tuberculum sellae; shallow horseshoe shaped fossa exhibits early deformation in the area. *b* The cranial base of a 5-year-old girl shows partly fused midsphenoid synchondrosis, large cartilaginous column of the widened spheno-occipital synchondrosis, resorbed tuberculum, extensive deformation of the shallowed fossa. In these congenital disorders the skeletal platybasia and extensive deformation of the fossa tend to increase with age.

in congenital disorders, it seems apparent that the role of change of the magnitude of skeletal platybasia also increases with age in nontreated patients. At the same time, the shape of fossa hypophysea assumes a more horseshoe-like shape. The tuberculum is much smaller or absent and the midsphenoidal synchondrosis, which is usually fused in the last month of the prenatal life under normal conditions, is not fused in those congenital disorders. As mentioned before, similar growth phenomena were observed in nonhuman primates. The exact mode of endochondral ossification and skeletal mechanism of cranial base flattening which we were able to establish reasonably accurately in nonhuman primates are still not clear in the case of the abnormal human skulls. However, striking resemblance in radiographic data suggests possible similarity of endochondral ossification of the skull base in congenital abnormalities and in nonhuman primates. Before any sure conclusions can be made about the mechanism of flexion in abnormal skulls, a detailed histologic study has to be made.

The characteristic phenomena of abnormal growth of the cranial base in mongolism, hypothyroidism or microcephaly revealed by resorption of the tuberculum sellae and delayed fusion of basicranial synchondroses may have different condrogenic and osteogenic mechanisms. The delayed fusion of basicranial synchondroses may be a result of arrested growth in the area. The cessation of growth may be due to the lack of cartilage formation which may be absorbed early and replaced by ossified tissue or to the lack of transformation of already existing cartilage into bone tissue. However, in both patterns of abnormal osteogenesis, typical for these growth centers, bipolar proliferation of cartilage is not present. Radiographic appearance of the sella turcica in mongolism and other congenital disorders characterized by general depression of endochondral bone formation is due to the wide open synchondrosis. This change of the fossa hypophysea and its pseudo enlargement is often erroneously interpreted to be the result of enlargement of the pituitary gland. However, in cases of the pituitary deficiency, one would rather have expected hypoplasia of the gland.

Summarizing our findings, it is hoped that they will be useful in elucidating the relationship that exists between chromosomal and skeletal abnormalities, and provide more information on skull base morphology in congenital disorders and the unique relationship between fossa hypophysea and its contents, the pituitary gland. Our study showed that the delayed fusion of the midsphenoid synchondrosis in postnatal life is essential to the mechanism of the flattening of the cranial base in nonhuman primates. The same growth phenomena, resulting in platybasia and changes in the shape of the sella turcica, are present in various congenital disorders, as can be seen in the diagrams.

Acknowledgements

The author would like to express special thanks to Prof. H. ZELLWEGER for his kind help in making the facilities at the Kinderspital available to her.

References

ASHTON, E. H.: Age changes in the basicranial axis of the anthropoidea. Proc. zool. Soc., Lond. *129:* 1–74 (1957).

BAUME, L. G.: Patterns of cephalofacial growth and development. Int. Dent. J. *18:* 489–513 (1968).

BJORK, A.: Cranial base development. Amer. J. Orthod. *41:* 198–225 (1955).

DAVIES, D. V. and DAVIES, F.: in GRAY Anatomy, 33rd ed. (Longmans & Green, New York 1962).

FORD, E. H. R.: Growth of the fetal skull. J. Anat., Lond. *90:* 63–72 (1956).

FORD, E. H. R.: Growth of the human cranial base. Amer. J. Orthod. *44:* 498–506 (1958).

HOYTE, D. A. N.: Experimental investigations of skull morphology and growth. Int. Rev. gen. exp. Zool. *2:* 345–407 (1966).

IRVIN, G. L.: Roentgen determination of the time of closure of the spheno-occipital synchondrosis. Radiology *75:* 450–453 (1960).

MICHEJDA, M.: Ontogenic changes of the cranial base in *Macaca mulatta.* Proc. 3rd Int. Congr. Primat., Zurich 1970, vol. 1, pp. 215–225 (Karger, Basel 1971).

MICHEJDA, M.: The role of basicranial synchondroses in flexure processes and ontogenetic development of the skull base. Amer. J. phys. Anthrop. *37:* 143–150 (1972).

MICHEJDA, M. and LAMEY, D.: Flexion and metric age changes of the cranial base in the *Macaca mulatta.* Folia primat. *34:* 133–141 (1971).

MOSS, M. L. and GREENBERG, S. N.: Postnatal growth of the human skull base. Angle Orthod. *25:* 77–84 (1955).

POWELL, T. V.: Laminographic study of closure of the spheno-occipital synchondrosis; thesis Chicago (1962).

SCOTT, J. H.: The cranial base. Amer. J. phys. Anthrop. *16:* 319–348 (1958).

SCOTT, J. H.: Factors determining skull form in primates. Symp. zool. Soc., Lond. *10:* 127–134 (1963).

SHAPIRO, R.: The normal skull, roentgen study, chap. 2, pp. 8–12 (Hoeber, New York 1960).

SICHER, H.: Oral anatomy, 4th ed. (Mosby, St. Louis 1965).

Author's address: Dr. MARIA MICHEJDA, University of Nebraska, Department of Pediatrics and Department of Orthodontics, *Lincoln, NB* (USA)

Medical Primatology 1972. Proc. 3rd Conf. exp. Med. Surg. Primates, Lyon 1972, part I, pp. 379–389 (Karger, Basel 1972)

The Nuclear DNA Content of Myocardial Cells of Monkeys as a Model for the Polyploidization in the Human Heart[1]

P. PFITZER and H. D. SCHULTE

Institute of Pathology and Clinical Surgery, University of Düsseldorf, Düsseldorf

Before discussing monkeys, some introductory remarks on human myocardial anatomy are necessary. Heart muscle cells of healthy adult human beings contain nuclei with different amounts of DNA. They have either the diploid quantity of 2 C or twice that amount (4 C) or even the fourfold quantity of 8 C [PFITZER, 1971a]. Thus, there are polyploid nuclei in the heart, a fact known since 1964 when SANDRITTER and SCOMAZZONI published their first cytophotometric studies on this subject. Their results were confirmed by investigations published by different laboratories [KOMPMANN *et al.*, 1966; PFITZER 1970 and 1971a; PFITZER and CAPURSO, 1970; FISCHER *et al.* 1970; EISENSTEIN and WIED, 1970; ADLER and HUECK, 1971; ADLER and SANDRITTER, 1971; ADLER, 1972; PFITZER, in press]. We used the UMSP/XD 50 and the SMP CARL ZEISS [for technical details, see PFITZER, 1971a].

The distribution of the different classes of myocardial nuclei is in general rather similar. In the left heart, the 2 C nuclei amount to 32% and the 4 C nuclei to 55% approximately; 10–15% belong to the 8 C or higher polyploid classes [PFITZER, 1971a]. In the histograms from the right heart the 2 C nuclei predominate, attaining about 60%. However, there are exceptions from this rule. In newborns and children younger than five years, both ventricles contain about 90% 2 C and 10% 4 C nuclei. The question at what age polyploidization occurs normally can best be answered by arranging our results according to the heart weights (fig. 1). It is evident that in the left ventricle a sharp rise of the percentage of 4 C nuclei takes place in hearts

1 Supported by the 'Sonderforschungsbereich Kardiologie Düsseldorf der Deutschen Forschungsgemeinschaft'.

weighing from 100 to 150 g. During further development no more significant changes can be detected. This means that polyploidization occurs at an age of about five to nine years, determined by the state of development. One of the problems in myocardial polyploidization is this timing for which we have no explanation.

Another important observation is made under pathological conditions. In all cases of hypertrophy a higher degree of polyploidization is attained which is in most cases significantly above normal. Thus, in a hypertrophic heart of 950 g due to hypertension, we find less than 1 % 2 C nuclei and a predominating group of 86 % 8 C and 16 C nuclei. In addition, there are some nuclei with 32 C and 64 C. The same situation can be found in the right ventricle. For example in a case of pulmonary hypertension 8 C and 16 C nuclei attained 72 % instead of the normal 3 %. As shown, polyploidization is far beyond normal in hypertrophic hearts of adults but what is happening in hypertrophic hearts of children, younger than five years? A case of a boy four months old with a so-called 'double outlet-chamber' can serve as example. While the histogram of the left ventricle is normal, the myocardium of the strained right ventricle contains nuclei with 8 C and 16 C in a heart weighing only 50 g [PFITZER, in press, fig. 3]. Polyploidization has taken place much earlier than under normal conditions and the same is true for other cardiac malformations.

By measuring the DNA content in a large number of normal and pathologic human hearts we know that the following factors are triggering or influencing polyploidization and DNA synthesis in myocardial cells: developmental state, severity and duration of strain, somatotropic hormone, genetic defects, differences in individual reaction. We know these factors, but we would like to know whether there are others and to have precise

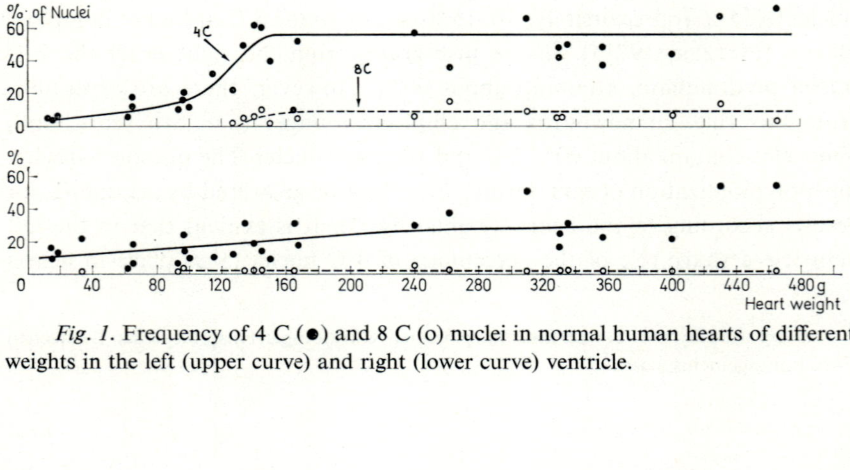

Fig. 1. Frequency of 4 C (●) and 8 C (o) nuclei in normal human hearts of different weights in the left (upper curve) and right (lower curve) ventricle.

knowledge about the mechanisms at work. We looked, therefore, for laboratory animals we could use for experiments. However, surprisingly neither in rats [MIRAKJAN and RUMYANTSEV, 1968; PFITZER and KUHN, 1970] nor in mice [PETERSEN and BASERGA, 1965], rabbits, cats or dogs [PFITZER, 1971d], a significant number of polyploid nuclei is present in the myocardial cells. The same is true for cattle [ADLER and BECKHOVE, 1971; PFITZER, 1971d] or the turkey which has a genuine hypertension [PFITZER, 1971b]. The only exception are pigs where approximately 4% of the nuclei are 6-, 8-, 12- or 16-ploid, but the proportion of the nuclear classes differs distinctly from that in human hearts [PFITZER, 1971c].

The only animals presenting conditions similar to or even identical with man are primates. We measured preparations from the left and right ventricle from 25 animals of eight species as listed in table I. The largest group is formed by 15 rhesus monkeys. The hearts of animals aged three to four years weigh between 7–10 g and their nuclei are not yet polyploidized (fig. 2). However, the histogram of the heart of an eight- to ten-year-old male (fig. 3) shows polyploidization identical to the normal human myocardium including the difference between the left and the right ventricle.

Heartweights of nine of the rhesus monkeys studied are known to us. If the results obtained from these hearts are arranged according to the percentage of the 4 C nuclei in the left ventricle (fig. 4), the hearts weighing 10–15 g contain about 10% 4 C nuclei. In the hearts of 23–25 g this nuclear class

Table I. Primates investigated

Species	Number of animals
Rhesus monkey *(Macaca mulatta)*	15
Chimpanzee *(Pan troglodytes)*	3
Gorilla *(Gorilla gorilla)*	1
Hamadryas baboon *(Papio hamadryas)*	2
White-handed gibbon *(Hylobates lar pileatus)*	1
Douc langur *(Pygathrix nemaeus)*	1
Black howler *(Alouatta caraya)*	1
Squirrel monkey *(Saimiri sciureus)*	1

attains 30%, and 55% in the 37.5-gram heart of the old male. The curve of the proportion of 8 C nuclei runs parallel to that of the 4 C values at a lower level. Polyploidization in the myocardium of rhesus monkeys most likely starts at a heart weight of about 20 g.

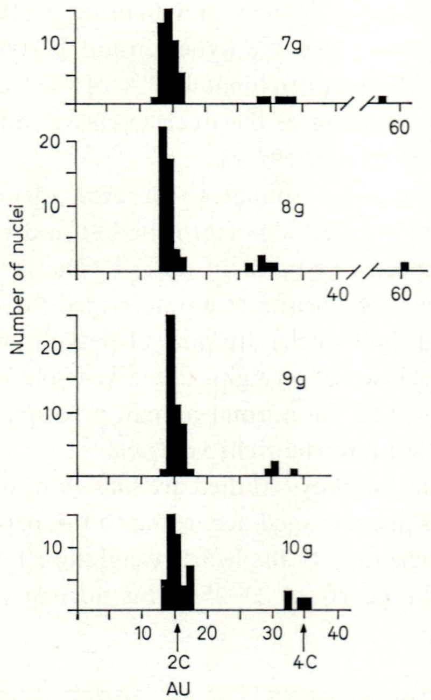

Fig. 2. Histogram of nuclear DNA content in myocardial cells of 4 rhesus monkeys three to four years old. AU= arbitrary units.

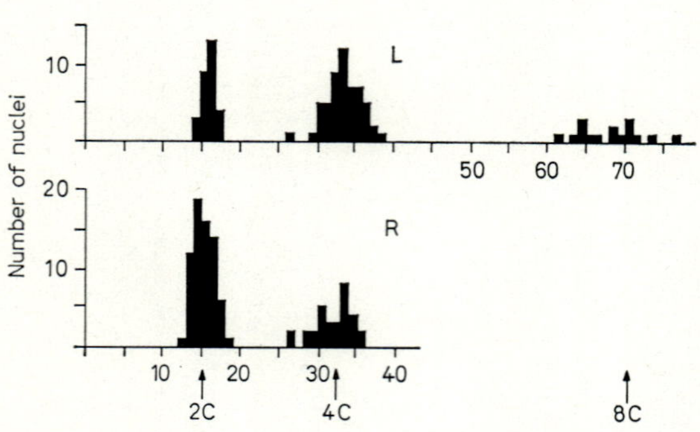

Fig. 3. Histogram of nuclear DNA content of the left (L) and right (R) ventricle of a rhesus monkey eight to ten years old (37.5 g).

The hearts of the other investigated species were either polyploidized or unpolyploidized. Since only one animal of each species up to now has been investigated, growth curves cannot be drawn. An exception is the small group of three chimpanzees, the hearts of which were kindly supplied by colleagues from the IRSAC-Institute in the Congo. The histograms from a heart of a seven-year-old male [PFITZER, 1972, fig. 2b] cannot be distinguished from those of a normal human heart or that of an adult rhesus male

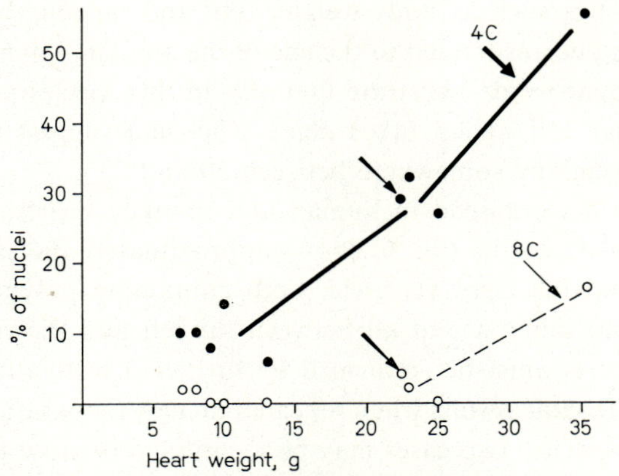

Fig. 4. Frequency of 4 C (●) and 8 C (o) nuclei in hearts of rhesus monkeys.

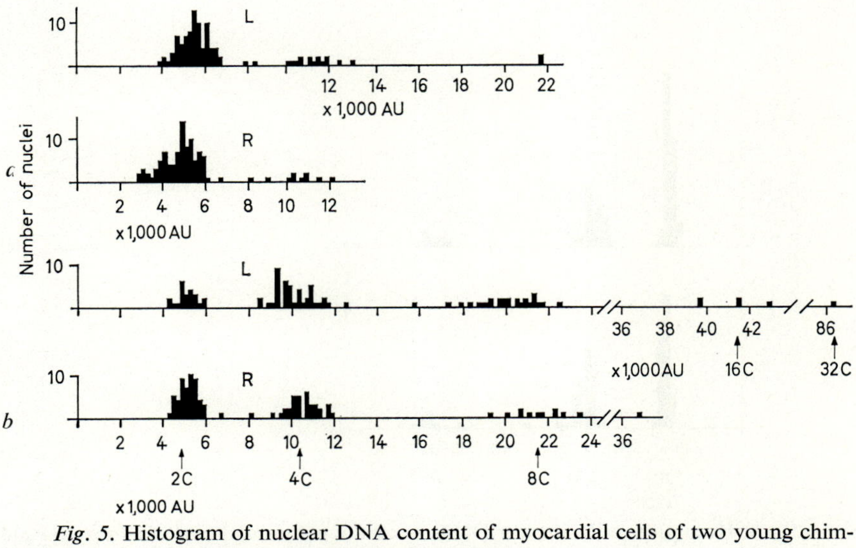

Fig. 5. Histogram of nuclear DNA content of myocardial cells of two young chimpanzees aged *a* 2.5 years (50 g) and *b* 1.5 years (76 g).

(fig. 3). The heart of a 2.5-year-old male chimpanzee weighs 50 g and shows no polyploidization (fig. 5a). Surprising are the histograms of a 1.5-year-old female (fig. 5b). Here polyploidization is even more pronounced than in the adult male. One-third of the nuclei in the left ventricle belongs to the 8 C, 16 C and 32 C classes, while polyploidization in the right ventricle is less pronounced though significant. The only explanation we see is the heart weight which attains 75 g instead of 50 g in the case of the 2.5-year-old male. The reason for this increase in weight is unknown. However, the other developmental characteristics such as body weight, foot and ear length, as well as dental development, corresponded to the age of the animal. Anyhow, these scarce data on chimpanzee demonstrate that also in this species polyploidization of myocardial cell nuclei takes place. The data suggest that it happens in the hearts weighing somewhere between 50 and 75 g.

In contrast to all results obtained in human and monkey hearts, the heart of a four-year-old male gorilla (fig. 6) shows approximately the same histograms for the left and the right ventricle, both containing polyploid nuclei. Whether there is no difference at all between the left and the right myocardium in gorilla hearts must be confirmed by further investigations. Our measurements gave the same results when repeated on new preparations.

From the remaining species, two cases may be mentioned because they show an increase in polyploid nuclei. A white-handed female gibbon *(Hylobates lar pileatus)* aged 9.5 years had 62% 8 C and 16 C nuclei in the left

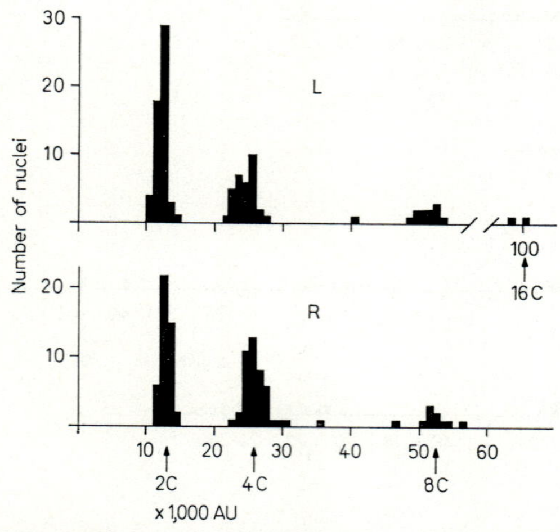

Fig. 6. Histogram of the nuclear DNA content of myocardial cells of a four-year-old male gorilla.

myocardium [PFITZER, 1972, fig. 3b], which corresponds to the findings in man suffering from hypertension. Since the blood pressure of the animal was not checked, we have no definite explanation. A 20-year-old female hamadryas baboon *(Papio hamadryas)* had 60% 4 C and 8 C nuclei in the right heart [PFITZER, 1972, fig. 2]. This case corresponds to the histograms of old people suffering from chronic emphysema and could thus be ex-

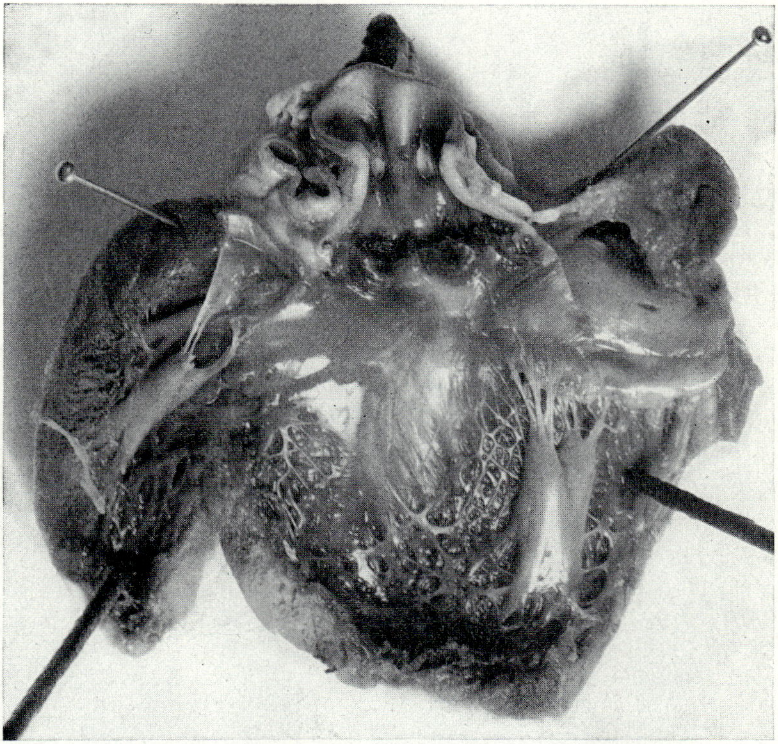

Fig. 7. Heart of a rhesus monkey with banding of the aorta ascendens.

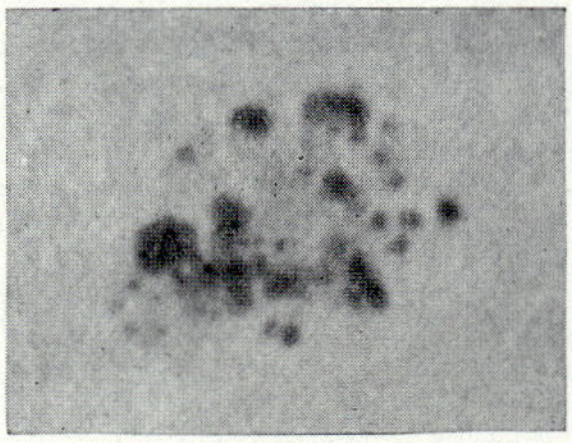

Fig. 8. Degenerative changes in myocardial nuclei with DNA in form of droplets. Feulgen stain. × 1,500.

plained. Whether the differences between these animals are due to hereditary differences between the species must be decided by further investigations; they may reveal important details for evaluating the reactions in human hearts under pathological conditions. At any rate, the most probable ex-

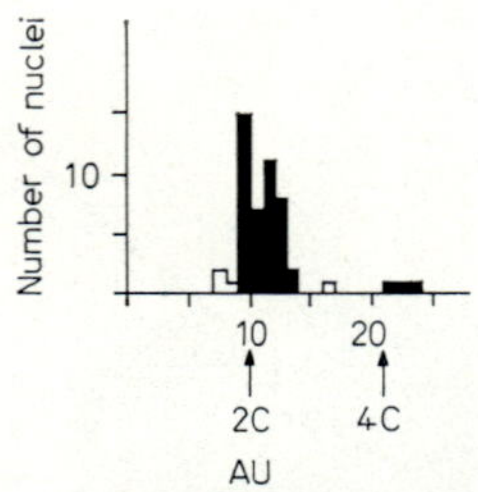

Fig. 9. Histogram of the nuclear DNA content of the left ventricle of a three-year-old rhesus monkey (13,7 g; 12 h postoperation) after extreme banding of the aorta. White intermediate values signify nuclei with DNA in form of droplets.

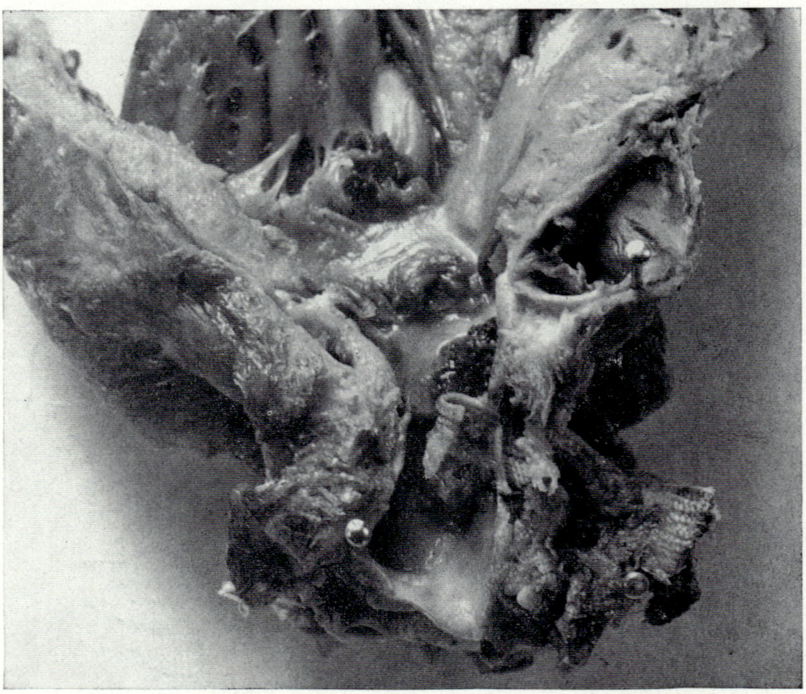

Fig. 10. Aorta of a rhesus monkey 30 days after banding operation, the banding material lying inside the vessel.

planation is that polyploidization can increase above the normal values in monkey hearts as in human myocardium.

In order to induce polyploidization and to study the preceding DNA synthesis we began operative experiments.[2] We performed banding operations of the aorta ascendens on eight rhesus monkeys. Six of the animals were about three years old and no polyploid nuclei were present at the time of the operation. We controlled this by examining a small piece of the myocardium taken during the operation from the top of the left ventricle (fig. 11). The degree of stenosis varied and was determined according to the intraventricular pressure during the operation. The maximum increase was 100% above the normal systolic pressure of 80 mm Hg in the left ventricle. Only one animal was exposed to this severe pressure increase and it died during the first 12 h due to heart failure. The degree of stenosis in this case can be seen in figure 7. Cytological signs of damage were nuclei containing

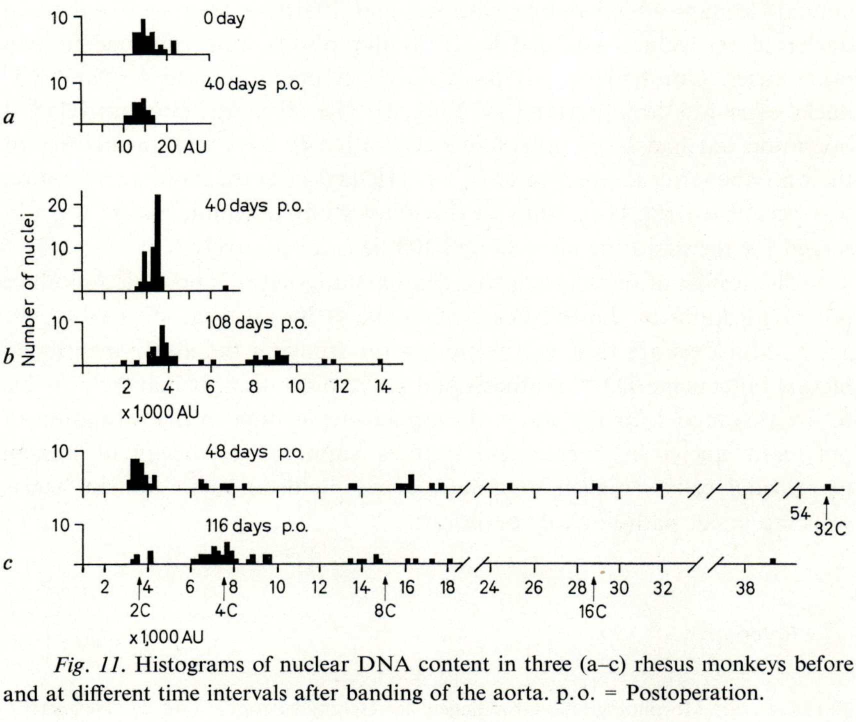

Fig. 11. Histograms of nuclear DNA content in three (a–c) rhesus monkeys before and at different time intervals after banding of the aorta. p.o. = Postoperation.

2 We thank Prof. VAN BEKKUM and Dr. BALNER for the facilities offered in their laboratories at the TNO Primate Center at Rijswijk.

the DNA as small droplets (fig. 8). The total amount of DNA can be reduced in such nuclei (fig. 9). This type of nuclear damage is never observed as *post mortem* change during the first days.

An increase in the ventricular pressure of 50–70% is tolerated over a prolonged period of time. Only one animal died 30 days after operation, because the banding material had cut through the aorta so that it was found inside the vessel at necropsy (fig. 10). However, there was no rupture with fatal hemorrhage and the animal died of heart failure.

Repeated thoracotomy allows one to control the myocardial development in the operated animals. We performed this operation 16 times on eight animals, four animals being operated on twice and two three times until now. At every occasion we took biopsy material for DNA measurements or electron microscopy. Figure 11a shows two histograms of the same animal, one at time of the banding operation and one 40 days later. In both no polyploidization is present. Figure 11b shows the histograms of another animal, 40 days after banding wtthout, and 108 days later with polyploid nuclei of the values 4 C and 8 C. Similar results were obtained in two other cases. One had no polyploid nuclei at operation and 4 C and 8 C nuclei after 116 days [PFITZER 1972, fig. 4]. The other was not controlled at operation but had some polyploid nuclei after 48 days and an increase of their number after another 68 days, i.e. 116 days after the banding operation was performed (fig.11c). Thus in this pilot study polyploidization was observed for the first time after 48 and 108 days, respectively.

The results of our experiments demonstrate that it is possible to induce polyploidization in the left heart of monkeys by artificial stenosis of the aorta. Monkeys are thus useful models for studying the above-mentioned factors influencing DNA synthesis and polyploidization. It may be possible to use this model for the study of mechanisms leading to the formation of polyploid nuclei by incomplete mitoses supposed to occur in human myocardial cells, or even to influence polyploidization in human hearts working under pathological conditions.

References

ADLER, C. P.: Morphologische Grundlagen der Herzhypertrophie und des Herzwachstums. Med. Welt *14:* 477–484 (1972).

ADLER, C. P. und BECKHOVE, P.: Postmortale DNS-Veränderungen im Herzmuskel. Beitr. path. Anat. *142:* 306–320 (1971).

ADLER, C. P. und HUECK, C.: Der DNS-Gehalt in wachsenden Menschenherzen. Verh. dtsch. Ges. Path. *55:* 464–470 (1971).

ADLER, C. P. und SANDRITTER, W.: Numerische Hyperplasie der Herzmuskelzellen bei Hypertrophie. Dtsch. med. Wschr. *96:* 1895–1897 (1971).

EISENSTEIN, R. and WIED, G. L.: Myocardial DNA and protein in maturing and hypertrophyed human hearts. Proc. Soc. exp. Biol. Med. *133:* 176–179 (1970).

FISCHER, B.; SCHLÜTER, G.; ADLER, C. P. und SANDRITTER, W.: Zytophotometrische DNS, Histon- und Nicht-Histonprotein-Bestimmungen an Zellkernen von menschlichen Herzen. Beitr. path. Anat. *141:* 238–260 (1970).

KOMPMANN, M.; PADDAGS, I., and SANDRITTER, W.: Feulgen cytophotometric DNA determination on human hearts. Arch. Path. *82:* 303–308 (1966).

MIRAKJAN, V.O. and RUMYANTSEV, P.P.: DNA synthesis in postnatal histogenesis of myocard and under its infarction hypertrophy and regeneration. USSR Academy of Science, 'Science' Leningrad Division, vol. 10, pp. 964–980 (1968).

PETERSEN, R.O. and BASERGA, R.: Nucleic acid and protein synthesis in cardiac muscle of growing and adult mice. Exp. Cell Res. *40:* 340–352 (1965).

PFITZER, P.: Der DNS-Gehalt der Zellkerne im Herzmuskel des Menschen; Habschr. Düsseldorf (1970).

PFITZER, P.: Nuclear DNA content of human myocardial cells. Curr. Top. Path. *54:* 125–168 (1971a).

PFITZER, P.: Der DNS-Gehalt der Zellkerne im Herzmuskel des Truthahns. Virchows Arch. B *8:* 175–178 (1971b).

PFITZER, P.: Polyploide Zellkerne im Herzmuskel des Schweines. Virch. Arch. B *9:* 180–186 (1971c).

PFITZER, P.: Polyploide Zellkerne im Herzmuskel von Tieren. Verh. dtsch. Ges. Path. *55:* 801 (1971d).

PFITZER, P.: Polyploide Zellkerne im Herzmuskel von Affen. Virchows Arch. B *10:* 268–274 (1972).

PFITZER, P.: Die karyologischen Grundlagen der Hypertrophie Verh. Ges. Kreislaufforsch. *38:* (im Druck).

PFITZER, P. und CAPURSO, A.: Der DNS-Gehalt der Zellkerne im Herzohr des Menschen. Virchows Arch. B *5:* 254–267 (1970).

PFITZER, P. und KUHN, H.: DNS-Gehalt und DNS-Synthese in den Zellkernen normaler und hypertrophierter Rattenherzen. Verh. dtsch. Ges. Path. *54:* 673 (1970).

SANDRITTER, W. and SCOMAZZONI, G.: Deoxyribonucleic acid content (Feulgen photometry) and dry weight (interference microscopy) of normal and hypertrophied heart muscle fibres. Nature, Lond. *202:* 100–101 (1964).

Authors' address: Dr. P. PFITZER and Dr. H. D. SCHULTE, Institute of Pathology, Clinical Surgery, University of Düsseldorf, *D–4000 Düsseldorf* (FRG)

Medical Primatology 1972. Proc. 3rd Conf. exp. Med. Surg. Primates, Lyon 1972, part I, pp. 390–397 (Karger, Basel 1972)

Homologues of the Human A-B-O Blood Groups in Nonhuman Primates[1]

W. W. Socha, A. S. Wiener and J. Moor-Jankowski

Laboratory for Experimental Medicine and Surgery in Primates (LEMSIP) of the New York University School of Medicine, New York, N.Y.

The four human blood groups O, A, B and AB are determined by the presence or absence of two agglutinable substances A and B on the red cells. There are two isoagglutinins, anti-**A** and anti-**B** and according to Landsteiner's rule, those and only those isoagglutinins are regularly present in the serum for which the corresponding agglutinogens are absent on the red cells. This explains the importance of the blood groups A-B-O when selecting donors for transfusion, for which purpose tens of millions of tests are being done each year. Consequently more is known about the human A-B-O blood groups than about any of the other blood group systems. When nonhuman primates became available for experimentation, it was natural, therefore, that one of the first investigations was the search for homologues of the human A-B-O blood groups in apes and monkeys, the study of which has thrown further light on their nature.

Apes

Chimpanzee *(Pan troglodytes)*

By now we have determined the A-B-O blood groups of several hundreds of chimpanzees, of which we have accurate records for 288 animals. Among them 37 were group O and 251 group A. Thus, our results, as well as the results of other investigators, indicate that of the four A-B-O blood groups, B and AB do not occur in these apes. As in man, when serum of a group O chimpanzee is mixed with the red cells of a group A chimpanzee, isoag-

1 Aided in whole by USPHS, NIH grant GM-12074-09.

glutination regularly occurs, but not in reverse combination. Landsteiner's rule of reciprocal relationship of agglutinogens and agglutinins holds in chimpanzees since, with few exceptions, the sera of group O chimpanzees have both anti-**A** and anti-**B**, while sera of group A chimpanzees have only anti-**B**, as can be demonstrated by testing chimpanzee serum against human red cells of groups O, A and B. Of the chimpanzees in our series, the salivas have been tested from more than a hundred animals, all of which proved to be secretors. As in human secretors, group O chimpanzee saliva contains only H, while group A saliva has both A and H. As for the distribution of the A-B-O blood groups in chimpanzees, the data cannot be pooled indiscriminately because of significant differences in their distribution in the few known subspecies of chimpanzees, comparable to the differences of the A-B-O blood groups in various populations of man. Thus, the frequency of group O in chimpanzee populations can range from as low as 9.5% to as high as 39.4%.

Aside from species-specific differences, it is difficult to distinguish serologically between chimpanzee and human blood of the same blood group unless more refined tests are done. For example, in tests with anti-**H** lectin group O chimpanzee red cells are agglutinated as in man but the reactions are often weaker, and chimpanzee group A red cells generally fail to agglutinate with this reagent irrespective of subgroup. Furthermore, in tests with anti-$\mathbf{A}_1$ reagents, chimpanzee group A red cells almost regularly react, but the reactions are usually weaker than for subgroup A_1 in man. In chimpanzee, as in man, subgroups of A appear to exist since about $^1/_{10}$ of group A chimpanzee red cells give markedly weaker reactions, though not as clearly negative as human subgroup A_2.

Only few of the rare dwarf chimpanzees *(Pan paniscus)* have so far been tested. They proved, like *Pan troglodytes,* to belong to groups A or O. Group A red cells of dwarf chimpanzee give, however, reactions indistinguishable from human subgroup A_1. These results and our simian-type blood group findings support the concept of *Pan paniscus* as a distinct species of chimpanzees.

Gibbons and Siamangs

In gibbons, isoagglutination occurs which resembles that occurring in man. However, the reactions define only three blood groups, homologues of the human blood groups A, B and AB; group O has so far not been found among more than 140 animals tested to date. The distribution of the A-B-O blood groups in *Hylobates lar lar,* which make up the bulk of gibbons tested so far, does not differ significantly from that required by the Hardy-Weinberg rule. Thus, the results for the A-B-O groups in gibbons can be explained on the basis of two allelic genes, *A* and *B* with gene *O* absent or, at most, extremely rare. With regard to other species and subspecies of gibbons, *H.l. pileatus, H. hoolock* and *H. concolor,* only a few animals of each have so far been tested. However, there is enough data to indicate a significant difference in the distribution of A-B-O blood groups in *H.l. pileatus* which were all group B, as compared with *H.l. lar.* Only two siamangs *Symphalangus brachytanites,* have been tested – both were group B.

Tests for the subgroups of A have been done in a series of 52 *H.l. lar,* and ten of them were found to be A_1, none was A_2, two were B, 20 were A_1B, and two were A_2B. The reactions of the red cells of group A_1 and A_1B gibbons with anti-**A** reagents were comparable to those of human red cells of the corresponding subgroup. Gibbon group A_2B cells gave reactions with anti-**A** resembling those of human A_2B cells. Thus, the subgroups of A in gibbon resemble the subgroups of A in man, and in this respect differ from chimpanzee group A. Gibbon red cells differ from human red cells, however, in that group B gibbon cells are regularly and strongly agglutinated by anti-**H** lectin like human group O cells. On the other hand, gibbon A and AB group red cells give negative or weak reactions with anti-**H**. Cross-matching of serum and red cells of gibbons gave the isoagglutination reactions expected from their blood groups. The salivas from 56 gibbons have also been tested by us, and all animals were found to be A-B-H secretors. All secreted the substance H in high titer together with the A-B blood group substances corresponding to their blood group.

Orangutans

Isoagglutination occurs in orangutans as in man, chimpanzees and gibbons, but tests to date have shown the presence of only three blood groups, homologous to human A, B and AB. Group O appears to be absent in this species as in gibbons. When our findings are combined with those published by other investigators, the total number of orangutans tested is 71; 41 group A, 14 group B, and 16 group AB. Application of the Hardy-Weinberg formula shows a marked deficit of group AB animals. This phenomenon could occur when two or more isolates are combined, contrasting with each other in frequencies of group A and B. There were no exceptions to Landsteiner's rule of reciprocal relationship between agglutinogens and agglutinins in a series of orangutans tested by us. The red cells of all group A orangutans tested with anti-$\mathbf{A}_1$ reagents gave reactions corresponding to human subgroup A_1. Orangutan red cells of all A-B-O groups gave negative reactions in tests with anti-**H** lectin. Salivas of 37 orangutans were tested and all but one proved to be secretors.

Gorillas

Because of rarity of this species only 25 gorillas have been tested so far. In contrast to other apes, the red cells of gorillas are devoid of A-B-H agglutinogens. Nevertheless their blood groups could be determined by testing the saliva for group-specific substances and the A-B-O group confirmed by testing the serum, after absorption with human group O cells, for anti-**A** and anti-**B** agglutinins. In this way 23 lowland gorillas *(Gorilla gorilla gorilla)* tested by us have all proved to be group B, while two mountain gorillas *(Gorilla g. berengei)* were both group A. In tests on salivas carried out by titration method, variations in inhibition titer for anti-**B** were observed, indicating the existence of subgroups B_1 and B_2.

Old World Monkeys

Baboons

Among the Old World monkeys, the most extensive studies have been carried out by us on baboons. By now we have typed more than 800 baboons.

The A-B-O blood grouping tests on baboons are carried out in the same way as for gorillas, namely, by testing saliva by the inhibition technique for the A-B-H group substances and the serum for anti-**A** and anti-**B** agglutinins after removal of nonspecific heteroagglutinins from the serum by absorption with human group O cells. Among all the baboons tested there were only three group O, the remainder were group A, B or AB. The distribution of the A-B-O blood groups is not the same in baboons of different species. These findings are again comparable to the variations in the distributions of A-B-O blood groups in different human populations.

Among the baboons of known species, group O has so far been found only in the Guinea baboons *(Papio papio)*. The gene frequencies in this species, estimated by the maximum likelihood method, proved to be $O = 10.6\%$, $A = 29.9\%$ and $B = 59.5\%$. The expected distribution of A-B-O blood groups, based on the theory of inheritance by triple allelic genes showed an excellent fit with the observed distribution. In the other three species of baboons, namely, yellow *(P. cynocephalus)*, olive *(P. anubis)* and Chacma *(P. ursinus)* baboons, although group O has not been so far observed, we believe the gene *O* is also present because the gene frequency analysis gives the best fit with the theoretical expectation when the presence of this gene is assumed. Only a few sacred baboons *(P. hamadryas)* have been, so far, tested by us.

Landsteiner's rule holds for baboons fairly well except that the reciprocal relationship is between agglutinins in serum and blood group substances in the saliva rather than on red cells. However, a few animals have been found with anti-**A** in their serum despite the presence of the group substance A in their saliva. The anti-**A** in these group A or AB animals is peculiar in that it is not inhibitable by secretor saliva from group A baboons or group A human beings in contrast to the anti-**A** from group B baboons or group B human beings. These results establish the existence of anti-**A** agglutinins of two kinds: one designated anti-$\mathbf{A}^c$ by WIENER which is reactive exclusively with the A substance on the red cells, while the other, designated anti-$\mathbf{A}^s$, is reactive for the group substances both in secretions and on red cells. Obviously, a group A or group AB baboon could produce anti-**A** antibodies only of the type anti-$\mathbf{A}^c$.

Baboon red cells fail to agglutinate in tests with human anti-**A** and anti-**B** sera and anti-**H** lectin. However, when tests were carried out with anti-**A** lectin (lima bean) agglutination resulted, but only in low titer, independent of animals A-B-O blood group as determined by tests on saliva. That the reaction was due to an **A**-like specificity could be proved by inhibition of anti-**A** lectin with group A or group AB saliva from man or baboon.

Macaques

There are as many as 12 species of macaques having numerous subspecies. Among these, so far, only a few species have been tested; our own studies have dealt principally with rhesus monkeys *(Macaca mulatta)* and crab-eating macaques *(M. fascicularis)*.

In macaques, as in baboons and other Old World monkeys, the red cells have no A-B-H agglutinogens demonstrable by the usual techniques, so that A-B-O grouping depends entirely on tests on saliva and serum. All the rhesus monkeys tested by us have reacted as group B, i.e., their saliva gave positive inhibition for B and H but not for A, while the serum regularly contained anti-**A** but lacked anti-**B**. There is some evidence which suggests the existence of subgroups of B in rhesus monkeys, such as we have found in gorillas. In contrast, in crab-eating macaques all four A-B-O groups have been found, though, as in baboons, group O is rare. Thus far, we have found no exceptions to Landsteiner's rule in this species. Of the other species of macaques we have tested only a few animals. All four stump-tailed macaques *(M. speciosa)* tested reacted as group B. In tests on saliva and serum of five pig-tailed macaques *(M. nemestrina)* all salivas inhibited anti-**H** but not anti-**A** nor anti-**B**; however, in only three of these monkeys were the expected anti-**A** and anti-**B** isoagglutinins found in their sera, while two had anti-**A** without anti-**B**.

Geladas *(Theropithecus gelada)*

Geladas are of interest because of their morphological resemblance to baboon. However, the human-type A-B-O groups in geladas differ sharply from those of baboons. All of the 18 geladas tested by us had saliva which inhibited anti-**H** but not anti-**A** or anti-**B** indicating that they belonged to group O, which is quite rare in baboons. However, the animals did not have the expected anti-**A** and anti-**B** in their sera; instead some animals had anti-**B** alone, others had anti-**A** alone, while still others have neither. This strongly suggests the possibility that the geladas tested actually had the blood groups A, B and AB but for some undetermined reason the group substance A and B could not be detected in their saliva. At any rate, these serological findings support the concept of geladas as a species distinct from baboons.

Other Old World Monkeys

All patas monkeys *(Erythrocebus patas)* tested proved to be group A since their saliva inhibits anti-**A** and anti-**H** but not anti-**B**, and their serum (absorbed with human group O cells) regularly agglutinated human group B but not group A cells. Differences in the anti-**A** inhibition titers of the saliva suggest the presence of subgroups of A in this species. Among 26 Celebes black apes *(Cynopithecus niger)* all but three reacted as group A, however, among the group A animals only a minority had the expected anti-**B** while the remainder had no group-specific agglutinins in their sera. Similar results were obtained in tests with saliva and serum of four drills *(Mandrillus leucopaeus)*. Only eight vervet monkeys *(Cercopithecus pygerythrus)* have been tested of which seven proved to be group A and one group B. In all those cases the serum tests for agglutinins gave the expected reactions.

New World Monkeys

Our experiments on New World monkeys have to date included 59 animals of which 31 were marmosets. Tests on salivas of various species of New World monkeys showed the existence of homologues of human A-B-O groups. The situation, however, is complicated by the presence on the red cells of all these monkeys of a B-like agglutinogen, as first pointed out by LANDSTEINER and MILLER. This practically precludes the presence of anti-**B** in the serum irrespective of the saliva A-B-O group. It is also noteworthy that in tests on saliva from marmosets only very weak or no inhibition was observed with anti-**H** lectin, though strong inhibition was obtained for anti-**A**.

Discussion

As has been demonstrated in this review, all apes and monkeys tested to date have homologues of human A-B-O blood groups. To be sure, A-like and B-like antigens are distributed throughout the animal kingdom but their occurrence as clear-cut blood groups appears to be characteristic of primates.

The anti-**A** and anti-**B** agglutinins in man probably result from inevitable exposure to microorganisms and foods having A-like and B-like specificities. The response to the exposure is canalized by the group-specific substances present in the individual's own body, so that when, for example, group specific substance A is present, there will be no response to the introduction of A-like antigen, while anti-**B** is readily produced. The fact that the Landsteiner's rule holds so well in apes and most species of monkeys adds support to this concept. An outstanding exception has been found in geladas, which have secretions reacting as group O but have sera usually devoid of anti-**A**, anti-**B** or both. This puzzling exception is presently under study. There are, in addition, certain apparent exceptions in New World monkeys which actually add further support instead of refuting the concept. For example, marmosets, which are all group A according to the reactions of their secretions, usually lack the expected anti-**B** in their sera. However, it turns out that all marmosets as well as all other New World monkeys tested by us to date, have a B-like agglutinogen on their cells, similar though not identical to the human agglutinogen B, so that anti-**B**, if it is formed at all, is necessarily different from the anti-**B** present in human group A serum. While the group substances found in the saliva of apes and monkeys appear to be homologues of the human A-B-H substances, the B-like agglutinogen of New World monkey red cells seems to be a heterophile antigen, since a similar B-like agglutinogen is found scattered throughout the animal and vegetable kingdom, in species as diverse as rabbits, dogs, seals and bacteria.

References

For complete references see

MOOR-JANKOWSKI, J. and WIENER, A. S.: Red cell antigens of primates; in FIENNES Pathology of simian primates, pp. 270–317 (Karger, Basel 1972).

WIENER, A. S. and MOOR-JANKOWSKI, J.: Blood groups on nonhuman primates and their relationship to the blood groups of man; in CHIARELLI Comparative genetics in monkeys, apes and man, pp. 71–95 (Academic Press, New York 1971).

Authors' address: Dr. W. W. SOCHA, Dr. A. S. WIENER and Dr. J. MOOR-JANKOWSKI, New York University Medical Center, 550 First Avenue, *New York, NY 10016* (USA)

Medical Primatology 1972. Proc. 3rd Conf. exp. Med. Surg. Primates, Lyon 1972, part I, pp. 398–401 (Karger, Basel 1972)

The Human and Chimpanzee Karyotypes

A Comparison of Their Banding Patterns

CATHERINE TURLEAU, J. DE GROUCHY and M. KLEIN

Clinique de Génétique Médicale, Hôpital des Enfants Malades Paris

Introduction

The primate closest to man is the chimpanzee. Their chromosome numbers differ by two: 2N = 46 in man, and 2N = 48 in the chimpanzee. The new banding techniques enable us to show the degree of identity between the karyotypes of both species.

Technique

Chromosome studies were made according to the controlled heating technique, developed by DUTRILLAUX and LEJEUNE [1971]. Two chimpanzees were studied: one male and one female.

Results

The chimpanzee chromosomes show banding patterns similar to those obtained with human chromosomes. Figure 1 shows a set of different metaphase chromosomes from chimpanzee and human cultures, in order to establish the points of identity and divergence between both karyotypes. This comparison shows the following:

A-Group Chromosomes (1–3)

Both chromosomes 1 show very similar banding patterns. The only obvious difference is the absence in 1(C) of the human subcentromeric constriction. This modifies the centromeric index of 1(C) as compared to 1(H).

No single chromosome in the chimpanzee karyotype matches the human 2. However, it is possible to match the human 2 with two acrocentrics from the chimpanzee, as shown in figure 1. The banding patterns would then coincide, with the exception of the distal end of 2p(H) and the centromeric region. Therefore, one cannot exclude a mechanism more complex than a simple Robertsonian fusion. Both chromosomes 3 are practically identical.

B-Group Chromosomes (4 and 5)

The B(H) appear more telocentric than the B(C). This difference may be accounted for by a pericentric inversion as indicated in figure 1. With

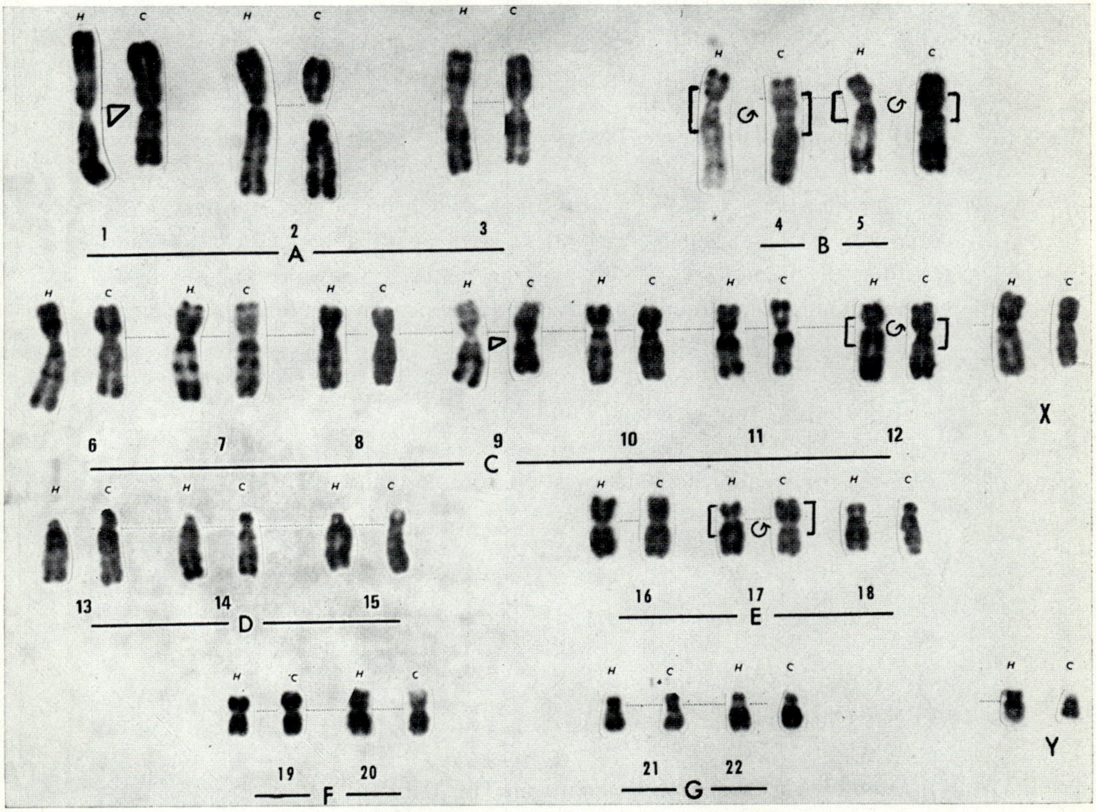

Fig. 1. A comparison of the human (H) and the chimpanzee (C) chromosomes.

this rearrangement both banding patterns match exactly. It is remarkable that both B-group chromosomes evolved in the same fashion.

C-Group Chromosomes (6–12)

Chromosomes 6–8, 10 and 11 have identical banding patterns in both species. Chromosome 9 differs in the same fashion as chromosome 1, i.e. 9(C) does not possess the subcentromeric constriction of 9(H). Chromosome 12 differs by its centromeric index. A pericentric inversion in the limits, shown in figure 1, allows an exact pairing of both banding patterns.

D-Group Chromosomes (13–15)

Chromosomes 13 and 14 show very good matching. Chromosome 15 is more difficult to match.

E-Group Chromosomes (16–18)

Chromosome 16 shows good homology. Chromosome 17 differs in the two species by its centromeric index. Here again a pericentric inversion restitutes the homology of both banding patterns. Chromosome 18 shows poor matching. The difference is difficult to explain by a simple mechanism.

F-Group Chromosomes (19 and 20)

Both the 19 and the 20 show good homology in both species.

G-Group Chromosomes (21 and 22)

Again the chromosomes are identical in both species.

Sex Chromosomes (X and Y)

The X of both species are identical. The Y differ by their size: the Y(C) is missing the distal part of the long arm of Y(H).

Discussion

The classification in figure 1 shows numerous points of identity between the human and the chimpanzee karyotypes. Without any speculation as to the time of evolutionary events, both karyotypes diverge today by a few major differences:

1) The presence in the chimpanzee karyotype of two acrocentrics instead of the human No. 2. The banding patterns show that if the latter results from a fusion of the former, the re-arrangement could be more complex than a simple Robertsonian fusion. This re-arrangement would account for the difference in chromosome numbers of both species.

2) Four chromosomes: the 4, 5, 12 and 17 differ by their centromeric index. This difference can be accounted for by pericentric inversions which allow perfect matching of the banding patterns. Pericentric inversions thus prove to be major steps in the course of evolution.

3) Two constrictions in the human karyotype, those of chromosomes 1 and 9, are not present in the chimpanzee karyotype.

4) Two chromosomes cannot be perfectly matched: numbers 15 and 18. This may be due in part to their poor banding patterns.

We suggest classifying the chimpanzee karyotype according to the human karyotype as shown in figure 1. The two acrocentrics corresponding to the human No. 2 may be called (2p) and (2q).

Reference

Dutrillaux, B. et Lejeune, J.: Sur une nouvelle technique d'analyse du caryotype humain. C.R. Acad. Sci. *272:* 2638–2640 (1971).

Authors' address: Dr. Catherine Turleau, Dr. Jean de Grouchy and Dr. Michel Klein, Clinique de Génétique Médicale, Hôpital des Enfants Malades, 149, rue de Sèvres, *F-75015 Paris* (France)

Index